THE PRE-INDUSTRIAL CITIES AND TECHNOLOGY READER

Cities and Technology, a series of three textbooks and three readers, explores one of the most fundamental changes in the history of human society: the transition from predominantly rural to urban ways of living. This series presents a new social history of technology, using primarily urban settings as a source of historical evidence and a focus for the interpretation of the historical relations of technology and society.

Drawing on perspectives and writings from across a number of disciplines involved in urban historical studies – including archaeology, urban history, historical geography and architectural history – the books in the series explore: how towns and cities have been shaped by applications of a range of technologies; and how such technological applications have been influenced by their contexts, including politics, economics, culture and the natural environment.

The *Pre-industrial Cities and Technology Reader* is designed to be used on its own or as a companion volume to the accompanying *Pre-industrial Cities*, Open University textbook, in the same series. Compiled as a reference source for students, this reader is divided into three main sections, presenting key readings on: Ancient Cities, Medieval and Early Modern Cities, and Pre-industrial Cities in China and Africa. Among the technologies discussed are: agricultural innovations such as the heavy plough, water transport, the medieval road revolution, the first urban public transport, aqueducts, building materials such as brick and Roman concrete, weaponry and fortifications, water clocks, street lighting and fire-fighting. Among the cities covered are: Uruk, Babylon, Thebes, Athens, Rome, Constantinople, Baghdad, Siena, Florence, Antwerp, London, Paris, Amsterdam, Mexico City, Hangzhou, Beijing and Hankou.

Colin Chant is Lecturer in the History of Science and Technology at the Open University.

THE PRE-INDUSTRIAL CITIES AND TECHNOLOGY READER

Edited by Colin Chant

 in association with

First published 1999
by Routledge
2 Park Square, Milton Park, Oxon, OX14 4RN

Simultaneously published in the USA and Canada
by Routledge
270 Madison Ave, New York, NY 10016

Routledge is an imprint of the Taylor & Francis Group, an informa business

Reprinted 2004, (twice), 2005, 2006, 2007, 2008 (twice), 2009

© 1999 The Open University for editorial matter and selection;
individual chapters the contributors

Typeset in Palatino by J&L Composition Ltd, Filey, North Yorkshire
Printed and bound in Great Britain by Bell & Bain Ltd., Glasgow

British Library Cataloguing in Publication Data
A catalogue record for this book is available from the British Library

Library of Congress Cataloging in Publication Data

The Pre-industrial cities and technology reader/edited by Colin Chant.
p. cm. - (Cities and technology)
1. Technology–Social aspects–History. I. Chant, Colin.
II. Series: Chant, Colin. Cities and technology.
T14.5.P74   1999
303.48′3–dc21
98-26901

ISBN-10: 0-415-20077-6 (hbk)
ISBN-10: 0-415-20078-4 (pbk)

ISBN-13: 978-0-415-20077-6 (hbk)
ISBN-13: 978-0-415-20078-3 (pbk)

CONTENTS

LIST OF FIGURES AND TABLES

Tables

INTRODUCTION

This reader is part of a series about the techno-logical dimension of one of the most fundamental changes in the history of human society: the transition from predominantly rural to urban ways of living. It is first and foremost a series in the social history of technology; the urban setting serves above all as a repository of historical evidence with which to interpret the historical relations of technology and society. The main focus, though not an exclusive one, is on the social relations of technology as exhibited in the physical form and fabric of towns and cities.

The main aims of the series are twofold. The first is to investigate the extent to which major changes in the physical form and fabric of towns and cities have been stimulated by technological developments (and conversely how far urban development has been constrained by the existing state of technology). The second aim is to explore within the urban setting the social origins and contexts of technology. The series draws upon a number of disciplines involved in urban historical studies (urban archaeology, urban history, urban historical geography, architectural history and environmental history) in order to correct any illusion of perspective that all major changes in urban form and fabric might be sufficiently explained by technological innovations. In brief, the series attempts to show not only how towns and cities have been shaped by applications of technology, but also how such applications have

been influenced by, for example, politics, economics, culture and the natural environment.

The wide chronological and geographical compass of the series serves to bring out the general features of urban form which differentiate particular civilizations and economic orders. Attention to these differences shows how civilizations and societies can be characterized both by their use of certain complexes of technologies, and also by the peculiar political, social and economic pathways through which the potentials of these technologies are channelled and shaped. Despite its wide sweep, the series does not sacrifice depth for breadth: studies of technologies in particular urban settings form the bulk of the three collections of existing texts in the series, this being the first.

This book is designed to stand on its own, but readers might note that there is a companion volume of original essays.[1] Readers should also note the policy that has guided the editing of the selected readings. With a general and undergraduate readership in mind, scholarly apparatus has been kept to a minimum. This means in many instances the omission of carefully compiled primary and secondary source references. Despite this pruning, in almost all cases, notes of sources have been retained where they would be expected at any level of scholarly writing, including undergraduate essays, in particular, where there is direct quotation, explicit reference to another author, or citation of statistical data. In practice, an unarguably consistent policy has been difficult to implement (among the exceptions are quotations from Mesopotamian inscriptions and Chinese manuscripts). In the end, as editor, I would encourage readers especially interested in any reading to follow up the reference to the original source which is always given.

In this series of textbooks and readers, authors have avoided what are now widely accepted as sexist expressions in their own writing. But where they occur in existing sources, they remain without editorial comment.

It should be made clear that in a collection like this, a product of that distinctive collaborative unit, the Open University Course Team, editorship signifies rather less individual input than in other such volumes with a single name on the cover. It means more that the named individual has taken on responsibility for corresponding with authors and publishers, and with preparing the handover manuscript; in this case, only in the Ancient Cities section has the selection and editing of the sources been entirely down to myself. And in all cases, the selection and editing has been subject to Course Team discussion and approval. The collaborative nature of the enterprise is particularly conspicuous in the case of the readings on China in Part 3, which were chosen and carefully edited by Arnold Pacey. I have also drawn heavily on his notes in the introduction to the section. I should like to record my gratitude to the many authors who have responded generously to my request that their work be made available in this edited form; some, of course, have been unavailable for comment. Finally, I wish to thank Denise Hall of the Open University Arts Faculty, for her part in the preparation of the manuscript, and Martin Barr at Routledge for saving me from a number of oversights that I should prefer to remain unquantified.

Note

1 C. Chant and D. Goodman (eds), *Pre-industrial Cities and Technology*, London, Routledge, 1999

Part 1
ANCIENT CITIES

The readings in this section cover a period of some four millennia, from the first mud-brick cities of Sumer in ancient Mesopotamia, to the great metropolis of imperial Rome during the first centuries of the Common Era.[1] The first, Harold Carter's review of monocausal theories about the origins of cities, presupposes in its opening discussion of surplus production the concept of a technology-driven Urban Revolution, first proposed by the archaeologist V. Gordon Childe.[2] In his careful diagnosis of the various weaknesses of monocausal explanations, Carter does not consider as such the technological determinism implicit in Childe's account; however, technological innovations are clearly fundamental to the agricultural surpluses discussed, and according to Karl Wittfogel's 'hydraulic' theory, it is irrigation technology on which urbanization hangs. Carter ends up counterbalancing technological and cultural explanations in his summary; his conclusion is by implication as critical of technological determinism as it is of all arguments making the historical phenomenon of urbanization dependent on some other variable – irrigation, trade, defence or religion. According to Carter all such arguments tend to be 'circular'; they presuppose the causal relationship they purport to prove, when the relationship could be the other way round. The only way out of these chicken-and-egg debates is to see cities as emerging from a complex mix of social, technological, economic and cultural changes.

The focus of the readings then changes from technologies that have been seen as necessary to the emergence of cities, to those deployed in the actual process of city-building. Apart from irrigation, the most distinctive technological attribute of ancient Mesopotamian urban civilization was building with mud-bricks, the subject of reading 2 by Roger Moorey. The author makes it clear that this technology was practised within a distinctive social and ideological context. This effort and expense necessary to build the great ziggurats, temples and defensive walls was reinforced by Mesopotamian religion; hence the ceremonies and rituals attending the making and laying of bricks for public buildings. In reading 3 from the *History* of the Greek travel writer Herodotus, there is another, more contemporary description of Mesopotamian mud-brick construction, as well as its irrigation farming and river transport, and further insights into the contexts of ancient building, not only in Mesopotamia, but also in Pharaonic Egypt. Herodotus's description of Babylon demonstrates the preoccupation of Babylon's rulers with security; to this end, the princess Nitocris is said to have diverted the Euphrates, created an artificial lake, and built a bridge across the Euphrates with removable wooden platforms. Herodotus's account needs to be treated with some scepticism by historians; he was writing in the fifth century, well after the time of Nebuchadnezzar, when Babylon was part of the Persian empire; there is some doubt as to whether Herodotus actually visited the city. He may therefore have been retailing some hyperbole and myth, and have been tempted to dress up his account for the delectation of the Athenian audiences to whom he would read from his work. His account of the building of the Great Pyramid shows how the cult of the dead ruler legitimized what seemed to Herodotus, from the viewpoint of democratic Athens, the years of 'oppression of the people' required by Pharaoh Cheops (Khufu) for the erection of his pyramid and causeway. Herodotus not only looked through Greek spectacles at the Egyptians' social arrangements, but at their technology too, as the appended notes indicate. Finally, he celebrates some of the engineering achievements of his fellow Greeks on the island of Samos, including Eupalinos's one-kilometre aqueduct through the base of a mountain. In reading 4 on the workers' town of Deir el-Medina in the vicinity of ancient Thebes, Rosalie David gives a more positive account of obligatory state work (corvée) in Pharaonic Egypt; the point is also made that part of the motivation of these craft specialists (apart from a bit of profit on the side), was entry to the afterlife on the Pharaoh's coat tails. David also tackles some earlier historical judgements that there were no truly urban settlements in Pharaonic Egypt. In the light of archaeological evidence that there were more walled settlements than was once thought, and also the evidence of communities of craft specialists provided by the workers' towns, the author places Egypt more firmly in the mainstream of urban civilization.

Readings 5, 6, and 7 deal with some of the principal technologies that enabled the Greeks, from the eighth century BCE, to build cities from Sicily through the eastern Mediterranean to the Black Sea coast. These include the means of moving heavy loads and lifting heavy blocks of stone that shaped Herodotus's perceptions of Egyptian building. In reading 5, Alison Burford convincingly rebuts the perennial propensity to read history backwards: in this case, to judge the efficiency of ancient transport by modern standards. Criticizing the ancients for their failure to harness the horse properly is to miss the point that under ancient conditions, the slow strength of the ox made it a more valuable working animal than the speedy horse. Similarly, in reading 7, J.J. Coulton argues that modern assumptions about the superiority of hoists based on compound pulley systems have led some archaeologists to look for its use by the Greeks to raise heavy building components before c.515 BCE. Coulton insists that before that date, an abundance of unskilled labour made the ramp a more rational choice for moving heavy blocks of stone. The introduction of the crane after that date is not a reflection of its greater efficiency – indeed, building with smaller stone blocks resulted – but of the growing scarcity of labour. But the arguments about the social relations of technology presented in this section are not all in the direction of the greater contextualization of technology. According to Alfred Burns in reading 6, the physical and even social character of its cities was shaped by the hydrogeography of Greece, with its strata of permeable limestone overlying impermeable rock, and the resultant technology of water supply.

The readings on ancient Rome begin with a selection from James C. Anderson's book on Roman architecture. This furnishes a good deal of the political, social, legal and economic context of Roman private and public building; in particular, he points out the increasing control over public building exercised by the imperial bureaucracy. The reading brings out some of the social complexity of the building industry of ancient Rome, including the role of women in managing the brick industry, and the considerable part played by free rather than slave labour. Reading 8 by Anderson is followed by excerpts from the two main primary sources on Roman technology: Vitruvius on architecture and Frontinus on water supply. Reading 9 from Vitruvius throws some light on the planning of colonies: he advised that Greek scientific principles should be applied to the selection of a healthy site, and that the internal layout should depend upon the prevailing winds; the religious dimension of Roman city-building is also well attested in his prescriptions about the location of temples. Of particular note for this series is the close connection detailed between the construction of fortifications and the weaponry of the time. Reading 10, taken from the report prepared in 97 CE by Frontinus, the *curator aquarum*, shows above all the concern of an imperial bureaucrat about private abuse of the public water supply, making it abundantly clear that within the ancient Roman context this was no neutral technology.

Readings 11 and 12, both taken from Neville Morley's book on ancient Rome and its hinterland, recall the theme introduced in the opening discussion of the origins of cities: the dependence of urban settlements on improvements in agricultural productivity. In the case of Rome, however, there were no further important agricultural innovations to explain its unprecedented growth; instead, it was able to feed its population through the conquest of fertile regions, and the organization of a Mediterranean trading system based mainly on water transport. Transport was one of the key technological considerations in Rome's food supply, as it is in the model of the German agriculturalist von Thünen (*The Isolated State*, 1826), who used transport costs to predict zones of agricultural production arranged in neat, concentric rings around a given town. Morley finds a suitably relaxed application of this model helpful in conceptualizing the changing relations between Rome and its *suburbium*. The relaxations include allowance for the differential costs of land, river and sea transport, variations in climate, geology and soil fertility, the demands of the elite both for unproductive uses of the countryside around Rome (the Campagna), such as gardens and tombs, and for the lion's share of the region's water resources.

Notes

1 In all original writing in this series of books, BCE (Before Common Era) and CE (Common Era) are used instead of BC and AD. However, authors' conventional usage has been retained in all previously published writing
2 For a discussion of Childe, see Chapter 1 of the companion textbook to this reader, C. Chant and D. Goodman, (eds), *Pre-industrial Cities and Technology*, London, Routledge, 1999

URBAN ORIGINS: A REVIEW OF THEORIES

by Harold Carter

Source: Harold Carter, *An Introduction to Urban Historical Geography*, London, Edward Arnold, 1983, pp. 3-9

Social scientists, including historians and archaeologists, have proposed four explanations for the emergence of towns:

1 Hydraulic theories or environmental bases to urbanism;
2 Economic theories or the growth of markets;
3 Military theories or growth about defensive strong-points;
4 Religious theories or growth about shrines.

Each of these can be reviewed in turn.

Hydraulic theories: the concept of a surplus

The basis of these theories rests on a fairly simple assertion which is well represented by Sir Leonard Woolley's opening to his discussion of 'The Urbanization of Society' in the UNESCO 'History of Mankind'. He writes:

> It is an axiom of economic history that real civilization can begin only in regions where the character of soil and climate makes surplus production possible and easy; only so is man relieved from the necessity of devoting all his energies and all his thought to the problem of mere survival, and only so is he able to procure from others by means of barter those things which minister to well-being and promote advance but are not naturally available in his own land; moreover, such conditions must prevail over an area large enough to maintain not merely a small group of indivi-duals but a population sufficiently numerous to encourage occupational specialization and social development. So does civilization begin.
>
> (Woolley, 1963, p. 414)

Here all the elements of the environmental theme are set out with the implication that specialization is a consequence of the ability to pay an income from surplus production and that from this not only social stratification follows, but the detachment of the specialist from tribe and kin makes residence the urban qualification. The surplus is the catalyst for that transformation which defines urbanism. To that conclusion can be added the view that the progress of agriculture, and the creation of a surplus, was largely a consequence of irrigation and that irrigation itself necessarily established a pattern of complex bureaucratic controls which brought into being an administrative system of an urban character; social change is in part the consequence of administrative necessity. From this two key concepts arise for discussion, for specialization and social stratification are but consequences. These are (i) the idea of a 'surplus' and (ii) the notion of a hydraulic society. [. . .]

The notion that due to progress in agriculture, possibly related to irrigation, a surplus accrued which was the foundation of all the changes which are subsumed in the concept 'urbanization', has come in for considerable criticism. This has been, essentially, because it is too simple an interpretation. [. . .]

This straightforward idea of the biological surplus carries little conviction for it is manifestly impossible to define in absolute terms a level below which life cannot be sustained and above which a disposable surplus is generated. There can be no absolute surplus and it follows that the idea can only be contemplated in relation to particular cultural contexts or as Lampard writes, 'natural endowment was less the "gift" of nature than a function of human resourcefulness and adaptive behaviour. In short Childe's celebrated "surplus" like the definitive city itself, was a societal product' (Lampard, 1965, p. 533). [. . .] This immediately robs the surplus of its primacy as a progenitor of those changes which are called 'urbanization'. 'Because the surplus is conceived of as individual and special to each and every society it is difficult or impossible to say anything of any great import about its specific role in either the emergence of urban forms or the functioning of urbanism in general' (Harvey, 1973, p. 220). [. . .]

The idea of the hydraulic society is closely related to an ecological interpretation of urbanization, that is, that it was a consequence of increasing size and density of population and that this was brought about by irrigation. [. . .]

The working out of the consequences of irrigation in greatest detail is to be found in Karl A. Wittfogel's book *Oriental Despotism*. Wittfogel, as the subtitle of his book – *A comparative study of total power* – implies, was more concerned with the state and political power than the city as such. In the index to his book the word 'urban', or its equivalents, such as 'town' and 'city', do not appear. Nevertheless, there are clear associations. Wittfogel's thesis proceeds from the basis that 'the characteristics of hydraulic economy are many but three are paramount. Hydraulic agriculture involves a specific type of division of labour. It intensifies cultivation. And it necessitates

cooperation on a large scale' (Wittfogel, 1957, p. 22). Intensification induces the appropriate population concentration, whilst cooperation leads to the need for a managerial and bureaucratic society. Power ultimately will reside in the managers and out of this, despotic control results. 'The effective management of these works involves an organizational web which covers either the whole, or at least the dynamic cores, of the country's population. In consequence, those who control this network are uniquely prepared to wield supreme political power' (Wittfogel, 1957, p. 27). The division of labour, the centralization of power and the administrative structure engender concentrated settlement so that the town emerges. So do all the associated characteristics. Thus Wittfogel maintains that the need for comprehensive works of defence arises as soon as irrigated agriculture is practised, for the whole system is fixed and unmovable and must be defined *in situ*. Again, 'a government apparatus capable of executing all these hydraulic and non-hydraulic works could easily be used in building palaces . . . monuments and tombs. It could be used wherever equalitarian conditions of a primitive tribal society yielded to . . . no longer tribal forms of autocracy' (Wittfogel, 1957, p. 45). Social transformation and monumental buildings are thus consequences of hydraulic agriculture. In short, urbanization itself can be viewed as following upon the development of irrigation. [. . .] To a large extent irrigation can be looked upon as creating the surpluses already discussed and the manipulating managers can be identified as the alienators of the surplus value. [. . .]

It is evident that dependability of production, as well as diversity of resources, are conditions which favour urban settlement. To those conditions irrigated agriculture must have made a clear contribution. There is, however, no agreement on the extent to which the irrigation schemes demanded an extensive bureaucracy. Moreover, there are critical areas such as Mesopotamia and Egypt where there is no evidence of the major irrigation schemes demanded to produce urbanization, as distinct from minor, small-scale, adaptive projects. Adams argues, 'there is nothing to suggest that the rise of dynastic authority in southern Mesopotamia was linked to the administrative requirements of a major canal system' (Adams, 1966, p. 68) and, again, that where Egypt is concerned there are no records of the construction of irrigation other than at a local level or reference to administrative posts connected with irrigation (Butzer, 1976). But even if all such evidence is discounted the whole argument may well be circular for possibly only the nucleated and organized forces of the city could have initiated and carried out the great irrigation schemes which are a consequence, not the creator, of urbanization. Nevertheless, in any argument for urbanization based on surplus production the impact of irrigation is an essential factor to be considered.

Economic theories: the city as market place

Two closely interrelated interpretations can be given to economic theories of urban origins. The first can be called 'mercantile' for it views the city as the product of long-distance trade, whilst the second can be called 'market' for it interprets the city as the centre created by a region to focus its internal processes of exchange. In this consideration no attempt will be made to treat these separately.

The Egyptian hieroglyph for a town was a cross within a circle and this is often seen as symbolizing two dominant functions of the earliest towns. The cross represents the meeting of routes at the market place, while the circle stands for the defensive walls, so that the city as a protected market place is overtly set down. [. . .]

A main exponent of the economic origins of cities is Jane Jacobs who forcibly argues the case which she epitomizes as, 'cities first – rural development later' (Jacobs, 1969, pp. 3–48). By this interpretation agricultural intensification is taken as being a consequence of city growth and not the other way round. 'Both in the past and today, then, the separation commonly made, dividing city commerce and industry from rural agriculture, is artificial and imaginary. The two do not come down different lines of descent. Rural work – whether that work is manufacturing . . . or growing food – is city work transplanted' (Jacobs, 1969, p. 18). In order to illustrate the process of urban origins Jacobs sets up an imaginary city called New Obsidian which deals at first in that stone. 'Since at least 9000 BC', she asserts, 'and possibly earlier the trading of the local obsidian [a volcanic glass from which cutting implements were made] had taken place by custom in the territory of a neighbouring hunting group who had become regular customers for the obsidian and, subsequently, go-betweens in the trade with more distant hunting peoples. It is the settlement of this group that has become the little city of New Obsidian' (Jacobs, 1969, p. 19). This trading post once established generates specialization and once more all the changes associated with urbanization are underway. Critical amongst the items traded is food and this must be non-perishable. It therefore comes in the form of live animals and seed and those responsible for looking after these supplies eventually turn to domesticating the animals and obtaining their own crops from the sown seed; agriculture is created by urban demand. The market place becomes a key feature in the urban layout; 'in the barter space, the two worlds (that of the townsfolk and the alien traders) meet. The square is thus the only "open space" in the city itself, left open originally because what has since become a busy meeting and trading spot was at first a space of separation, deliberately kept empty' (Jacobs, 1969, p. 23).

Jane Jacobs tries to provide a real-world

parallel to her New Obsidian by quoting Catal Huyuk, a Neolithic town in Anatolia whose origins date to 7560 BC. This was excavated by James Mellaart (Mellaart, 1967) and one of the conclusions which Jacobs draws from his evidence is that 'Civilization came directly – without a break – from the hunting life'. That evidence is mainly in the art forms which are strongly related to the Upper Palaeolithic tradition of naturalistic painting carrying the implication that part of the population of Catal Huyuk was of Upper Palaeolithic stock, that is, the old hunting, food-gathering stock preserving the remains of an Upper Palaeolithic heritage. 'Catal Huyuk had a valuable resource and a trade in that resource to be sure, but it had something else valuable and more wondrous. It had a creative local economy. It is this that sets the city apart from a mere trading post with access to a mine. The people of Catal Huyuk had added one kind of work after another into their own local city economy' (Jacobs, 1969, p. 35). It was this which produced the almost explosive development in the arts and crafts which Mellaart records. Mellaart is not drawn into the speculative conclusions which Jane Jacobs derives however and concludes 'it would be premature to speculate further about the ancestry of Catal Huyuk' (Mellaart, 1967, p. 226).

There are at least three reasons why the market origins of towns must be regarded with some scepticism. The first is that in contrast to modern free trading principles, economic relations and trading in the early towns seem to have been controlled by treaty arrangements carried out by 'traders by status' who formed part of the urban bureaucracy. '"Prices" took the form of equivalencies established by authority of custom, statute or proclamation.' Given these conditions then no market place is necessary and there is evidence that this was the case. 'Babylonia, as a matter of fact, possessed neither market places nor a functioning market system of any description' (Polanyi, 1957, p. 116). Jacobs's scenario of the emergence of towns of the New Obsidian variety, therefore, seems to have no basis in fact and, indeed, the large open spaces which she describes as the locations of markets never existed. They are features of much later times set back by her into the remote past.

The second reason for disbelief in the market as the creator of towns is the simple fact that in more recent times both the centre of long-distance trade, the fair, and that of local exchange, the market, have been carried on without bringing permanent settlement into being. Fairs in Britain were often held on open sites remote from actual villages or towns. [. . .]

The third reason is, to some extent, a reiteration of the second. [. . .] If the market need not generate permanent settlement, still less is it likely to precipitate all those complex changes in social and economic structure which are the concomitants of urbanization. Although trade was an important element in the early city its intensification must be seen as an offshoot of urbanism rather than a creator of it.

Military theories: the city as strong-point

It has been noted that the circular element in the Egyptian hieroglyph for a town symbolized a wall or a set of external defences. Accordingly it can be contended that the origin of cities lay in the need for people to gather together in search of protection. Once that agglomeration had been brought about under the oversight of a military caste, then the other changes characterizing urbanization took place. In support it is adduced that defences appear at the same time as towns. Thus Mellaart notes that 'the need for defence may be the original reason for the peculiar way in which the people of Catal Huyuk constructed dwellings without doorways, and with the sole entry through the roof'. Those who were responsible for the layout of the city did not build a solid wall but fringed the settlement with an unbroken line of houses and store-rooms where access was only through the roofs. 'The efficiency of the defence system is obvious and, whatever discomfort it involved for the inhabitants of the city, there is no evidence for any sack or massacre during the 800 years of the existence of Catal Huyuk' (Mellaart, 1967, p. 69). Again, Kathleen Kenyon writing of 'the first Jericho', the pre-pottery Neolithic settlement, records, 'The first settlement is quite clearly on the scale, not of a village, but of a town. Its claim to a true civic status is established by the discovery . . . that it possessed a massive defensive wall' (Kenyon, 1957, pp. 65-6). The equating of a defensive wall with civic status need not necessarily have implications for origin, but the discovery of a town wall with 'its foundations resting on bedrock' certainly does. The wall is associated with a great rock-cut ditch and a massive tower. 'We have, therefore, indications that the elaborate system of defences has a very long history. The earlier stages have not yet been fully traced. But the earliest . . . with the nucleus of the tower and a free-standing stone wall must in itself have been sufficiently magnificent. . . . It belongs to the earliest phase in the history of the town. . . . We have now reached bedrock in several places without finding any suggestion of a yet earlier phase' (Kenyon, 1957, p. 72).[1]

The evidence quoted above suggests that from the earliest identifiable times in the oldest known towns there were elaborate and strong defences, providing grounds for the belief that military necessity might have been a cause for the origin of towns. Even Wheatley from his strongly committed viewpoint is forced to admit that, 'Warfare may often have made a significant contribution to the *intensification* of urban development by inducing a concentration of settlement for purposes of defence and by stimulating craft specialization' (Wheatley, 1971, pp. 298-9). The critical difficulty is that much of the evidence is capable of a comple-

tely circular form of interpretation. From the viewpoint of the believer in the economic surplus, formal and institutionalized warfare is itself the creation of the so-called urban revolution. Tribal raiding and intergroup fighting obviously go back to pre-urban times, but as occasional and sporadic affairs; organized war of the sort which characterized Babylonia, for example, only developed with a city based civilization. [. . .] This militarism follows as a result of those classes who had alienated the economic surplus in the cities, initiating policies of external aggrandizement. War in this sense is not a cause of socioeconomic change but rather is generated by it. If 'protection' is needed, then protection from whom? And if the conflicts were not of a different order why were not towns generated at an earlier date? If the answer is that a certain size and agglomeration of population were needed, then the catalytic role of warfare and defence is lost over against the forces which led to demographic growth and ecological organization. Moreover, although defences appear in the earliest of towns, not all early towns have defences. But there is a great danger of this becoming a 'chicken and egg' type of dispute to which no answer of any certainty can be provided.

It must be accepted that some settlements were from the beginning defensive strongpoints and, accordingly, the close agglomeration and the institutional organization which are a necessary part of elaborate communal defences possibly played a role in the emergence of towns.

Religious theories: the city as temple

[. . .] 'We must if we are to explain the growth, spread and decline of cities, comment upon the city as a mechanism by which a society's rulers can consolidate and maintain their power and, more important, the essentiality of a well developed power structure for the formation and perpetuation of urban centres' (Sjoberg, 1960, p. 67). This is a view closely parallel to that of Wittfogel (see above p. 8) but within the context of the argument of a sacred basis to the town it is fundamental that the power structure developed and became ecologically effective in the hands of the priesthood.

In a more objective sense the process of agglomeration is related to the designation by tribal groups of sacred territories administered by the priesthood. The priests, therefore, in their role of ensuring the safety and security of the people – not by walls against physical enemies, but by rites against the menaces of nature – became the first group to be detached, the first specialized sector of the population. The sacerdotal elite in this way operated as disposers of 'surplus' produce brought as offerings. Change was instigated, not as Jane Jacobs would have it through a free market, but through the concentration of the surplus in priestly hands. Moreover, the dominance of the sacred territory not only brought people to it and bound them to it so that actual physical agglomeration took place, but the same process meant that citizenship, the crux of belonging, was defined by a specific religious territorial allegiance. C.J. Gadd records of the cities of Babylonia, 'It has already been observed as peculiar to this period that it had developed a form of human government which seemed to reproduce upon earth exactly the hierarchy of heaven, so that it was sometimes hardly clear whether gods or men were the acting parties' (Gadd, 1962, p. 46). It is little wonder that cities were so often thought of as being the creations of the gods themselves.

One further line of evidence can be put forward. There is a clear association of shrines and temples with excavated cities. Inevitably no inductive evidence of this order can ever be complete but the predominant role of religious buildings in city layouts cannot be easily put aside. Again, Catal Huyuk as one of the earliest cities can be cited and there, of the 139 living rooms excavated in levels 2 to 10, as many as 40 probably served Neolithic religion.

There can be little doubt that religion played a significant part in that process of transformation which brought cities into being. What is less convincing is why the particular metamorphosis should have taken place at a point or period in time, since belief in some other world and cult practices do not seem to begin with urbanism. Moreover, if markets do not produce permanent settlement neither do shrines – 'the tribal market functioned as the capital of the tribe, especially when the nearby shrine was that of the patron saint of the tribe. Neither market nor shrine, however, was surrounded by either an agglomeration of dwellings, or by civic, religious or commercial buildings' (Fogg, 1939, pp. 322–3). Wheatley, however, argues that the period of urban beginnings was marked by a 'growing distinction, in contrast with tribal religion, between gods and men, a distinction which necessitated the elaboration of a communication system mediated through worship and sacrifice' (Wheatley, 1971, p. 319). The mediators in this way became the alienators, alienating people not from any material surplus but from direct relations with the other-world forces and, in political terms, from each other. Nevertheless, for this to take place, a progression is postulated from terminal food collecting, through primary village farming efficiency to developed village farming efficiency, from which condition the dispersed ceremonial centre emerges and by a process akin to synoecism the compact city develops. That is, ecological and technological progression becomes an essential basis. Again the argument can become circular, for it can be proposed that any one group alienating and expropriating a surplus would exploit the irrationality of the expropriated in order to legitimize the process. The masses accept social transformation because belief is exploited and the early citizen is reconciled to the situation in the same way as the immanence of the next world was used in the nineteenth century to persuade people to accept the inequalities and the miseries of this. Perhaps such a view suggests the relationship postulated between the protestant ethic and the rise of capitalism. But such a notion is a very long way from the proposal of a linear and direct relationship between religion and industrial urbanism. It is equally inappropriate to suggest a relationship of such a character in the context of urban origins.

A summary of theories of urban origins

It is possible to initiate a discussion of the various theories on urban origins by considering, very briefly, the Industrial Revolution and the cities it brought into being. If it does nothing more it will indicate how complex these major socioeconomic transformations are. In school textbooks it was traditional to present the Industrial Revolution through an account of technical innovations, discoveries which 'revolutionized' the modes of production. This has long since been abandoned for it is a narration of the correlates of change rather than its initiator. Again, towns were built quite simply and directly about mine, mill and factory and yet to call these the 'causes' of towns and cities is true only in an elementary way. In order to trace those forces which transformed industry and created towns it would be necessary to review the whole pattern of economic and social change in the preceding centuries. Yet this, too, depends on what is meant by 'origin'. In a direct sense the factory was the town's origin. So it is with the beginnings of urbanism, and much of the debate seems simplistic in the extreme as single, direct causes are sought. Certainly monumental buildings, even defensive walls and temples, are somewhat akin to mines and factories. They are critical indicators of the changes taking place, and of the agglomeration seed, yet in themselves were no more than evidence of large-scale, deeper-lying processes. Undoubtedly the critical factor in the beginning of towns was the gradual progression of technology and a growing density of settled or village population, perhaps lifted to a critical take-off point by the intensity of cultivation made possible by irrigation. Even so in that progression, even without Childe's notion of an urban revolution, a catalyst metamorphosed the old kin-structured, tribal organization into a class-based territorial one. That catalyst was probably the intricately related role of temple, fortress and market place. Perhaps the best summary is Wheatley's:

> It is doubtful if a single autonomous, causative factor will ever be identified in the nexus of social, economic and political transformations which resulted in the emergence of urban forms . . . whatever structural changes in social organization were induced by commerce, warfare or technology, they needed to be validated by some instrument of authority if they were to achieve institutionalized permanence.
>
> (Wheatley, 1971, p. 318)

At this point it is tempting to suggest that the notion of an 'urban revolution' might well be the root cause of much of the difficulty and obscurantism which has characterized the debate on urban origins. The whole topic still rests under the shadow of Childe's proposition for the concept of a single cause is inexorably linked with the idea of a revolution. The analogy with the so-called Industrial Revolution, which has already been made, might well be appropriate for there, too, the view of a revolution simply caused by technological change has long been abandoned. Industrial settlement there certainly was but one wonders whether the causes of its origin, in those simple terms, is a sensible question to ask. Perhaps the time has come to look critically at the concept of an urban revolution and to consider the city as emergent from a longer period of social and economic change and cultural adaptation in which an elaborate complex of factors was mingled.

Note

1 Kenyon's inference that Jericho's first wall was defensive has been challenged. A more recent archaeological interpretation is that it protected the settlement against flooding

References

ADAMS, R.McC. (1966) *The Evolution of Urban Society*, Chicago, Aldine Pub. Co.

BUTZER, K.W. (1976) *Early Hydraulic Civilization in Egypt: A Study in Cultural Ecology*, Chicago, University of Chicago Press

FOGG, W. (1939) Tribal Markets in Spanish Morocco, *Journal Royal African Society* 38, pp. 322–6

GADD, C.J. (1962) 'The Cities of Babylon' in EDWARDS, I.E.S. et al. (eds), *The Cambridge Ancient History*, Cambridge, Cambridge University Press, Vol. 1, Ch. 13, p. 34

HARVEY, D. (1973) *Social Justice and The City*, London, Edward Arnold

JACOBS, J. (1969) *The Economy of Cities*, New York, Random House

KENYON, K. (1957) *Digging up Jericho*, London, Ernest Benn

LAMPARD, E.E. (1965) 'The History of Cities in Economically-Advanced Areas', *Economic Development and Cultural Change*, 3, p. 92

MELLAART, J. (1967) *Catal Huyuk. A Neolithic Town in Anatolia*, London, Thames & Hudson

POLANYI, K. (1957) 'Marketless Trading in Hammurabi's Time', in POLANYI, K. et al. (eds) *Trade and Market in the Early Empires*, New York, Free Press

SJOBERG, G. (1960) *The Pre-industrial City*, New York, Free Press

WHEATLEY, P. (1971) *The Pivot of the Four Quarters*, Edinburgh, Edinburgh University Press

WITTFOGEL, K. (1957) *Oriental Despotism. A Comparative Study of Total Power*, New Haven, Conn., Yale University Press

WOOLLEY, L. (1963) 'The Urbanization of Society', in HAWKES, J. and WOOLLEY, L. (eds) *History of Mankind*, Vol. 1, *Prehistory and the Beginnings of Civilization*, Paris, UNESCO, Pt. 2, Ch. 3

2

BRICKS AND BRICKMAKING IN MUD AND CLAY

by P.R.S. Moorey

Source: P.R.S. Moorey, *Ancient Mesopotamian Materials and Industries*, Oxford, Clarendon Press, 1994, pp. 302, 304–7

Mesopotamia is a land of mud and mudbrick architecture. From the earliest phase of settlement in the region it was appreciated that the soils available there were generally excellent for building purposes. They provide a very acceptable substitute for stone as a building material when moulded and exposed to the intense summer heat of the plain (it can reach 50°C in Babylonia in the shade) for an appropriate length of time, or better still when kiln-fired. Once bricks had been developed, it became general practice to build the mass of a building in sun-dried bricks, whilst facing the lower courses and paving the floors with kiln-fired bricks. In a country short of wood for fuel baked bricks were a luxury, commonly used only where necessary to protect the unfired from erosion by wind or water. There may, however, have been a ritual preference for sun-dried mudbrick in some temple construction from early times.

This preference for brick is not surprising in Babylonia where good building stone is generally absent, but its equal popularity in Assyria is remarkable, since suitable building stones were available close to major settlements or within reach by water transport. 'With stone so plentiful at no great distance, its comparatively sparse use in buildings at Dur Sharrukin [Khorsabad] can be attributed only to the architects' failure to recognize its possibilities or their

desire to stick fast to tradition' (Loud and Altman, 1938, p. 15). It should also be said that timber was more readily available in the north to fire kilns whilst in the south, where fuel was scarce, *terre pisé* [rammed earth] and sun-dried bricks had distinct economic advantages. Mudbrick is very adaptable and easy to use. It is also remarkably durable when properly protected against the weather. [. . .]

The basic limitations of an architecture of *terre pisé* had a profound long-term effect on the builders of ancient Mesopotamia. The laws of gravity and the quality of the workmanship in foundation setting and in ramming technique determine the relationship between height and width in packed earth walls. *Terre pisé* tends to be unstable. Certain fundamental inhibitions survived the introduction of pre-dried and standardized bricks, which made walls lighter and thus capable of being taken higher so long as points and lines of stress were appropriately treated. The real key lay in the proper use of mortar and of kiln-fired bricks:

> The Mesopotamian architect, who used bricks profusely, was always hiding them behind the thick mud facing applied to all walls. He failed to realize that the use of a different type of mortar to set bricks would permit him to increase the height of his walls without making them so thick as to endanger their durability. Eventually . . . mortar was used in combination with fired bricks,

and the technology of arches and domes was transposed from the level of stone block architecture to that of brick architecture. The width of the rooms, hitherto restricted by the span of the roof timbers, or expanded by means of a forest of columns, then increased. A new technique was created, based on the interplay of weight and support, stress and counterstress, structure and fill, and the heavy mud-faced and gaudily painted walls and the massive, piled-up temple towers were replaced, after little more than a millennium, by scintillating walls in enameled and intricately patterned bricks, and by slender towers and graceful domes.

(Oppenheim, 1977, p. 325)

[. . .]

The preferred brick manufacturing month was the 'third' (May–June), immediately after the spring rains, when water would be plentiful and the whole summer lay ahead, if necessary, for drying. Chaff or straw was easily available at this time. The July–August period was characterized as a time of building, when the dryness of the ground would have facilitated foundation-laying. The association of the fire-god with building may arise from this conjunction of intense heat and construction. [. . .]

Mudbricks were commonly produced in rectangular wooden moulds, open at the top and bottom, usually singly, but sometimes in twos or threes. Almost any soil may be used as the medium, though one with a greater clay content is more satisfactory. Generally, red-brown bricks have derived their colour from fresh soil taken from agricultural land, whilst grey bricks are those for which occupational debris was used. Red bricks, however, often appear with grey mortar, perhaps coloured by ash, since settlement debris may have given a stronger bond. Some form of tempering was always necessary to avoid warping and cracking. Chopped straw or dung was most commonly used. It has been calculated that 100 bricks require about 60 kg. of straw (i.e. $\frac{1}{8}$ hectare of barley). The resistance of sun-dried mudbricks to fracture decreases with the decay of the

straw bonding. Pulverized sherds and other mineral matter were sometimes employed. The lime content of many clays in Iraq made them particularly suitable for the manufacture of durable mudbricks. [. . .] When kings were involved in formal ceremonies at the start of a building project, tools of ivory and equipment of precious woods are mentioned:

I (Assurbanipal) made bricks for the (Gula Temple in Babylon) out of cuttings of aromatic plants, in a mould of ebony and *musukannu*-wood. I had (them) wield the hoes, and I saw to the correct laying of the foundations.

As the basic material and temper are puddled with water to a suitable consistency for putting into moulds by hand, the presence of plenty of water is a decisive factor in the location of brickfields. They are commonly, and presumably always were, in or adjacent to a cultivated field beside a canal or river. The brickmakers move regularly so as not to dig too large a cavity in any one cultivated area. The surplus mud is cleaned off the top of the mould by hand. The mould is removed and the brick dried in the sun for a period of time appropriate to the season, with regular turning. [. . .]

A liquid form of the same mixture used for bricks generally serves both for *mortar* and for wall *plaster*.

The making and laying of bricks for public buildings, especially temples, is known from textual sources to have been accompanied by ceremonies and ritual to propitiate the gods, including a specific brick god, and to create the most favourable circumstances, especially for the crucial process of making the first brick. For each new project unbaked mudbricks had to be freshly made, as they cannot be salvaged from old buildings. Written evidence indicates that such rituals attended work on simple houses as well.

As Delougaz (1933, pp. 6–7) pointed out, it is by no means easy to estimate the degree of standardization in sun-dried mudbrick manufacture from observation of the results: 'brickmaking does not require any special technical knowledge, so that practically every villager does it occasionally. Of course there are some men in every village specially skilled in the making and handling of mudbricks.' Delougaz recorded that his best brickmaker in the Diyala [a tributary of the river Tigris] in the early 1930s held a record for making almost three thousand a day. 'If we attribute approximately the same number to brickmakers in antiquity, we can understand why rather large sections of buildings are made of bricks of identical size, not through any intentional standardization of size, but simply because one brickmaker turned out thousands of bricks from his own particular frame. Therefore only in large buildings, where of necessity many brickmakers had to be employed, do we find that bricks of different sizes and proportions were used during the same period.' Whatever the rate may have been this was, in general, a simple but slow method of production. [. . .]

In structures, especially large public buildings, the proportions of sun-dried and kiln-fired bricks varied. In general, often to a surprising degree, sun-dried were preferred to kiln-fired. It appears that for both court and temple, as well as for private persons, the high cost of fuel made kiln-fired bricks so expensive to manufacture that their use was minimized. Moisture is a great threat to mudbrick. In southern Iraq, where the ground water is extremely saline, rising damp is a persistent danger. Salts, drawn up into a wall by capillary action, recrystallize as they dry causing the brickwork to crumble. Thus whoever could afford it employed baked brick and bitumen for damp-courses. Elsewhere kiln-fired bricks were used only in areas requiring protection from water, such as water

basins and drainage channels, or where resistance to wear, as in thresholds, entrance halls, or courtyards, was particularly needed. They were also used for door-sockets in houses; stone being preferred for this in major public buildings. In some contexts where bitumen might be anticipated, baked brick served, as in the early second-millennium BC houses at Ur. In ziggurats open channels were regularly provided in the core, where courses of sun-dried brick generally alternated at set intervals with layers of reeds to create a ventilation system that counteracted the effect of infiltrating humidity and thus conserved the integrity of the brickwork. [. . .]

The emergence of the widely distributed 'tripartite' plan [entrance/vestibule; reception area; private rooms] for houses and temples in the Ubaid period reinforces the argument that new levels of social organization now affected the builder's craft across the whole of Mesopotamia. Fairly elaborate systems of bricklaying had already been evolved by the time of the emergence in the archaeological record of the first major monumental buildings in the south (Eridu, Uqair, and Uruk), with an interplay of projecting piers and wall recesses that probably owes much to earlier building there in reeds and palmwood.

References

DELOUGAZ, P (1933) *Plano-convex Bricks and the Methods of their Employment*, Studies in Ancient Oriental Civilization, 7, Oriental Institute Publications, Chicago

LOUD, G. and ALTMAN, C.B. (1938) *Khorsabad II: The Citadel and the Town*, Oriental Institute Publications 40, Chicago

OPPENHEIM, A.L. (1977) *Ancient Mesopotamia*, rev., 2nd edn, Chicago, University of Chicago Press

3

HISTORY

by Herodotus

Source: Herodotus, *The History of Herodotus*, in *Great Books of the Western World*, 6: *Herodotus, Thucydides*, Encyclopaedia Brittanica, 1952, pp. 40-4, 75-6, 102

Book I

178. Assyria possesses a vast number of great cities, whereof the most renowned and strongest at this time was Babylon, whither, after the fall of Nineveh, the seat of government had been removed. The following is a description of the place: – The city stands on a broad plain, and is an exact square, a hundred and twenty furlongs in length each way, so that the entire circuit is four hundred and eighty furlongs. While such is its size, in magnificence there is no other city that approaches to it. It is surrounded, in the first place, by a broad and deep moat, full of water, behind which rises a wall fifty royal cubits in width, and two hundred in height. (The royal cubit is longer by three fingers' breadth than the common cubit.)

179. And here I may not omit to tell the use to which the mould dug out of the great moat was turned, nor the manner wherein the wall was wrought. As fast as they dug the moat the soil which they got from the cutting was made into bricks, and when a sufficient number were completed they baked the bricks in kilns. Then they set to building, and began with bricking the borders of the moat, after which they proceeded to construct the wall itself, using throughout for their cement hot bitumen, and interposing a layer of wattled reeds at every thirtieth course of the bricks. On the top, along the edges of the wall, they constructed buildings of a single chamber facing one another, leaving between them room for a four-horse chariot to turn. In the circuit of the wall are a hundred gates, all of brass, with brazen lintels and side-posts. The bitumen used in the work was brought to Babylon from the Is, a small stream which flows into the Euphrates at the point where the city of the same name stands, eight days' journey from Babylon. Lumps of bitumen are found in great abundance in this river.

180. The city is divided into two portions by the river which runs through the midst of it. This river is the Euphrates, a broad, deep, swift stream, which rises in Armenia, and empties itself into the Erythræan sea[1]. The city wall is brought down on both sides to the edge of the stream: thence, from the corners of the wall, there is carried along each bank of the river a fence of burnt bricks. The houses are mostly three and four stories high; the streets all run in straight lines, not only those parallel to the river, but also the cross streets which lead down to the water-side. At the river end of these cross streets are low gates in the fence that skirts the stream, which are, like the great gates in the outer wall, of brass, and open on the water.

181. The outer wall is the main defence of the city. There is, however, a second inner wall, of less thickness than the first, but very little inferior to it in strength. The centre of

each division of the town was occupied by a fortress. In the one stood the palace of the kings, surrounded by a wall of great strength and size: in the other was the sacred precinct of Jupiter Belus [Bel, chief deity of the later Babylonian pantheon], a square enclosure two furlongs each way, with gates of solid brass; which was also remaining in my time. In the middle of the precinct there was a tower of solid masonry, a furlong in length and breadth, upon which was raised a second tower, and on that a third, and so on up to eight. The ascent to the top is on the outside, by a path which winds round all the towers. When one is about half-way up, one finds a resting-place and seats, where persons are wont to sit some time on their way to the summit. On the topmost tower there is a spacious temple, and inside the temple stands a couch of unusual size, richly adorned, with a golden table by its side. There is no statue of any kind set up in the place, nor is the chamber occupied of nights by any one but a single native woman, who, as the Chaldæans, the priests of this god, affirm, is chosen for himself by the deity out of all the women of the land. [. . .]

184. Many sovereigns have ruled over this city of Babylon, and lent their aid to the building of its walls and the adornment of its temples, of whom I shall make mention in my Assyrian history. Among them two were women. Of these, the earlier, called Semiramis [Sammuramat, queen of Assyria during the late ninth century BCE], held the throne five generations before the later princess. She raised certain embankments well worthy of inspection, in the plain near Babylon, to control the river, which, till then, used to overflow, and flood the whole country round about.

185. The later of the two queens, whose name was Nitocris [identity uncertain; possibly wife of Nebuchadnezzar II], a wiser princess than her predecessor, not only left behind her, as memorials of her occupancy of the throne, the works which I shall presently describe, but

also, observing the great power and restless enterprise of the Medes, who had taken so large a number of cities, and among them Nineveh, and expecting to be attacked in her turn, made all possible exertions to increase the defences of her empire. And first, whereas the river Euphrates, which traverses the city, ran formerly with a straight course to Babylon, she, by certain excavations which she made at some distance up the stream, rendered it so winding that it comes three several times in sight of the same village, a village in Assyria, which is called Ardericca; and to this day, they who would go from our sea to Babylon, on descending to the river touch three times, and on three different days, at this very place. She also made an embankment along each side of the Euphrates, wonderful both for breadth and height, and dug a basin for a lake a great way above Babylon, close alongside of the stream, which was sunk everywhere to the point where they came to water, and was of such breadth that the whole circuit measured four hundred and twenty furlongs. The soil dug out of this basin was made use of in the embankments along the waterside. When the excavation was finished, she had stones brought, and bordered with them the entire margin of the reservoir. These two things were done, the river made to wind, and the lake excavated, that the stream might be slacker by reason of the number of curves, and the voyage be rendered circuitous, and that at the end of the voyage it might be necessary to skirt the lake and so make a long round. [. . .]

186. While the soil from the excavation was being thus used for the defence of the city, Nitocris engaged also in another undertaking, a mere by-work compared with those we have already mentioned. The city, as I said, was divided by the river into two distinct portions. Under the former kings, if a man wanted to pass from one of these divisions to the other, he had to cross in a boat; which must, it seems

to me, have been very troublesome. Accordingly, while she was digging the lake, Nitocris bethought herself of turning it to a use which should at once remove this inconvenience, and enable her to leave another monument of her reign over Babylon. She gave orders for the hewing of immense blocks of stone, and when they were ready and the basin was excavated, she turned the entire stream of the Euphrates into the cutting, and thus for a time, while the basin was filling, the natural channel of the river was left dry. Forthwith she set to work, and in the first place lined the banks of the stream within the city with quays of burnt brick, and also bricked the landing-places opposite the river-gates, adopting throughout the same fashion of brickwork which had been used in the town wall; after which, with the materials which had been prepared, she built, as near the middle of the town as possible, a stone bridge, the blocks whereof were bound together with iron and lead. In the daytime square wooden platforms were laid along from pier to pier, on which the inhabitants crossed the stream; but at night they were withdrawn, to prevent people passing from side to side in the dark to commit robberies. When the river had filled the cutting, and the bridge was finished, the Euphrates was turned back again into its ancient bed; and thus the basin, transformed suddenly into a lake, was seen to answer the purpose for which it was made, and the inhabitants, by help of the basin, obtained the advantage of a bridge. [. . .]

193. But little rain falls in Assyria, enough, however, to make the corn begin to sprout, after which the plant is nourished and the ears formed by means of irrigation from the river. For the river does not, as in Egypt, overflow the corn-lands of its own accord, but is spread over them by the hand, or by the help of engines. The whole of Babylonia is, like Egypt, intersected with canals. The largest of them all, which runs towards the winter sun, and is impassable except in boats, is carried from the Euphrates into another stream, called the Tigris, the river upon which the town of Nineveh formerly stood. Of all the countries that we know there is none which is so fruitful in grain. It makes no pretension indeed of growing the fig, the olive, the vine, or any other tree of the kind; but in grain it is so fruitful as to yield commonly two-hundred-fold, and when the production is the greatest, even three-hundred-fold. The blade of the wheat-plant and barley-plant is often four fingers in breadth. As for the millet and the sesame, I shall not say to what height they grow, though within my own knowledge; for I am not ignorant that what I have already written concerning the fruitfulness of Babylonia must seem incredible to those who have never visited the country. The only oil they use is made from the sesame-plant. Palm-trees grow in great numbers over the whole of the flat country, mostly of the kind which bears fruit, and this fruit supplies them with bread, wine, and honey. They are cultivated like the fig-tree in all respects, among others in this. The natives tie the fruit of the male-palms, as they are called by the Greeks, to the branches of the date-bearing palm, to let the gall-fly enter the dates and ripen them, and to prevent the fruit from falling off. The male-palms, like the wild fig-trees, have usually the gall-fly in their fruit.

194. But that which surprises me most in the land, after the city itself, I will now proceed to mention. The boats which come down the river to Babylon are circular, and made of skins. The frames, which are of willow, are cut in the country of the Armenians above Assyria, and on these, which serve for hulls, a covering of skins is stretched outside, and thus the boats are made, without either stem or stern, quite round like a shield. They are then entirely filled with straw, and their cargo is put on board, after which they are suffered to float down the stream. Their chief freight is wine, stored in casks made of the wood of the palm-tree. They are managed by two men who stand

upright in them, each plying an oar, one pulling and the other pushing. The boats are of various sizes, some larger, some smaller; the biggest reach as high as five thousand talents' burthen. Each vessel has a live ass on board; those of larger size have more than one. When they reach Babylon, the cargo is landed and offered for sale; after which the men break up their boats, sell the straw and the frames, and loading their asses with the skins, set off on their way back to Armenia. The current is too strong to allow a boat to return upstream, for which reason they make their boats of skins rather than wood. On their return to Armenia they build fresh boats for the next voyage. [. . .]

Book II

124. Till the death of Rhampsinitus [probably one of the Pharaohs named Rameses], the priests said, Egypt was excellently governed, and flourished greatly; but after him Cheops succeeded to the throne, and plunged into all manner of wickedness. He closed the temples, and forbade the Egyptians to offer sacrifice, compelling them instead to labour, one and all, in his service. Some were required to drag blocks of stone down to the Nile from the quarries in the Arabian range of hills; others received the blocks after they had been conveyed in boats across the river, and drew them to the range of hills called the Libyan. A hundred thousand men laboured constantly, and were relieved every three months by a fresh lot. It took ten years' oppression of the people to make the causeway for the conveyance of the stones,[2] a work not much inferior, in my judgment, to the pyramid itself. This causeway is five furlongs in length, ten fathoms wide, and in height, at the highest part, eight fathoms. It is built of polished stone, and is covered with carvings of animals. To make it took ten years, as I said - or rather to make the causeway, the works on the mound where the pyramid

stands, and the underground chambers, which Cheops intended as vaults for his own use: these last were built on a sort of island, surrounded by water introduced from the Nile by a canal. The pyramid itself was twenty years in building. It is a square, eight hundred feet each way, and the height the same, built entirely of polished stone, fitted together with the utmost care. The stones of which it is composed are none of them less than thirty feet in length.

125. The pyramid was built in steps, battlement-wise, as it is called, or, according to others, altar-wise. After laying the stones for the base, they raised the remaining stones to their places by means of machines formed of short wooden planks[3]. The first machine raised them from the ground to the top of the first step. On this there was another machine, which received the stone upon its arrival, and conveyed it to the second step, whence a third machine advanced it still higher. Either they had as many machines as there were steps in the pyramid, or possibly they had but a single machine, which, being easily moved, was transferred from tier to tier as the stone rose - both accounts are given, and therefore I mention both. The upper portion of the pyramid was finished first, then the middle, and finally the part which was lowest and nearest the ground. There is an inscription in Egyptian characters on the pyramid which records the quantity of radishes, onions, and garlic consumed by the labourers who constructed it; and I perfectly well remember that the interpreter who read the writing to me said that the money expended in this way was 1600 talents of silver. If this then is a true record, what a vast sum must have been spent on the iron tools[4] used in the work, and on the feeding and clothing of the labourers, considering the length of time the work lasted, which has already been stated, and the additional time - no small space, I imagine - which must have been occupied by the quarrying of the stones,

their conveyance, and the formation of the underground apartments.

[. . .]

Book III

60. I have dwelt the longer on the affairs of the Samians, because three of the greatest works in all Greece were made by them. One is a tunnel, under a hill one hundred and fifty fathoms high, carried entirely through the base of the hill, with a mouth at either end. The length of the cutting is seven furlongs - the height and width are each eight feet. Along the whole course there is a second cutting, twenty cubits deep and three feet broad, whereby water is brought, through pipes, from an abundant source into the city. The architect of this tunnel was Eupalinus, son of Naustrophus, a Megarian. Such is the first of their great works; the second is a mole in the sea, which goes all round the harbour, near twenty fathoms deep, and in length above two furlongs. The third is a tem-

ple; the largest of all the temples known to us, whereof Rhœcus, son of Phileus, a Samian, was first architect. Because of these works I have dwelt the longer on the affairs of Samos.

Notes

1 Literally, the 'Red Sea', though the term as used by Herodotus covers a lot more of the waters of the Indian Ocean than it does now; in this instance, we should say the Arabian (or Persian) Gulf

2 Although most Egyptologists are agreed that ramps were constructed for dragging blocks of stone, Herodotus is wrong to identify as such a ramp the monumental causeway joining a temple in the valley with the pyramid (Alan B. Lloyd, 'Herodotus on Egyptian Buildings; a test case', in A. Powell, (ed.), *The Greek World*, London, Routledge, 1995, p. 276)

3 His supposition that stones were raised into place by wooden machines was probably inspired by the Greeks' use of cranes for such tasks; there is no evidence that the ancient Egyptians used them (ibid., p. 277). See, however, Reading 7, pp. 44–5, for an alternative explanation.

4 His reference to the Egyptian masons' use of iron tools is anachronistic; their metal tools would have been made of copper (ibid., pp. 276–7)

4

DEIR EL-MEDINA

by A.R. David

Source: A.R. David, *The Pyramid Builders of Ancient Egypt: a modern investigation of the Pharaohs' workforce*, London, Routledge and Kegan Paul, 1986, pp. 56-9, 62, 64-7, 75-7

Much of our knowledge of ancient Egypt is derived from tombs and temples - from the scenes carved and painted on the walls, and from the artefacts buried in the tombs. These sites, built of stone to last for eternity, are well preserved and in many cases have not been built over in more recent times. However, the towns - or settlement sites, as they are called - should also play a major role in any consideration of Egypt's society and civilisation. Because they were built of mud-brick, and in many cases have successive levels of occupation, they have survived less well and have therefore not received the same degree of attention as the funerary and religious monuments. Nevertheless, they are of vital importance in providing a more complete picture of life in Egypt.

Some archaeologists have suggested that true urban development never existed on a widespread scale in ancient Egypt; that because of environmental, political and religious systems, the walled city with different building levels and continuous settlement was not found throughout the country. It has been argued that, because a stable centralised monarchy had been established by King Menes at the beginning of the 1st Dynasty, subsequently there was no real political need for walled towns, and that the natural barriers provided by the deserts and mountains protected Egypt from most external threats. Thus, the need for true towns was limited to areas where products

entered Egypt or along the east-west trade route involving the Red Sea and the oases in the Western Desert. In between, it is claimed, there was only a string of small 'harbours' along the Nile, in place of substantial town development. Each nome or district also had its modest centre, where the offices and houses for the administrators and officials were situated. The royal capital, or centre of government, was moved by different kings from one site to another; kings also had numbers of residences scattered around Egypt to which they paid occasional visits. In addition to housing the officials, all these urban developments also attracted craftsmen, traders and farmers to supply food for the townspeople. However, most of Egypt's resources were directed towards the construction of temples, tombs, and especially the king's mortuary complex, rather than to the towns, and this state of non-urbanisation persisted until the New Kingdom. By contrast, in Mesopotamia, urbanism was highly developed from the earliest times, and the city-state persisted as the most important element of the society.

The disappearance of the mud-brick towns in Egypt, which have been lost either because they have disappeared under the alluvial mud of the inundation or because the bricks have been removed by successive generations of locals for use as fertiliser, has not facilitated a correct assessment of the number and size of real towns in Egypt in the earlier periods. However, sufficient evidence exists to allow the concept of non-urbanism to be strongly contested by other archaeologists. It has been claimed that, as well as the major sites such as Memphis, even in the Old Kingdom, at sites in Upper Egypt such as Edfu, Abydos and Thebes, it is evident that there were walled towns of various sizes and types, and that towns existed at most or all of the places known from other sources to be administrative centres. These towns were occupied by officials with local duties, agricultural workers, and craftsmen and did not merely exist as the result of socio-economic conditions. In some cases they were specifically created to house personnel associated with temples or other monuments, at the behest of the government. This alternative concept of urbanism in Egypt presents a view of a country with an ordinary pattern of town development, rather than the sparse townships previously suggested. However, although they were obviously an important element in the society, the existence and development of settlement sites is a subject which does not find adequate coverage in the surviving inscriptional sources on which much of our understanding of the society is based.

Although the quantity, importance and spread of settlement sites in Egypt is disputed, it is apparent that two main types of urban development occurred. One was the natural, unplanned growth which evolved from the conditions of the predynastic villages. The second was a planned growth; certain towns were initiated for specific reasons in particular areas; they continued for the duration of the project, and were finally abandoned. Because their location was dependent upon the site of the project, they were not natural choices for continuing occupation and therefore were not levelled down for re-settlement. Some of these towns were built to house the royal workmen engaged on the building, decoration and maintenance of the king's funerary com-

plex, and although they are not the oldest settlements, they are particularly important and also relatively well preserved.

In the Old Kingdom, there would certainly have existed large workforces to build the pyramids, but no details remain of how these men were controlled and managed, nor where they were housed. Most of the labourforce was made up of conscripted peasants. In theory, every Egyptian was liable to perform corvée-duty and was required to work for the state for a certain number of days each year. The wealthier evaded the duty by providing substitutes or paying their way out of the obligation, so it was the peasants who effectively supplied this labour. At first, their duties consisted of building and maintaining the network of irrigation systems, but since the land was annually covered with water for several months because of the inundation, the peasants were later gainfully employed during this period on the construction of tombs (especially the royal tomb or pyramid), and on the temples. This not only provided them with food (they were paid in kind) at a time when agricultural work was impossible, but also acted as a focus for the use of Egypt's manpower and resources. It has been suggested that the early use of the labourforce in this way – when they were engaged in building the king's pyramid during the Old Kingdom – provided a strong, unifying factor which enabled Egypt to develop as a powerful centralised state. Each pyramid project acted as a political, social and economic focus, and also, in the Old Kingdom, as a potent religious force, since at that time it was believed that individual eternity could only be attained through the king's own ascension to the heavens after death. Thus, the labourers and craftsmen sought to ensure their own eternity through service in constructing the royal tomb. The peasants were not slaves in the strict sense of the word, although their freedom and choice of action in terms of their place and type of work was strictly limited by social and economic factors.

Although most of the workforce were conscripted peasants, even in the Old Kingdom there would have been professional craftsmen and architects responsible for the more detailed work on the funerary complex, and these would have been housed near the pyramid sites. However, although Petrie maintained that he had uncovered some traces near Giza of the barracks in which the workforce was housed, no complete workmen's town of the Old Kingdom has yet been discovered. The earliest example of a purpose-built royal workmen's town which has so far been revealed is Kahun [situated by the al-Fayyum oasis in the Western Desert]. However, two other towns of a similar type have been excavated – the site known as Deir el-Medina, built to house the workmen engaged in building and decorating the royal tombs during the New Kingdom at Thebes, and that at Tell el-Amarna, the capital city with its associated rock tombs built by the heretic Pharaoh, Akhenaten, at the end of the 18th Dynasty.

Although these towns were constructed at different periods, they shared a common functional purpose, and certain physical and environmental characteristics.

The three towns were conceived and built to a predetermined plan and none grew out of any previous random settlement. Each was enclosed by a thick brick wall, designed to confine the workmen and their families to a certain area. It is evident that the sites, all on the desert edge, were chosen because they were near to the worksite, but also because, isolated and surrounded by hills, they could be guarded. The inhabitants, after all, had knowledge of the position and structure of the royal tomb, and it was essential that such knowledge was retained as effectively as possible within an enclosed community. Planned within boundary walls, there is some evidence both at Amarna and at Deir el-Medina (occupied over a much

longer period) that random growth of houses eventually occurred outside the walls. It is also noteworthy that even the proximity of a good water supply was not considered essential to these town sites, the requirements of isolation and security being greater. It is evident in all the towns that they were built to conform to a definite plan, with walls dividing them in a north/south orientation. Constructed for speed and efficiency, they are distinguished by regular rows of terraced houses for the workforce, and at Kahun and Deir el-Medina, some officials were also resident. At Amarna, there were 74 houses, at Kahun 100, and 140 at Deir el-Medina. [. . .]

The community of workmen at Deir el-Medina was involved in the construction and decoration of the kings' tombs. Together with their wives and families, they occupied the town for some 450 years, from the beginning of the 18th Dynasty to the end of the 20th Dynasty. [. . .]

In its heyday, the town supported a thriving community, and the evidence which is preserved to us in the houses, the nearby tombs, and the papyri, ostraca [sherds of pottery or limestone used for writing on], stelae [inscribed slabs or building surfaces] and inscriptions provide an unparalleled source of detailed information. Unlike Kahun or Amarna, the excavations here have yielded large quantities of written material – official records, private letters, and literary texts – making it possible to identify the names of the workmen, their wives and their families, the houses they occupied, their legal transactions, and their working conditions. In addition, the artists' sketches on limestone flakes and potsherds which have been found in large quantities here provide a vivid insight into their everyday concerns and show us something of their humour. [. . .]

The physical arrangement of the village of Deir el-Medina closely resembled the situation at Amarna and at Kahun. Situated on the west bank near Thebes, it was hidden from the river; the only approach was along a narrow road which ran in a north to south direction through the valley, and this must have been the route taken by servants bringing supplies to the village and by members of the community who went to trade goods with those living in the cultivation. However, like Kahun and Amarna, its isolation afforded ease of surveillance, and it was also near to the worksite – considerations more important than proximity to a good water supply. The town had no well and the water had to be brought from the river which was about 1.61 km away. Special parties went to fetch the water supplies on donkeys, and it was then kept in a large tank outside the north gate, under the watchful eye of a special guard. From this, women drew their rations and kept them in large pots at the entrance to the house, to be used for domestic requirements.

The original town was enclosed inside a thick mud-brick wall. There is little information about the layout of the town within this wall during the 18th Dynasty, although it seems that not all the area was taken up with buildings at that time. This first village was destroyed by fire, perhaps during the Amarna Period, but under the restoration of King Horemheb, when the Court had already returned to Thebes and there was a return to earlier traditions, Deir el-Medina expanded. New houses were built, the damaged ones restored, and in the 19th Dynasty, the community reached its zenith. Then, the village occupied an area some 132 metres long and 50 metres wide. The area within the wall contained seventy houses, but other houses, numbering between forty and fifty, grew up outside the wall. Although there were over 400 years of occupation, there was no evidence of continuous levels of building within the enclosure walls and the floor levels of the houses were never raised. The expansion in numbers of the community was dealt with by erecting houses for them outside the walls, some of which were built amongst and

over early tombs. This was perhaps the result of official policy not to extend the village itself. Certainly, the enclosure wall, originally conceived as a means of isolating and perhaps defending the community, eventually became a distinction between two social levels, with the descendants of the original inhabitants living inside the walls and considering themselves a superior category, while the less privileged remained outside.

The original village was bisected by a main north to south street; with expansion in the number of houses, further side roads were created. The houses were arranged in blocks and the village presented a rectangular appearance, within its enclosure wall. The houses were all terraces; the earliest ones had no foundations and were built entirely of mud-brick, but the later ones, built on rubble, had basements of stone or brick and stone walls topped with mud-brick. Stone was also used for some of the thresholds. The houses had just one storey, and the roofs were always flat, constructed of wooden beams and matting, with small holes left to let in some light. Wood (from the date-palm, sycamore, acacia and carob trees) was readily available here, unlike Kahun, where the mud-brick roofs were vaulted. At Deir el-Medina, stone or wood was also used for door-frames, and the small natural light supply was doubtless augmented by leaving the wooden doors ajar. Hieroglyphic texts, painted in red, often occur on the door jambs and lintels and make it possible to identify some of the occupants. It seems that the houses were originally assigned to individual tenants by the government, but that, over the years, their families took on the tenancies on a hereditary basis.

Generally, the houses were quite cramped and dark. They all opened directly on to the street, and followed a basic pattern, although variations reflected some difference in status and wealth. The more affluent might plaster and whitewash their outside walls and paint their doors red but the floors were all made of earth. An average house had four rooms. An entrance hall led off the street, and here there was a brick structure which resembled a four-poster bed, often decorated with painted figures of women and the household god, Bes. This has been variously described as a 'birth-bed', or an altar. Here, also, there were niches in the walls to contain painted stelae, offering tables and ancestral busts, and the area probably acted as a household chapel.

A second room led off this; this was the main living room and it was higher than the first, with one or more columns to support the roof. Here, there was also a low brick platform, probably used as a sitting or sleeping divan. Small windows set high in the walls provided this room with some light. In a few houses, the archaeologists discovered child burials beneath this main room, a custom also encountered at Kahun.

One or two small rooms were entered from the main room, which were probably used as sleeping and storage quarters. At the back of the house, a kitchen was situated, consisting of a walled, open area, with storage bins, a small brick or pottery oven in which to bake bread, an open hearth and an area for grinding the grain supplied as payment to the workmen. A staircase to the roof led from the first, second or fourth rooms, and cellars under some of the houses were used to store possessions.

The walls of the rooms were either decorated with frescoes or whitewashed; the columns, window and door surrounds were coloured, and blue and yellow were favourite choices. Although Deir el-Medina has not provided the wealth of domestic artefacts which Petrie uncovered at Kahun, since most of the furniture had disappeared from the village, nevertheless, scenes on the walls of the workmen's tombs show the type of furniture which would have been used in the houses. Furniture was sparse and functional, and would have included stools, tables, and headrests for sleep-ing, and chests, boxes, baskets and jars to contain the family's possessions.

The accommodation of the workmen was not, however, spartan, and reflected their comparative affluence as highly skilled craftsmen in a society that valued their importance as builders of tombs and artificers of funerary equipment. [. . .]

The selection of the site of the royal tomb was the responsibility of a royal commission, headed by the vizier. The various phases of constructing and decorating the tomb were carried out in close concert, but there was clear specialisation of the various crafts, with one phase following another in each area of the tomb. The men were supplied with tools by the state, and we have seen that certain posts carried the responsibility of handing these out as required, and attending to those that needed repair. In addition to the copper tools, wicks and oil or fat were supplied to provide artificial lighting. Salt, it has been suggested, was added to prevent the wicks from smoking.

The construction of the royal tomb was obviously a major undertaking, commenced as soon as a ruler ascended the throne. If the king's tomb was finished in good time, then the workforce was employed on tombs in the Valley of the Queens, and in some cases, on those of favoured courtiers. However, in most cases, the nobles had to obtain craftsmen from elsewhere and doubtless employed some of those artisans from Deir el-Medina who were unable to find a place on the royal gang.

The workmen were also much concerned with their own burial places. Finds at Deir el-Medina indicate that many of the inhabitants were prosperous and had aspiring social ambitions. [. . .] There is never any indication that they were slaves, or even serfs, and although their total freedom of habitation and movement was evidently restricted, there is no indication that there was any strict regulation of their domestic lives or religious practices.

Indeed, they had considerable opportunities

to augment their income and attain a comfortable standard of living. They accepted commissions for private work, and such was the demand for tomb equipment and funerary inscriptions that they could regularly increase their income. They held privileged positions, and some were able to acquire their own expensive metal tools, to own land and livestock, and slaves.

Their relative affluence is nowhere better illustrated than in their tombs which they built in the nearby cliffs. They devoted much of their spare time to these, which provide a valuable source of information, for, although the tombs were usually plundered, the walls, beautifully decorated with scenes of the deceased in the underworld in the company of the gods, provide information and frequently give the name and position held by the workman.

The tombs were built on a basic plan, with a small open courtyard and a vaulted chapel surmounted by a brick pyramid topped with a pyramidion [like the pyramidal apex of an obelisk]. At a time when the kings had abandoned pyramids in favour of rock-cut tombs, the wealthier necropolis workmen were incorporating miniature pyramids in their funerary monuments. A shaft in the courtyard led to an underground passage and a vaulted burial chamber. Stelae were set on the side of the mud-brick walls, and in the courtyard there was a large stela which commemorated the deceased and depicted his funeral.

The tomb equipment included coffins, canopic jars [used to hold the entrails of embalmed bodies], ushabtis [figurines of the deceased], statuettes, furniture, clothing, jewellery, tools, pottery and stone vessels. The villagers made their own tomb equipment as well as that for sale to private buyers. Amongst themselves, they paid each other for required items. They also acted as embalmers and priests at each other's funerals. The tombs that have survived well date mostly to the early Ramesside period [19th and 20th Dynasties, during which rulers frequently used the 'birth name' Rameses], when conditions were stable, but tombs were used at different periods for family burials and passed on in Wills.

5

HEAVY TRANSPORT IN CLASSICAL ANTIQUITY

by A. Burford

Source: A. Burford, 'Heavy Transport in Classical Antiquity', *Economic History Review*, 2nd series, Vol. 18, 1960, pp. 1–12, 16–18

[I]

The question of moving heavy loads bears on some aspects of the economic structure of ancient Greek society, and on the way in which it was able to answer mechanical problems. So far this topic has been virtually untouched.

Lefebvre des Noëttes' discussion of ancient harness dismissed heavy transport as a minor issue. It was a rare activity, with little bearing on general conditions and available methods, and an inefficient one.[1] His argument, henceforth the orthodox view, has recently been restated in the *History of Technology*.[2] It maintains that the kind of harness in use in Europe until the tenth century AD only allowed the animal to exert a part of his total strength, so that the maximum load which could be pulled by one yoke of animals was about 1100 lb.; and that there was no practicable means known of increasing power by multiple yoking in file. The argument is based on a wide range of archaeological and literary evidence. Taking together the predominance of horse-chariots in decoration, and literary evidence for low loads as a general rule, the orthodox view asserts that ancient transport was permanently hampered by people's inability to remedy a drastic impediment. This lay in the tendency of the throat-and-girth harness to ride up against the horse's windpipe, choking it as

soon as it tried to exert its full strength. The harness failed to fulfil its function, which, as Lefebvre des Noëttes says, is to 'permit the complete utilisation of the force of all the animals, so that they may work as a team'.[3] The result of this inefficiency was to reduce the power achieved to about one third of that of modern draught animals.

As for multiple yoking, the evidence is scanty, and concerned with extraordinary instances; and on the strength of this Lefebvre des Noëttes declares that 'if the Greeks had known of a method of multiple yoking, other people would have been obliged to adopt it too.'[4]

These reasons are given for the lack of development: the ancient world was uninventive and not interested in improvement, and the use of slave-power rendered technological experiment unnecessary. Lefebvre des Noëttes says that slavery, 'a fatal consequence of the lack of motive power, was the bane of ancient society'.[5] In effect, heavy transport was impeded by bad harness and bad roads; and this in turn impeded the development of better roads and better harness. If there were to be any progress, then one factor must improve first. R.J. Forbes maintains that 'at each stage of technological development the availability of prime movers, such as a type of machinery which supplies motive power for other tools or machinery, is the keystone'. But there was no such development because 'manpower was always readily available in antiquity', and the reason why 'harnessed animals did not largely take over the part played by human labour' is the 'insufficient knowledge of animal anatomy, which caused the ancients to use ox-harness for donkeys, mules and horses too with disastrous effect'.[6] [. . .]

[II]

The chief objections to the orthodox view are several. First, it ignores almost entirely, as I shall point out more fully below, the inscribed building-accounts which yield the most important evidence we have of heavy transport. And secondly it does not start at the right end of the problem: one should ask, not what limitations were imposed by prevailing transport conditions, but what demands were made upon the means available, i.e. what was in fact transported. My argument runs upon these lines. First, heavy transport was undertaken much more frequently than the orthodox view maintains. Secondly, it achieved its purpose – that is to say, ancient transport though inefficient by modern standards was effective by any standard. Thirdly, draught-animals and not men provided the motive power. The orthodox view is right in maintaining that ancient harness was unsuited to the horse – but this is not relevant to a study of heavy transport because, as I shall demonstrate, the ox was the working animal, not the horse. It is also right to say that no change in method took place throughout antiquity, but to hold this up as a radical fault in the ancient system shows a basic misconception of the nature of technological development, and of the ancients' use of the methods to hand.

What transport did people require?

No city was self-sufficient. All supplies, of food and materials for ship-building, house-building, and industry – such as wood, stone,

wool, metal, and potter's clay – had to be brought in either from the surrounding countryside, or from overseas. Land transport was always necessary in the first case, and often so in the second, since many cities lay some miles inland from their ports. For example, Argos and Corinth are about 5 miles from the coast, and Athens is 7 miles from the Piraeus. We know too that Athens depended on corn imported from Euboea and cities on the Black Sea. Some of this was brought not to the Piraeus but to harbours on the north-east coast of Attica, and came to Athens along the Decelea road, a distance of at least 30 miles. Land communications were obviously good enough, in some places at least, for freight to go considerable distances, in considerable bulk. [. . .] This transport would all of it have been carried on by ox-cart, pack-animal, and porter (for short distances).

Heavy transport did not go on every day, but it must have been required occasionally for normal business, such as bringing heavy timbers into the shipyards. And it was essential to one widespread activity in Greece, public works. This involved moving considerable quantities of building-stone for distances of anything up to 25 miles, and it was an activity undertaken by both large and small cities from the sixth century BC onwards.

The building programme of one small city alone shows that the occasions when heavy transport was necessary were not so rare as to be unique in the experience of a whole generation. At the beginning of the fourth century BC, Epidaurus [city in north-east of Peloponnese] found itself involved in a long-term building scheme. The healing-cult of Asklepios had become increasingly popular towards the end of the fifth century throughout Greece, so that the Epidaurians thought it worthwhile to enlarge and enrich the sanctuary which was in their territory. We know from the building accounts that foreign workmen (e.g. Argive [from Argos], Corinthian, Parian [from the island of Paros], Athenian) contracted to work there, and that various materials such as wood and stone were imported. In this place alone people had to deal with building problems including transport for at least a century, during which time they built five temples, a theatre, and various houses for the reception of pilgrims.

Not only was building material moved in large quantities: loads which could not be divided into units small enough for one yoke to move *were moved somehow*. The orthodox view, with the modern idea of a standard of efficiency in mind, dismisses the Greek method of increasing power as unrealistic and uneconomical. Yet the fact that a different harnessing-system was not developed shows that there was no pressing need for improvement. R.J. Forbes says that 'the keystone of technological development is the availability of prime movers' (i.e. the inventions themselves). But the keystone is, surely, the need of a new technological device, for technological development is a practical answer to some particular demand, not a mystical accident.

The orthodox view seems to gain support from a considerable weight of archaeological evidence. But first it is an ill-balanced, prejudiced selection, and secondly, it simply is not in the nature of this kind of evidence – e.g. fine-painted pottery, stone-reliefs from temple- and treasury-friezes and gravestones, and terracotta and bronze models (intended for toys or votive offerings) – to give us information of industrial or heavy transport. The horse was the most decorative, and, as we shall see, the most socially acceptable animal to portray. It is from a few vases only that we know that the same harness was used on mules and donkeys. What most of the pictorial evidence tells us is that horses were harnessed with throat-and-girth harness, two or four abreast, to light, two-wheeled chariots; that they were not shod (a recognized impediment, on which both Thucydides and Xenophon remark); and

that this kind of harness remained unchanged until after the end of the Roman Empire. Pictorial evidence for ox-transport is almost non-existent. I know of only the Tourah relief, carved in an Egyptian quarry, which shows three yoke of oxen harnessed more or less in file in order to move a block of stone.

Lefebvre des Noëttes argues from this evidence the predominant and inefficient use of the horse, the non-existence of a sensible way of harnessing in file, and the relative unimportance of the ox. Of the available literary evidence, he quotes, in the first place, Xenophon's description of the Persian Cyrus' experiment. He harnessed eight yoke of oxen to a siege-tower, and the load per yoke was reckoned at 15 talents. Xenophon quotes the normal load for one yoke at 25 talents, which is about 1100 lb., to explain his comment on the ease with which the animals moved the tower.[7] This is supposed to suggest both that multiple harnessing wasted power, and that the generally-accepted load was about one fifth of that carried by modern draught-animals. The second piece of evidence is a passage in the Theodosian Code which is supposed to corroborate Xenophon's observation. The Code, published in 438 AD, includes edicts against abuse of the Roman Imperial services, of which one was transport. Maximum loads were prescribed for each type of cart, and for a heavy one-yoke cart it was 1500 Roman pounds, or about 1100 lb.

Another piece of evidence is Diodorus' description of Alexander's funeral car, which is quoted in full to prove that, because the load involved was ridiculously small in relation to the number of animals harnessed, the harness must have been inefficient. But it is ludicrous even to imagine that one could derive any useful information from the description of a purely spectacular array of beasts, in four ranks 16 abreast, pulling one coffin. Finally, Vitruvius, in his book on Greek and Roman architecture, refers to an invention in the sixth century BC for moving heavy blocks, of a wooden frame on wheels to which two yoke of oxen were harnessed (abreast, not in file).[8]

The pictorial evidence is mostly concerned with social and military subjects. The literary evidence is only interested in industrial transport in special instances, or where the evidence consists of domestic or civic records, such as the Theodosian Code or the Mycenaean Tablets.

But there is other evidence not used by Lefebvre des Noëttes. In Plutarch's life of Lycurgus, the legendary reformer of the Spartan state is said to have devalued the iron currency to such an extent that one yoke of oxen was required to move ten minae-worth. If one reckons by the so-called Pheidonian standard (i.e. the archaic ratio of iron to silver), ten minae of iron work out at about 3300 lb.[9] Cato, in his agricultural treatise, describes how an oil-mill, probably weighing about 3500 lb., was taken 25 miles by three yoke of oxen.[10] And Pliny, writing in the first century AD, refers to a Gallic plough drawn by two or three yoke in file.[11] We also have abundant evidence for the use of the ox with the plough. Yoke of oxen were modelled in bronze and terra-cotta, rather less frequently than horses and chariots, but often enough to show that it was oxen which performed the heavy work of ploughing. This is borne out by the fact that, in collections of votive ornaments, the ox usually outnumbers the horse, and that plough-oxen even appear on gravestones of different date, suggesting that oxen were used continuously throughout antiquity for farm-work.

The most important evidence of all consists of building accounts which come from all over the Greek world during a period of about three hundred years. Heavy transport is a recurring item. But this evidence is ignored by the orthodox view, which is yet further invalidated by the fact that, while there is no direct evidence for the use of manpower in transporting building material, we have it stated explicitly in these accounts that oxen were employed for this purpose. The accounts are permanent records, inscribed on large stone slabs (only partly preserved), of expenses incurred in the construction of temples, fortifications and other civic amenities. Materials, the work done, and the labour employed, are listed not in columns as we know accounts but in a narrative form. In some cases they give the names of the workmen employed. But the content of the inscriptions varies from place to place, and from year to year. Items are recorded in working order, and sometimes in minute detail, so that a comprehensive picture can be obtained of the problems which arose.

The earliest known accounts are Athenian and date from c. 450 BC: they give a very condensed summary of expenses for quarrying and transport. Then the remaining fragments of the accounts for the Parthenon and the Propylea [the monumental gateway to the Athenian Acropolis] between 448 and 432 BC, record the treasurers' summaries of receipt of income, and of what materials and work this was spent on. The Erechtheum [a temple on the Athenian Acropolis] accounts are quite different. Individual labourers are listed by name: they are skilled craftsmen, and the work, of finishing off wooden and marble ornament, is described in detail. At the beginning of the fourth century there are a few brief statements of money and labour contributed towards the rebuilding of Athens' Long Walls [enclosing the city and its port at the Piraeus] set into the walls themselves. Then at Epidaurus we have the full record of the building of one temple, and incomplete but more detailed records of works carried on later in the century. At Delphi work done year by year on the rebuilding of the Apollo temple is recorded with the minimum detail. There survive from the third quarter of the fourth century at Athens very detailed records of specifications and accounts for work in the sanctuary at Eleusis and in the Piraeus. There is a series of records from Delos, from about 315 to 250 BC, covering all kinds of expenses, with no distinction between large and small items. Other fourth-century accounts, like those from Epidaurus, have been found at Tegea, Nemea, Troezen, and Hermione in the Peloponnese. Apart from lists of private individuals' contributions to public works, the only other building accounts come from Didyma near Miletus in Asia Minor, in the early second century BC.

Inconsistent in form and spasmodic they may be, but they provide the most direct evidence we can have of working conditions. I have already pointed out that indivisible loads too heavy for one yoke to move were moved. Blocks weighing 2, 3, or 4 tons are commonplace on many building-sites, and some column-drums at Eleusis, by no means exceptionally large, weigh from 6.5 to 8 tons. We know from the inscription dismissed by Lefebvre des Noëttes that these blocks were moved a distance of 22 miles from the Pentelic quarries to Eleusis by teams of oxen, ranging in size from 19 to 37 yoke. (Lefebvre des Noëttes assumed that the column-drums only weighed 1 ton each.) Moreover, a survey of the kind of stone used in temples and other buildings shows that transport was by no means prohibitive to one's choice of material. For example, it seems likely that imported stone, possibly from Corinth, was used in the temple of Zeus at Nemea, about 20 miles inland from the Corinthian quarries. The inscriptions tell us that Corinthian stone was used in the sanctuary of Asklepios at Epidaurus; and, to reach the sanctuary, people had to take the stone first to the Corinthian port of Cenchreae, then ship it from there to Epidaurus, and carry it 7 miles inland from there to the sanctuary. It is clear from the accounts that the small towns of Hermione and Troezen, in the south of the Argolid [region of Peloponnese including Argos and Mycenae], also imported stone. In fact local stone was only used in places where it was known to be suitable, i.e. at Corinth. It is quite likely that

poros-limestone [granular variety of calcareous tufa, not necessarily porous], similar to the Corinthian stone, could have been quarried near Epidaurus. But the difficulty of transporting stone from the known source at Corinth was not so great as to oblige the Epidaurian building-commission to look for local material.

It is precisely this evidence which the orthodox view neglects; and unless one considers heavy transport in its context, together with the evidence, it can only appear exceptional and irrelevant.

[III]

The ox was the first animal to be yoked, and it retained this significance throughout antiquity. [. . .] The ox was domesticated long before the horse, and even in areas where the horse had his natural habitat cattle predominated. It may be argued that oxen, as well as sheep and pigs, were kept in far greater numbers simply because they were more edible than horses. But the greater distinction between the ox and the horse is one of slow strength as opposed to speed. When it came to the point of harnessing animal-power for agricultural work, as in Sumeria, the ox was not only the most obvious draught-animal to hand, but also adequately if not eminently suited to working with a yoke and plough – and so it continued to be.

In those parts of the world to which the horse was not native – the Near Eastern and Mediterranean countries – it only appeared after the art of riding had been usefully developed, and consciously imported. No horse-remains have been found in Crete before c. 1700–1600 BC, and none in the Early Helladic settlements (pre-1900 BC) of mainland Greece. The horse, whose essential quality is speed, has always possessed glamour, an aristocratic mystique. Its prestige value has never, certainly in antiquity, been outweighed by its utility. When war chariots were introduced in the Near East, people were probably as much impressed by their psychological effect as by any practical advantage they afforded. The late use of horses is also indicated by the written evidence. There is no mention of them in Sumerian texts c. 3000 BC, the earliest occurs in Assyrian texts c. 1700 BC. The Mycenaean tablets also speak for a late date – it is said of ideograms that 'some commodities which are themselves innovations in LM II' (i.e. well after 1600 BC) 'such as horses, chariots . . . require new symbols'.[12] That the horse did not replace the ox in farmwork is suggested by the fact that oxen outnumber horses in agricultural contexts. The tablets only mention horses in connexion with war-chariots, but oxen are specified as 'working oxen'. [. . .]

The orthodox view is right in assuming that the horse was choked by its unsuitable harness. But the horse chariot was not intended to be an efficient and sensible means of transport. The aim was to have a fine show of horses rearing and struggling, and drawing a ridiculously light load as fast as possible. In any case the horse would have been too precious, too lightly built, and too nervous for heavy work. I have found only one example of a working-horse, in a list of contributions for building in the third century BC at Callatis, on the Black Sea.

So that it is the ox whose capabilities must be considered with regard to harness and the power available. We have already seen that Xenophon and the Theodosian Code establish a maximum load of 1100 lb.; but Plutarch suggests a normal load about three times as heavy, and Cato's oil-mill produces a load of about 1100 lb. per yoke – but here some form of multiple yoking was used, and this, according to the orthodox view, should have reduced the maximum power of each yoke. So that what should have been a lower effort achieved the same as that claimed by the orthodox view for a maximum effort.

Which is right? And if one figure is right does this mean that the others are wrong? Evidence

for the speed of working oxen provides about the same degree of variation. Pliny says that a fair day's work for one yoke is to replough one and a half acres with a nine-inch furrow, which comes to about 11 miles.[13] Cato allows expenses for six day-wages, that is, presumably, two days' wages to each of the three yoke-drivers; in which case, if the whole journey was 25 miles long, then 12 or 13 miles were covered each day. But it took two and a half to three days to take the column-drums 22 miles from Mt. Pentelikon to Eleusis, though the delay may have been due to some other now unknown and unimaginable complication.

It is thus futile to try to make out a standard rate of efficiency for ancient transport. First, the evidence does not allow it, and secondly it is much more likely that people adapted the means available to the demands of the moment. It was to no one's advantage to work according to some abstract standard of speed or load, nor were commentators interested in any but the particular instance they had in mind. Thus Xenophon's comment refers to an experiment with army baggage-animals, so that the standard he quotes may be a military one; while the Theodosian Code seeks to impose a limit on loads for the protection of state-roads and transports. Neither refers to private or business methods. [. . .]

[IV]

The practical importance of the ox is relevant to the whole question of harness. The adequacy or otherwise of ancient harness depended on the animal harnessed. The yoke is peculiarly suited to the ox, as it is not to the horse, since the horse's neck provides no ridge of backbone and muscle for the yoke to rest against when the animal pulls forward, so that the harness slips back and drags the throat-strap up in front, off the horse's shoulders onto its windpipe. Successful harness for the horse was invented when its anatomy received special study – and this can only have happened when the ox was too rare or too slow for the needs of the moment. The Chinese invented the breast-strap harness as early as the fourth century BC. [. . .]

The Chinese also developed the prototype of the modern horse-collar, from necessity – this time to meet a threat to security. Dr Needham suggests that this development took place on the edge of the Gobi desert in northwest China, where in the fifth century AD people required speedy and dependable transport in the face of nomad attacks.[14] [. . .]

Other people in antiquity, e.g. the Greeks, did not have to solve this particular problem; they were not faced by any difficulty which could be overcome only by inventing different harness. As V. Gordon Childe said, 'technological progress depends not only on an accumulation of useful knowledge, but also on a multiplication of wants'.[15]

When did the Greeks find it necessary to employ devices for heavy transport? Occasionally, as I have suggested, for moving heavy ship's timbers, for oil-mills, grindstones, and military equipment such as battering-rams and catapults (in the Hellenistic period); but more frequently for building materials. Another fairly regular demand on heavy transport would have been made to move ships along the Diolkos [slipway for passage of ships across land] at the Isthmus of Corinth. Not much is known about this, except for a few yards of the paved roadway itself. This has grooves about a yard apart cut in it to accommodate cartwheels. Ships could thus be wheeled across on a cradle of some kind and so avoid the long voyage round the Peloponnese when travelling east or west. [. . .]

Scholars have often assumed that column-drums were rolled along behind animals whose harness was fastened to pivots in the central cuttings on the horizontal surfaces of the drum. In most cases the lifting-bosses projecting on the round surface would have been rather an

impediment, besides which the drum would have chipped, and on a downhill slope would probably have broken the animals' legs. Nor would one have moved *squared* blocks in this manner. Rollers provide a simple means of moving large blocks for very short distances. Obviously this method was used on the site and on quaysides – payment was made at Epidaurus for two kinds of 'rolling', on and off, where a quayside context seems likely. But the Eleusis-drums were moved on waggons; wood and rope were brought up from the Piraeus ship-yards, four coils of rope were cut into lengths for traces, axle-blocks were purchased, and heavy beams, of the kind which bore the weight of a ship's anchor at the bows, were also provided. So that something larger and stronger than an ordinary waggon came of it. Where such weights (7 or 8 tons) were involved, something had to be done about the road-surfaces; iron (i.e. tools) was supplied for 'road-making', which probably meant filling in holes and removing boulders. The Greeks could make good roads when occasion demanded, as is shown in the Panathenaic Way in the Athenian agora, the paved track down from the Pentelic quarries, the Sacred Way at Delphi, the Diolkos, and in the cities of Selinus and Acragas in Sicily. The Didyma accounts record roadwork at the quarries; Plutarch mentions roadmakers among the other workers on Pericles' scheme, and they also appear in the Propylea accounts; and Diodorus says that roadmenders went with Alexander's funeral-car on its way from Babylon to Egypt. But despite this evidence for roadmaking, it should be pointed out that the Eleusis-drums were moved in August, when there was least likelihood of mud on the roads. [. . .]

Transport-contractors never formed joint-stock companies. There is no evidence that draught-animals were maintained either publicly or privately for heavy transport alone, and there is no likelihood that this was so. For normal purposes there were porters, packmules and donkeys owned or rented on a small scale by slaves and freemen. The Eleusis inscription shows that two, three, or four men worked together to transport a bulk of material, which suggests that no one had more than the minimum resources. It was not these professional, small-time carriers who undertook transport contracts for public works, because they had neither the resources nor the social position necessary for organizing transport on a large scale. Obviously Eudemus of Plataea was a big man, socially speaking; and it is possible that the social prestige gained by this kind of work was more important than the financial profit, if any. Even at Corinth, there would only have been occasional need of transport on a large scale, although the quarries supplied stone for other places. [. . .]

I have suggested that the social standing of the organizers, i.e. the contractors, was fairly high. This would accord with the idea that heavy transport was organized on a more or less voluntary basis; and that the owners were willing to join in transporting building-stone for their own city's temple or defences, or for the sanctuary of a cult in which they felt a particular interest. Cities usually had to depend on foreign skilled labour to some extent, but the transport of material could be done by any citizen with time to spare, under the direction of a contractor or official with some experience of loading heavy blocks on to carts. Contributing money towards the cost of public works became a common practice in Greek cities from the fourth century onward. People felt themselves honoured by being inscribed as minor benefactors of their city; and sometimes contributions were made in kind - e.g. labour, materials, or draught-animals. So that contributing service as a haulier could be considered a part of this social practice. But of course it was service for payment; though the pay would not have been regarded by the farmers as a major source of income, but as a bonus. [. . .]

There were accidents - blocks did fall off carts on the way - yet the system worked,

the means were adequate. Nobody would have understood the distinction between efficiency and inefficiency made by modern critics of ancient devices. Why did harness remain unaltered from the time the first ox-yoke was used in Sumer c. 3500 BC until the introduction of breast-strap harness (invented in China c. 330 BC) into Europe about the sixth century AD? The answer is, not that ancient society was slothful, uninventive, and slave-ridden, but that the harness then in vogue was perfectly adequate and went on being so.[16] People knew certain ways of doing necessary jobs. They could, therefore, see no reason for other methods to come into existence.

Notes

1 R.J.E.C. Lefebvre des Noëttes, *Le cheval de selle à travers les âges: contribution à l'histoire de l'esclavage*, Paris, A. Picard, 1931

2 Vol. 2, *The Mediterranean Civilisations and the Middle Ages*, Oxford, Clarendon Press, 1956, C. Singer, E.J. Holmyard and T.I. Williams (eds). See especially R.G. Goodchild, section 14, 'Roads and land-travel', and E.M. Jope, section 15, 'Vehicles and harness'

3 Ibid. Introduction

4 Ibid. p. 74

5 Ibid. p. 174

6 *Studies in Ancient Technology*, Vol. 2, Leiden, E.J. Brill, 1955, pp. 78 ff.

7 Xen. *Cyr.* VI, i, 52 ff.

8 *De architectura* X, ii, 11–12

9 Plut[arch], *Lycurgus*, ix, 1

10 *De agricultura*, xxiii, 3

11 Pliny, *Natural History*, XVIII, 173

12 M. Ventris and J. Chadwick, *Documents in Mycenean Greek*, Cambridge, Cambridge University Press, 1956, p. 42

13 Pliny, *Natural History*, XVIII, 178

14 *Science and Civilisation in China*, IV, Cambridge, Cambridge University Press

15 'Magic, craftsmanship, and science', *Frazer lecture* Liverpool, University Press, 1950, p. 9

16 As I have shown, the availability of manpower has no bearing on heavy transport, because power was supplied by oxen. So that the question whether or not manpower was slave or free has no relevance here; and I would suggest that social status has nothing to do with technological invention, or the lack of it

6

ANCIENT GREEK WATER SUPPLY

by Alfred Burns

Source: Alfred Burns, 'Ancient Greek Water Supply and City Planning: a study of Syracuse and Acragas', *Technology and Culture*, Vol. 15, 1974, pp. 404-11

[. . .] Although the water systems of Syracuse and Acragas [in Sicily] are not as spectacular engineering feats as the tunnel of Eupalinus or the later syphon of Pergamon, they are very impressive accomplishments because of the magnitude of the systems and the quality of their workmanship. The fact that miles of aqueducts are still in existence, some of them still in working order, testifies to the skill of their builders. They are, however, in no way unique. Anyone visiting the Sicilian works and water installations of the same period in such widely separated places as Megara, Samos, Lindos, and Athens will notice the many common features. The uniformity of the techniques, materials, and dimensions from place to place indicates the existence of a well-developed profession with established traditions, standards, and methods. That many of these are later reflected in Vitruvius's specifications (*De Architectura* 8.6.1) suggests that the Romans had learned this art, as so many others, from the Greeks either directly or by way of the Etruscans.

A salient characteristic of the Greek installations that has been noted repeatedly is that they are constructed entirely underground. It is generally assumed that protection of the supply from enemies was the reason. But since no less effort was expended when the entire installations were contained within the fortifications of the city walls (e.g., in Acragas), protection from pollution must have been a primary reason. We know that the Greeks believed that epidemics were spread by wind and water (e.g., Hippocrates *Airs, Waters and Places*; Aristotle *Pol.* 7. 1330a39-1330b11-14); it was natural for them to avoid exposing their drinking water to the open air. Obviously, the water also was thus kept cooler and more palatable. We must not forget that in the semi-arid Greek lands cool springs were identified with nymphs and accorded divine honors. For the Greek, as Pindar says, 'water is the best thing' (*Ol.* 1.1).

Even the reservoirs were built underground with ceilings consisting of heavy stone slabs supported by many pillars. Typical structures are the reservoir of Samos with fifteen sturdy pillars and the one in Acragas with forty-nine arranged in seven rows of seven each. Both are the starting points of long aqueducts carrying the water to its desired destination. In most cities this destination was a centrally located fountain house on or near the agora. The availability of an adequate source of good water was one of the paramount considerations in the choice of the site for a city to be founded and one of the factors determining its layout. In older cities that had grown gradually from early settlements, the fountain house, like the acropolis, was a focal point in the spontaneous development as well as in conscious replanning during times of growth and prosperity.

The age of the tyrants was such a time; thus,

it is not surprising that the 6th century BC marked the beginning of some city planning, development of the architectural styles, and the inception of major civil engineering projects – such as jetties and harbors, drainage and water-supply systems, enlargement of market places, and construction of public buildings, including fountain houses. In addition to the obvious utilitarian necessity, the fountain house in or near the civic center of the polis played an important part in its social life, for as the agora and the gymnasium were the meeting places for the men, the fountain house was the informal social center for the women. Thus, in both the spontaneous layout of the 'old style city' and in the planned 'new, that is Hippodamian' city with its rectangular streets (Aristotle *Pol.* 7. 1330b24–25), the fountain house is a crucial, well-planned feature. It is often combined with the water reservoir, especially where the fountain is close to the source of the water. That source is almost invariably in limestone rock. The typical fountain house contains a distribution tank with the water running from several spouts. The distribution tank is fed by reservoirs or cisterns constructed behind it. Wherever possible, the fountain house is built with its back side against the rock, and the reservoirs are excavated within the rock. These reservoirs may be fed either by a spring in the rock or by an aqueduct bringing the water from a more distant spring.

Examples of such installations are Pirene and Glauke in Corinth, the Nympheum in Syracuse, the fountain house at Ialysos, and the town fountain in Lindos [Ialysos and Lindos were cities on the island of Rhodes]. Of course, where no such rock exists, the fountain house with the reservoirs may be found constructed from masonry above ground. The fountain house of this kind in Megara, with its thirty-five pillars in five rows, is interesting because the reservoir was divided by a center wall into two halves, each with separate inlets and outlets, so that the two parts could be emptied and cleaned independently without interrupting the water supply. Although the water supply system of Megara is one of the oldest, the present form of the fountain house is the result of rebuilding in Hellenistic-Roman times. (Since many of the installations described here were repaired, enlarged, and rebuilt repeatedly, it is generally difficult, if not impossible, to determine the exact age of many of their features.) Most of the fountain houses seem to have started as natural or man-made caves that were successively enlarged; the outer rock faces were cut smooth and colonnades were added.

The aqueducts feeding the reservoirs or drawing from them are laid in subterranean passages that are usually large enough to allow a man of slight build to walk through upright; dimensions vary from 70 cm to 1 m wide by 1.50–2.00 m high; although typically 1.80 m high, they can on rare occasions be as low as 1.30. In soft ground the passages are masonry lined and covered by stone slabs. Where the soil is rocky, they are tunneled through. Manhole-type vertical shafts to permit access are provided at regular intervals varying from 20 to 40 m. The galleries through hard rock were dug by sinking the shafts first and then connecting them by digging from both sides toward each other. The shafts, too, are quite uniform in most locations, either round with a diameter of 65–90 cm, or rectangular, with sides of 90 cm by 1–1.50 m. The 90-cm width seems preferred because it is the most comfortable for 'chimney climbing,' that is, holding on to a rope while bracing one's legs against the sides. In many locations (e.g., Rhodes and Sicily), vertical rows of footholds at 24-cm intervals are carved into both sides to facilitate this type of climbing. The surface openings of the shafts are covered by heavy stone slabs. The rows of openings, whether covered or not, are quite conspicuous in areas with little topsoil and scarce groundcover, such as, for instance, the plateau above Syracuse. And, indeed, Thucydides tells us that the Athenians during their siege of that city were able to find and interrupt some of the aqueducts supplying Syracuse (*Hist.* 6,101). In the underground channels the water was led through conduits of baked clay very similar to modern sewer pipes.

The pipes used in various areas are again of very similar type, manufactured on the potter's wheel with tongue-and-bell fittings, that is, a narrower socket on the end of one pipe fitting into the flared or flanged or sleevelike end of the next pipe. The joints are held together and sealed by a cement generally consisting of lime and sand; an admixture of bitumen occurs on Cyprus, and the use of lead appears in some conduit lines in Athens. The length of the pipe sections varies from 0.65 to 0.95 m, outside diameters from 0.15 to 0.25 m, and wall thickness from 0.015 to 0.025 m. Many pipe sections have a round or oval hole cut into the upper wall and covered by a lid. These holes are usually referred to as 'cleaning holes' and no doubt were used for this purpose. From the peculiar locations of the holes, however, it seems that their primary purpose must have been to permit reaching into the pipe to apply the cement from the inside. One finds two different arrangements; in some locations (e.g., in Athens and Acragas), the hole is found in each pipe close to the same end where it is easiest to reach the joint; in others (Samos, Priene), the holes are in the center and only in every other pipe. Since from the middle it was possible to reach both joints, the holes obviously were needed only in alternating pipes. This makes it quite clear that the holes were required for the original sealing process.

In addition to all these similarities, however, the most important common feature of Greek aqueduct systems is the method of obtaining the water supply. Influenced by the rediscovery of the tunnel of Eupalinus on Samos, the early archaeologists (e.g., Dörpfeld, Schubring) thought that all the newly discovered tunnels

had served the same purpose, namely, to transport water through an obstacle. Consequently, their logical reaction when they found a tunnel opening was to look for the other end or some river or other likely copious source of water. As some aqueducts were indeed designed for the transport of water, this attempt was sometimes successful, especially since such aqueducts are usually found close to the surface following the contour line of the terrain almost horizontally with only minimal descent. Even where the channels are tunneled through hard rock, their course can be easily followed by the manhole-like shaft openings at regular intervals. But in most cases when a tunnel was found driven head-on into the side of a mountain, the attempt to locate the other exit had to end in frustration because there is no other exit.

The geological configuration prevalent in most of the ancient Greek lands on the rim and the islands of the Aegean is a permeable limestone cap superimposed on an older impermeable layer. Consequently, there is a general shortage of surface water. The water seeps through the limestone and collects at its base, where the limestone meets the impermeable strata. This seam is also where most springs and trickles of water are found, often issuing from a natural cave or fissure. These sources are often the locations of early fountain houses and the later, more elaborate *nymphea*, which, as we have seen above, are usually built against and into the living rock. When the natural flow was insufficient, tunnels were driven into the mountain, not to get through the mountain, but to find and collect the seeping waters within. Tunnels of this nature usually are winding and turning as the diggers followed some natural trickle. The galleries divide into many strands, which often branch again. Frequently there is a rectangular gutter-like channel carved in the center of the tunnel floor coated with a sealing cement to collect the water dripping from walls and ceiling and trickling from the side galleries, which are sometimes only a few meters deep, but often branch out in many directions; generally, two ledges are left along the walls on both sides of the water channel to allow walking without stepping into the water. The tunnel of Lindos on Rhodes is of this kind, a veritable labyrinth that feeds the town fountain and is still the city's only water supply. As we have seen, most aqueduct tunnels at Syracuse and Acragas also belong in this category. [. . .]

From the foregoing, I believe, it can be readily seen how the physical, and to some extent the social, development of the Greek cities was inextricably bound up with the hydrogeography and the resulting technology. The early city typically developed at the foot of a calcareous massif that provided the two essentials: a defensible site for the acropolis and a source of water at its base. The vicinity of the spring with the fountain house determined the location of the agora with its cluster of public and religious buildings. The rest of the city would spread around this natural civic center. Beginning with the 7th century BC and gaining momentum in the 6th when the growth of the old cities and the founding of new ones favored a more conscious planning process, the pattern remained essentially the same, although the population increase in many cases made it necessary to tap more distant formations of aquifer rock and to bring the water to pre-existing or new fountain locations by means of aqueducts.

From the late 5th century onward, with the advent of the fully planned 'Hippodamian' city, the fountain house or *nympheum* became an integral part of the architectural composition which framed the agora (e.g., at Miletus and Ephesus). In Rhodes the aqueducts and drainage channels were planned and built in conformity with the rectangular grid of the street plan apparently before any surface construction was begun.

At no time, however, was a step taken toward a distribution of water to private dwellings. Distribution always remained limited to a few outlets at focal points of the city. Some private houses did have wells or cisterns to capture the rainwater. Well or cistern water, however, was generally not used for drinking but for other purposes only (Aristotle *Pol.* 7. 1330b15–17). Also, restrictions were placed on the drilling of wells by private citizens (Plutarch *Solon* 23, 6). The restrictions on well digging and the limited number of fountains dispensing running water may have been essentially conservation measures. I suspect there was no popular pressure for wider distribution, as the women would not have wanted to forego their daily walk to the fountain where they could meet their friends and exchange gossip. The gatherings at the fountain must have been an important part of their meager social life. Later on, when prosperity increased, the availability of slaves obviated the need for wider distribution.

From the consistency of the techniques in most hydraulic installations, it can be seen that the building of aqueducts, sometime during the 6th century BC, had become a well-developed craft, the domain of a specialized profession with its own standardized methods and skills in their application. On the whole, there is no evidence that these methods required or included any sophisticated mathematical procedures other than straightforward horizontal and vertical measurements that could be performed with ropes, stakes, and plumb lines.

If we were to look for any deviations from the norm, we would have to consider the aqueduct of Theagenes at Megara more primitive because of its below-normal height of only 0.47 m, probably because it was among the earliest such installations since Mycenaean times. The one project, however, that is unique is the tunnel of Eupalinus on Samos. Although the rest of the system conforms to the general pattern of 6th- and 5th-century aqueducts, the tunnel itself, as Herodotus (*Hist.* 3.60) points out, is certainly an outstanding engineering feat without parallel among contemporary works. Not only are the dimensions of 2 m high by 2 m wide, with an additional lower channel for the conduit line, more than twice the normal, but also its 1-km length, drilled from opposite sides through a mountain for an almost perfect link-up without guidance by any intermediate shafts, was not equaled anywhere in antiquity, even in late Roman times. [. . .]

LIFTING IN EARLY GREEK ARCHITECTURE

by J.J. Coulton

Source: J.J. Coulton, 'Lifting in Early Greek Architecture', *Journal of Hellenic Studies*, Vol. 94, 1974, pp. 1, 7–14, 16–17

In the standard handbooks on the techniques of Greek architecture, the problem of lifting heavy architectural members is considered mainly in terms of the various cranes and hoists based on compound pulley systems which are described by Vitruvius and Hero of Alexandria.[1] It is assumed that the same basic method was employed also in the archaic period, and that the use of an earth ramp by Chersiphron to raise the architraves of the temple of Artemis at Ephesos[2] in the mid-sixth century was exceptional. If this is true, it is a matter of some interest in the history of technology. [. . .]

After *c.* 515 BC, the situation seems fairly clear; cranes were in common use. But positive evidence suggesting the use of a crane before that date is scanty. [. . .] Pairs of U-shaped holes in the early temples of Apollo and Athena at Delphi could be used with a hoist, but since another explanation is possible, and since other U-shaped holes were definitely not used with a crane, we cannot place much confidence there. [. . .]

The two substantial cases where there appears to be evidence of lifting by crane in an early building are the temple of Athena at Assos and the first temple of Hera ('Basilica') at Paestum. At Assos there are U-shaped channels at each end of some cornice blocks, while others have unusual cuttings which are interpreted as sockets to take iron hooks attached to a hoist. At Paestum the frieze backers of the first temple of Hera also have U-shaped channels at each end.

If we wish to get rid of these 'exceptions', we must suppose either that the U-shaped channels and other cuttings were in these cases used with levers rather than with cranes; or that these two temples are considerably later than is often thought; or that the temples were indeed begun early, but that they were under construction for a long time and that the blocks suggesting the use of a crane were not set until after *c.* 515 BC. The first way out is rather an unfair argument; the second is perhaps possible, since there is no concrete evidence for the date of either temple and some scholars have suggested that the temple at Assos belongs to the late sixth century BC; but the third seems the most reasonable. At Assos the cornice would of course be the latest main element of the exterior, and although the alternating sizes of its mutules [ornamental features of a Doric cornice] may seem an early feature, it is one which derives necessarily from the sizes of the triglyphs and metopes [standard decorative components of a Doric frieze], which in turn must have been defined by the time the architrave was put in place – perhaps long before the cornice. At Paestum a firm *terminus ante quem* [finishing-point] for work on the first temple of Hera must be the start of the second temple of Hera ('Poseidon') in the mid-fifth century. It is uncertain how long before that the frieze backers were set in place.

It may be, however, that none of these explanations is required, and that we should rather take the evidence from Assos and Paestum at its face value, as indicating that cranes were already used by Greek builders in the third or even the second quarter of the sixth century BC. A decision on this point must await the conclusion of the argument. If in the meantime, however, the temples at Assos and Paestum are left aside, it appears that there is nothing to compel belief in the use of cranes in Greek architecture before *c.* 515 BC, and the scantiness of the evidence that can be taken as suggesting it in itself indicates that the use of a crane cannot have been common practice. We are simply faced with the fact that heavy blocks were raised to the required level by *some* means. The question of how this was done can perhaps be more usefully looked at from the other end. Instead of seeing how far the classical Greek methods can be traced back, let us consider how the Greeks may have learnt to handle heavy weights, and how far the methods they are likely to have adopted initially can be traced forward.

In Greece itself the problem of lifting heavy loads did not seriously arise before the second quarter of the seventh century BC. Before that date there was no monumental sculpture and the stones used in building were not normally larger than could be lifted by two men, so that there was no need for any special techniques. In about the middle of the seventh century the Greeks began to produce both sculpture and architecture involving large pieces of stone, and since statues were normally set on bases, they too, like architectural blocks would need to be raised to the required height. Temple A at Prinias [in central Crete] (*c.* 630 BC ?) provides one of the earliest examples of large stone blocks in both architecture and sculpture, the largest frieze block weighing about half a ton.

The architrave blocks of the temple of Artemis at Kerkyra [on Corfu] (*c.* 590–80) would have weighed 5 or 6 tons, but such loads seem insignificant beside the Colossos of the Naxians at Delos, which must have weighed over 20 tons itself, and stood on a base weighing over 30 tons. So within little more than 50 years the Greeks had learnt to handle blocks more than 100 times heavier than those they had been used to previously. It is unlikely that their efforts were undertaken in complete isolation from the rest of the world and that in this short time they invented a completely new way of dealing with heavy weights.

Large stone blocks were used to some extent in the architecture of the Levant and Assyria, but most commonly for orthostates [upright stones], thresholds and column bases, all positions where the blocks would not have to be raised to any extent. It is true that the first evidence for the use of the pulley comes from an Assyrian relief, but it shows just a simple pulley used to haul up a bucket of water. There is no evidence that the system was further developed, and if not, it would not be suitable for the kind of loads we are dealing with. A series of reliefs from the palace of Sennacherib (705–681 BC) shows the various stages in the transportation and setting up of a colossal winged bull, which must be imagined as weighing 40–50 tons; and the job is done with levers, a sledge and rollers, and an artificial ramp. It is hardly surprising that no crane is used, for it is doubtful, as we shall see, whether a simple crane could lift such a weight; but there is no sign of a block and tackle or a winch, both of which would be useful in hauling such a load. Neither here, nor in the colossal statues actually preserved do the Assyrians show any desire to piece a colossus together from smaller blocks of stone that could be handled by a crane, the method adopted for the monument of Antiochos I of Commagene on Nemrud Dagh [a sanctuary in eastern Anatolia].

The method used by Sennacherib is exactly that familiar in Egypt, and Egypt was of course the chief home of the arts of large scale sculpture and large scale architecture in stone; it can surely be no coincidence that large scale stone sculpture and architecture began to appear in Greece at precisely the time when close contact with Egypt was resumed. A brief examination shows that although Greek borrowing from Egypt in the field of architectural form was comparatively limited, the technical similarities between Greek and Egyptian architecture are extremely close, covering virtually all aspects from the methods of quarrying stone to the dressing down of the building in the final stages. Among these similarities are two of the features we have already looked at, the U-shaped hole and the projecting boss. They occur more rarely in Pharaonic architecture than in Greece, but the fact that they occur at all confirms the arguments based on the evidence of Greek architecture, that both these devices are to be associated with the use of wooden levers rather than a crane or hoist. For it is virtually certain that Egyptian architects of the third and second millennia BC did not use any kind of pulley system.

The method most clearly used by the Egyptians for raising heavy blocks of stone was the ramp. A temporary ramp of earth and mud-brick was built against the wall under construction, its height being raised as work progressed, and the stone blocks were hauled up the ramp on rollers and then levered into position. A hypostyle hall would be filled with earth as it was built, so that the architrave blocks could be hauled up the ramp, too.

Besides the ramp, the Egyptians used the lever to some extent, although it is uncertain whether they ever used it to lift a block more than a few centimetres. They may also have used what has been called a rocker; that is, a kind of sledge with runners shaped like segments of a circle so that a block loaded on to it could be rocked backwards and forwards like a rocking-chair. By judiciously placing pieces of wood beneath the runners as the sledge is rocked, it is possible to raise it gradually (see Figure 7.1). Although rockers of this type were certainly used by Egyptian builders, there is some doubt about whether they were used for lifting. The main positive evidence is in Herodotus' description of the construction of the Pyramid of Cheops, where he says that in the last phase the stones were raised by means of 'a device made with short pieces of wood'.[3] It seems unlikely that this is a fairy story, yet these terms can hardly be applied to a ramp or a lever or a crane. They do suit the rocker technique, however, for in addition to the rocker itself being made out of fairly short pieces of wood, a vital part is played by the short pieces of wood placed beneath the runners, for the feasibility of the system depends

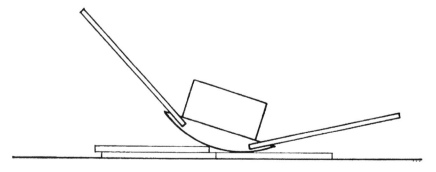

Figure 7.1 Rocker, possibly used for lifting architectural blocks

on the stability of the platform which they form.

Thus ramps, levers, and perhaps rockers are the lifting devices for heavy loads which the Greeks could have learnt of when they first took to monumental building. We have already seen reason to believe in the use of levers in early Greek architecture, although in Greece, as in Egypt, it is uncertain whether they were ever used for extensive lifting. For the ramp we have the well-known story in Pliny's *Natural History*[4] of how Chersiphron raised the architraves of the temple of Artemis at Ephesos by means of a ramp. This story has been disbelieved, and even where it has not been rejected, it has generally been taken as an unusual departure from normal Greek practice. It probably would have been unusual in the fifth century, but it is much less certain that it was unusual in the sixth century. As we shall see, builders in the archaic period, like Egyptian builders, handled and raised blocks of stone much heavier than could be raised with a simple crane or hoist. The architrave blocks of the temple of Artemis must have been among the heavier blocks lifted in the sixth century, but they would not have been in a class of their own. For the use of the rocker in Greek architecture we have no evidence, and short of a direct reference to one it is difficult to imagine what evidence could ever be found. If it was used, it had gone out of use by the time of Herodotus, for it is fairly clear that he does not understand exactly what was the device that the Egyptians had told him about.

Much of the favour with which the hoist based on a compound pulley system is regarded by modern archaeologists is perhaps due to the fact that it is the normal method used today for raising heavy building components. It is doubtful, however, if its advantages were so striking in the sixth century BC, and it may be worthwhile to compare the two systems from the point of view of a sixth-century builder. As far as mechanical advantage is concerned, the ramp is superior to the compound pulley system, for with the latter a theoretical advantage of 1:6, using 6 pulley wheels, is the useful maximum if natural fibre ropes are used, and this is the largest number of pulleys specified by Vitruvius. On the other hand there is evidence of ramps with slopes of 1:8 and 1:12 in Egypt, giving a corresponding mechanical advantage. In terms of power input, the two systems are comparable if, with the pulley system, the power is applied directly by men pulling a rope; for in both cases the power input can easily be increased by setting more men to pull. But in this case, in addition to offering less mechanical advantage, the pulley system offers less control, for it would be hard to lower a load gently, and there would be no easy way of applying a brake if the men stumbled.

If the compound pulley system is combined with the use of a winch, the mechanical advantage is much increased, perhaps ten times or more; but there is a limit to the amount of power that can be harnessed to a winch with a horizontal axle. It would be easy enough to set two men to work at each end of the winch axle, but hard to increase the number of men efficiently employed much beyond that number. A pair of men working a winch cannot apply a continuous force of more than about 50 kg, so that the power input would be only about 100 kg. A reasonable axle diameter for the winch would be 0.15 m, with handles turning through a circle about 1.50 m in diameter, so giving a mechanical advantage of 1:10. If this were used in association with a six-wheel pulley system, and worked by four men, the maximum load would be $10 \times 6 \times 100$ kg = 6 tons. Friction would reduce this maximum in practice, but different dimensions in the winch could increase it somewhat, and of course if two such hoists were used together, they could lift double the load. But it is hard to see how a simple winch and hoist could raise the loads of 20 to 40 tons that were not avoided in the archaic period. To raise very heavy loads,

Vitruvius[5] advises either what Drachmann[6] calls a geared winch (which increases the mechanical advantage, so allowing the power input to remain small) or a treadwheel attached to the winch axle (which perhaps increases the mechanical advantage somewhat, but more obviously allows the power input to be substantially increased). By these means weights of 20–30 tons could be lifted, but it seems highly unlikely that the Greeks, beginners in the field, had devised such refinements as these in the mid-sixth century BC. We shall see that this limit in the lifting capacity of the winch and pulley hoist, in contrast to the unlimited capacity of the ramp, may be of some significance in the argument.

The advantages of the winch and pulley hoist do not lie in its greater efficiency as a means of lifting loads. The most obvious advantage over the ramp is the much smaller amount of preliminary work involved. A hoist may require a good deal of highly skilled labour in its construction, but once made, it can be used repeatedly, and moved from one place to another. The ramp requires a considerable expenditure of effort to reach a height of several metres, and almost the same amount of effort is needed to remove it; and it has to be constructed afresh on each occasion it is required. Another advantage of the hoist is that if the load is within its capacity, then it can be handled by fewer men working over a longer time, and probably with more skill, than the men pulling a block up a ramp on rollers. Thus the chief difference between the performance of the hoist and the winch is that the former involves more skilled labour working over a longer time, while the latter involves the use of a large body of unskilled labour working perhaps over a shorter time. This characteristic of the ramp was not seen as a disadvantage by the rulers of Egypt and Assyria, who had a plentiful supply of unskilled labour, and it is doubtful whether their builders would have used a compound pulley hoist, even if they had known

about it, for it would have placed an unwelcome limit on the size of stone they could employ. But in Greece, particularly by the end of the sixth century, political, social and economic conditions made it difficult to bring together a large body of unskilled labour for a limited period, and it was preferable to use methods which allowed fewer men, but professionals, to work full time. [. . .]

The very rapid growth in the ability to handle heavy weights in the late seventh–early sixth century has already been noted, with the suggestion that this implies a technique learnt from others, not developed in Greece. Almost as striking is the drop in the weights handled during the fifth and much of the fourth century. There were at least thirteen blocks weighing over 20 tons used during the sixth century, and some of those listed as just one do in fact constitute a whole series of blocks, as with the architraves of the temple of Artemis at Ephesos and Temples FS and GT at Selinous or the column shafts of the temples of Apollo at Syracuse and at Corinth. Yet in the fifth century no blocks weighing over twenty tons were used, and only two series, the architrave blocks of the temple of Zeus at Olympia and Temple ER at Selinous [in Sicily], weigh much over 12 tons. The smaller scale of the buildings in the fifth century, and the increasing interest in detail and refinement rather than sheer size, explain this drop in block size to some extent, but there was also a positive effort to keep down the weight of individual blocks. [. . .]

It is therefore hard to avoid the conclusion that the builders of the fifth century BC felt themselves bound by some limitation in the lifting of heavy weights which had not been felt by sixth century builders. It is only with the rebuilding of the colossal Ionic temples in Asia Minor, from the mid-fourth century onwards, that the weights lifted grow again to match those lifted in the archaic period, and by that time there is no reason why the more sophisticated types of winch described by Vitruvius should not have been developed. Nevertheless, the columns of the temple of Apollo at Didyma are built up from quite low drums, and the architrave (which was not in fact set until the Roman period) consists of two rows of blocks side by side, the heavier of them weighing just over twenty tons each. The feeling that unlimited lifting power was available comes only in the Imperial period, with its fashion for monolithic column shafts. [. . .]

The increase in the size of blocks lifted in the Hellenistic period, and particularly under the Roman Empire, is probably due not to a reversion to the use of ramps, but to the development of more sophisticated cranes, such as those described by Vitruvius and illustrated on Roman reliefs.

Notes

1 Vitruvius 10.2, Hero, *Mechanica* 3.2–5
2 Pliny, *Nat. Hist.* 36.14
3 Herodotus 2.125
4 Pliny, *Nat. Hist.* 36.14, 21, 95–6
5 Vitruvius 10.2.5–7
6 A.G. Drachmann, *The Mechanical Technology of Greek and Roman Antiquity* Copenhagen, Muncksgaard, 1963, p. 204; see pp. 146–7

8

THE ORGANIZATION AND SUPPLY OF ROMAN BUILDING

by James C. Anderson, Jr.

Source: James C. Anderson, Jr., *Roman Architecture and Society*, Baltimore, Johns Hopkins University Press, 1997, pp. 68-9, 76-7, 79-80, 82-6, 88-91, 119-27, 156-61, 164-5

Building contracts

Any sort of building or construction requires that the labor essential to it, skilled and unskilled, and the materials necessary for it be organized before the actual work begins. This coordination of organization and supply was regularly handled in the ancient Roman world by forming a legally binding contract between the landowner who wanted something built on property he owned and a builder who had or could provide the needed expertise. Since a contract is concluded between the proprietor and the builder, it can be set up in such a way that, for each specific job, all the various essential elements can be provided for, and the employer can choose how much or how little specific control he wants to exercise over the project, frequently allowing the employer to transfer all real control (and responsibility) onto the builder who has agreed to organize the job and who must then assume the burden of accountability, while the landowner takes no, or very little, part in the project.

In the Roman world, at least well into the second century AD and probably much later, private contractors continued to furnish the majority of the labor and supplies necessary for public and private building, although from the late first century AD onward the increasing size of the Imperial bureaucracy – especially the creation of the office called *opera Caesaris*, probably during the reign of Domitian – meant that public building came more and more under governmental administrative control. But the importance of the contract system continued; prior to that time it constituted the only means by which architectural projects were realized. [. . .]

Cicero provides an excellent example of the type of wealthy landowner from the Roman upper classes who was most likely to let contracts for private construction. His projects concentrated on housing, tombs, commercial property, and maintenance or repair work. We have direct attestation in his letters of building in all four categories.[1] Certain points can be noted. While Cicero purchased and sold a number of houses and villas during his lifetime, he was apparently quite scrupulous about maintaining and repairing them; he did similar service for his brother Quintus's personal property while Quintus was with Caesar in Gaul. The story was quite different with rental and commercial properties, such as *tabernae* and *insulae*, that Cicero owned. He was interested solely in the profits to be earned from those and farmed out all maintenance and repair on rental property to be taken care of by others. We know that he owned, or in one case co-owned, apartment buildings (*insulae*) in Rome on the Argiletum and on the Aventine hill, that he rented out shops (*tabernae*) in Puteoli, and that both were excellent, if not trouble-free, sources of income for him.

What is abundantly clear is that the brothers Cicero, while they probably maintained a small number of semi- or minimally skilled workers able to undertake maintenance and repair jobs on their permanent staffs, turned to *redemptores* [contractors] and *architecti* whenever they indulged in larger projects or those that required any kind of specialized skill. Labor costs money, of course, and skilled labor of the kind needed to build a new house or totally refurbish a villa or build a top-quality funeral monument required contracting with private builders; it could not have been cost efficient, or even conceivable, for the Ciceros to have supported such skilled builders permanently (i.e., as slaves), since they did not engage in sufficiently extensive building projects to justify the expense. [. . .]

Public building during the Republic

We are told by Polybius that, by the second century BC, expenditure for public construction, including new projects and the maintenance of existing structures, was the biggest single drain on the Roman treasury.[2] But public building by contract is attested at least two centuries earlier and probably can be seen as far back as 435 BC, when the censors gave their approval of the newly built *villa publica* [gathering-place in the Campus Martius], which certainly implies approval of its construction by the magistrates (*probatio operis*) and, in turn, strongly suggests that they had let a contract for its construction, which they were seeing completed in the required manner.[3] Indeed, there is reason to suppose that public contracting, for all sorts of necessities in addition to construction, may have been in existence as far back as the fifth century or even the regal period of Roman history [. . .]

The near cataclysm and associated expenses of the Second Punic War caused an understandable hiatus in magisterial building projects at Rome, and the period of recovery after the defeat of Hannibal in 202–201 BC seems to have continued into the following decade. But in 196 BC, Livy provides his first specific reference to a building contract let in that year by the aediles for the construction of a temple.[4] In 193 the curule aediles [magistrates whose responsibilities included public buildings] built two *porticus* [colonnaded walkways], both called Aemilia, for utilitarian purposes as well as a huge wholesale market, called *Emporium*, next to the Tiber. The next year, the curule aediles built another porticus, which was paid for with revenue from fines they had levied.

By the 180s BC the economic recovery was complete and the censors returned to public building in an extremely important way beginning in 184 BC when M. Porcius Cato and L. Valerius Flaccus held the magistracy and spent large amounts of money on new and repair work to drains, the water supply, and the roads, as well as the construction of the first basilica, the Basilica Porcia, in Rome. While the evidence indicates that there was some senatorial opposition to these schemes, it was much less than that exerted in the fourth century BC against Appius Claudius Caecus's program of construction and probably reflects a generally more positive attitude toward censorial proposals aimed at coping with the problems brought on by the city's growth. [. . .]

In 179 BC, the censors M. Fulvius Nobilior and M. Aemilius Lepidus enjoyed the allocation of an entire year's *vectigal*, or tax revenue, for public building projects, which included the usual provisions for maintenance and repair of the aqueducts, the road system, harbors, drains, and sanitation but also allowed the construction of the Basilica Aemilia in the Forum and was supposed to permit the construction of a new aqueduct. The last item was blocked by M. Licinius Crassus, who would not give right of way for it to cross land that he owned. [. . .]

The censorial system of contracting for public building had serious disadvantages. Nonetheless it worked quite well, at least through the second century BC. The problems included the fact that the censorship was a unique magistracy within the governmental system of Republican Rome, which was otherwise primarily made up of annually elected offices. But the censors were elected for a term of eighteen months only once every five years. Eighteen months was hardly sufficient time for the holders of the office to plan, contract for, and see to completion any kind of major building project, and special exceptions and arrangements had to be made repeatedly in order for an important building to be completed according to the contract. Even more problematical, however, must have been the utter lack of any kind of professional staff to advise on, supervise, or carry out public works. This left the censors, generally aristocratic amateurs, at the mercy of contractors who seem quite often to have been untrustworthy. This was a problem that continued into the first century BC and was solved only under the early Empire. [. . .]

Nonetheless, the censorial system was surprisingly effective at least through the second century BC [. . .] Especially in the second century, it was to a large degree censorial contracting that resulted in the design and execution of new and very advanced works of architecture, new and exciting developments in construction technique (most important, of course, the discovery of the bonding properties of *pulvis puteolanum*, or pozzolana sand, which led to the development of *opus caementicium* [concrete]), and the greatly extended and improved amenities necessary to a rapidly expanding city. [. . .]

As in so many other areas of Roman life, the first century BC proved a crucial but crisis-oriented period of substantial change in the system of public building. While the censorial system remained intact, the rise to power of the great Roman *condottieri* [leaders of mercenaries] – Marius, Sulla, Pompey, and Caesar – combined with the tremendous new technical possibilities that had been opened up to Roman architects and builders by the advances in techniques and design made during the second half of the second century BC led to a major change in the methods and control of public building. It was Sulla who combined, during the dictatorship in Rome, the necessary power, resources, and ambition and took advantage of the opportunity to undertake the first massive public building program in the capital. He indulged in all manner of grandiose projects including the rebuilding of the temple of Jupiter Capitolinus, the repaving of the Forum and (perhaps) rebuilding of the Curia [Senate-house] and the Rostra [Speaker's platform], and the creation of the Tabularium [Record-office] [. . .] This use of public works for propaganda purposes and political prestige presaged what would happen soon thereafter; public building would become a kind of power play between the participants in the struggle for supremacy of the last decades of the Republic.

Not only did Sulla not overhaul the general system and functioning of public works administration, he rendered it even less efficient since no censors were appointed for a decade after his death and their absence rendered the old system utterly unusable. Other annual magistrates, with other duties to handle in addition, found themselves swamped under more maintenance contracts than they could possibly manage, and so extortion and fraud ran riot. To keep the system from collapsing completely the whole spirit of the Republican system had to be abrogated and curators appointed to handle public works for long periods of time. Even so, public works were badly managed throughout the first two-thirds of the first century BC, and by the time of Octavian's rise to power, their condition was desperate. [. . .]

Public building during the Empire

[. . .] Octavian realized that he needed to go far beyond glorious architectural creations: the functioning fabric of the city was deteriorating alarmingly, and the system for maintaining and repairing it was in hopeless disarray. His efforts to convince some of the wealthier senators to put some of their own money, or spoils of war (*manubiae*), into the repair of roads, for instance, met with little success.

As early as 33 BC, it had become apparent to Octavian that he would have to reorganize the public works administration, and in that year he began by persuading Marcus Agrippa to assume the office of aedile, in which he was to take on a vast responsibility for every facet of public works and public building in the city. The aedileship, like the censorship, had fallen largely out of use in the decades preceding 33 BC, and very little of the office's traditional *cura urbis* [literally, 'care of the city'] had been exercised, much to the city's detriment. Agrippa's program as aedile would have been unthinkably broad at any previous point in Roman history. Agrippa undertook repairs and long-deferred maintenance of the roads, buildings of every sort, and the water system, while at the same time reorganizing the services that were responsible for them. He repaired the three republican aqueducts – the Appia, Anio Vetus, and Marcia – and began the new Aqua Julia. He also made detailed arrangements for the maintenance of the aqueducts in the future and continued to take direct responsibility for them after 33 BC. To carry out this last task, Agrippa had to reorganize the whole water operation, and he did so by setting up a permanent staff of stonemasons and other skilled or semiskilled workmen, who provided the core of an efficient department responsible for this essential public work. For these and other building projects, Agrippa maintained a large, professional, highly organized staff of advisers, supervisors, architects, and agents, which at

his death in 12 BC may have numbered as many as 240 workers involved with the aqueducts alone. This staff must have provided the model not only for the Imperial water board but also for what was to become in later reigns the permanent construction board called *opera Caearis*.

While Agrippa may well have provided the model for the later Imperial public works administration, it is quite clear that Augustus (as Octavian became in 27 BC) did not attempt to set up any kind of single central authority to handle these tasks. That was a reform for a later reign. What he did do was twofold: he removed his program of new building projects completely from contact with those officials responsible for maintenance and repair of public works (for which he devised a permanent bureaucracy) and initiated a policy of setting up permanent boards supervised by *curatores* which would have the responsibility in perpetuity for maintaining and repairing essential public works. [. . .]

Augustus's reformed system of administration remained essentially unchanged well into the last years of the first century AD. Control of public works passed slowly but seemingly inexorably out of private control into the Imperial administration. [. . .] We know that, beginning with the reign of Claudius (AD 41–54), all the various boards for public works administration were operating officially *ex auctoritate Caesaris* [on the authority of the emperor], no longer even attempting to preserve the myth of functioning *ex auctoritate senatus* [on the authority of the Senate], as Augustus had maintained. Their finances were provided not from the nominally senate-controlled *aerarium* [public treasury] but directly from the Imperial *fiscus* [Imperial treasury]. More and more, every form of public works, whether new or not, came under the control of Claudius's staff, probably in part because of the increase in private and quasi-private building undertaken by the Julio-Claudians. As more and more buildings were put up by the emperors and their relations, they seem to have needed and maintained an ever growing staff of architects and other skilled builders who were kept completely distinct from builders working on public construction projects for the official boards. Their main charge was to maintain and repair the Imperial family's own possessions. After the fire of AD 64, Nero assumed responsibility for rebuilding the city, apparently making use of his own architects and planners. While his plan for altering the whole face of the city seems never to have been completed, it did add to the power and influence of the emperor's building staff and put them in positions of supervisory authority over much of the city's architecture.

The accession of the Flavians seems to have led to ever greater concentration of public works administration in Imperial hands. This development led eventually to the establishment of a massive public works department called the *opera Caesaris*, which became the dominating official presence in Roman public architecture from that time forward. By Trajan's reign it had become customary for all public building projects to be scrutinized by the emperor's staff, which now controlled almost every aspect of architecture and city planning and was even exercising supervisory rights over maintenance and repair. The emperors wanted such control over building, in part, for propaganda purposes, given how effectively architecture and public building had served to promote the policies of Augustus and Domitian. [. . .]

Supply of manpower to the building trade

What evidence we have for how manpower was supplied to the building industry in the Roman world is concentrated almost exclusively on the city of Rome itself. While this might give cause for suspicion – that the situation in what was by far the largest city of antiquity might be unique rather than exemplary –

recent investigations of what little evidence there is outside Rome have, instead, tended to corroborate the evidence from the city on how unskilled labor was supplied, not only to the building trade but in general. Thus it seems reasonable to analyze the situation in the city and allow it to serve as a general model for the Roman world.

To understand the nature of the supply of manpower to the building trade in Rome, it is necessary first to review quickly what we know of the makeup of the city's population. There are notoriously few reliable pieces of evidence with which to determine the actual number of people in the ancient city at any point during its existence: such evidence simply has not been preserved, and what little there is is open to dispute and interpretation. For Italy as a whole, information is even more sketchy, and for the Empire the evidence is effectively nonexistent. It is now widely assumed that a figure between 750,000 and 1,000,000 is not too far wrong for the city in the age of Augustus, based on the known numbers of male citizens who are attested as recipients of the grain dole for the years 46, 45, 44, 29, 24, 23, 12, 5, and 2 BC and for AD, 14 and 37. Whether the Imperial city grew beyond such a figure is at best a tortured subject, and one not really susceptible of proof. For our purposes, however, the simple realization that by the latter half of the first century Rome was far and away the largest urban concentration of people in the ancient Mediterranean world and that she must have remained so well into the Empire will at least permit us to analyze the nature of the unskilled labor supply that was available there.

We are told over and over again in literary sources for the late Republic and early Empire that the countryfolk of Italy tended to drift to the city in large numbers and that this constituted a significant societal problem in the ancient city. [. . .]

The phenomenon of Italian peasant migration into Rome, and the resulting problems of massive urban unemployment and reliance on public dole, continued at least well into the reign of Augustus, if not later [. . .] The question then becomes: aside from reliance on the grain dole, which was hardly sufficient to support one man and utterly insufficient for a family, how did this burgeoning mass of largely unskilled immigrant labor survive? Rome's great size and power resulted, after all, from military and subsequent political conquest. The city never possessed any sort of industry that, in a modern sense, could provide widespread employment, with the single exception of the very trade and transport that supplied it with food and raw materials. Thus, as Brunt has pointed out,[5] the very necessity of supplying Rome's artificially swollen population must to a large degree have created a demand for cheap and unskilled labor to service that supplying. Clearly there would have been a demand for muscle at docks, on transport from the docks to the city (both up the river from Ostia and, before the mid first century AD if not later, on coastal routes to Ostia from Puteoli and other ports on the western coast of Italy), at the unloading sites such as the Emporium in the city, and to the *macella* and other wholesale and retail markets throughout the city. Thus the supply of transport could have made use of a certain proportion of the unskilled labor available in Rome, and it is easy then to hypothesize other areas of demand for muscle and sweat that would have stemmed from that industry. By the same token, it seems reasonable to assume that substantial numbers of unskilled laborers might have come to be employed in the building trade, from its initial expansion in the second century on well into the second century AD at least. We have seen from our survey of the history of building in Rome that not only did public construction increase dramatically from *c.* 181 BC onward, but that private construction in and around the city also was on the rise throughout the central centuries of Roman history. Wealthy citizens,

like Marcus and Quintus Cicero, constantly seem to have been engaged in building projects, both new and reconstruction, both in the city and in the nearby countryside. At the same time, it is reasonable to assume that the continuing increase of the city's population from the Italian peasantry provided an ever increasing demand in and around the city for humble dwellings, tenements, and *tabernae* and that the private building trade supplied these, at least in part, by providing employment for unskilled laborers. Thus demand mandated supply, and supply struggled to keep up with demand.

It cannot be denied that Rome's was an economy supported and propelled to a great degree by slave labor. We know that slaves were employed in every capacity in Rome and throughout the Roman world, in skilled work as well as in the most gruelling of manual tasks. But the evidence for the supplying of unskilled labor in the building industry through gangs of slave workmen is practically nonexistent. Such an arrangement is mentioned only in situations in which a *redemptor* could expect regular or constant employment that would justify the maintenance and feeding of such a servile company - attested to us, as we have seen, in the case of contractors who were engaged on the maintenance and repair of the water system during the Republic and who were required by law to maintain the necessary slave labor force in order to be permitted to bid for the aqueduct contracts[6] - and permit the contractor to make a profit even with the relatively heavy expense of maintaining his slaves. In most industries, and particularly in building and construction, the demand must have fluctuated with a number of variables: the season of the year (Frontinus says construction should be undertaken only between April and November and halted when the weather was too hot),[7] the state of the private economy or of the money available for public expenditure on building, and the very nature of a construction industry itself (in which the completion of the building also terminates the demand for the labor that built it). We know that the ancient Roman landowner did not maintain a sufficient slave *familia* on his estate(s) to cope with the seasonal harvests or with major operations beyond normal maintenance of the property[8] but hired a contractor who could provide, as well as skilled workers and (sometimes) materials, the necessary manpower. While we are told that some building contractors did maintain some number of permanent slaves, there can be little question that these regular workers would not have sufficed, and were not intended to, during any period of heavy demand for construction (a 'building boom') and may well have been proficient in particular skills that justified their maintenance by the contractor even when there was no immediate demand. [. . .]

While building certainly requires a pool of skilled workers who can deal with the relatively technical tasks – from the plans and advice of the architect through the specific skills required of the woodworkers, stonemasons, plasterers, painters, and so forth – these seem often to have been either slaves of particular contractors or of a *familia* involved in the building trade, whose special skills were recognized and respected, often leading to their manumission and continuance in the building industry as freed specialists. These were not the men who provided the sheer hard labor of construction. It is important to remember that Rome was an almost completely nonmechanized society where heavy labor required sheer muscle and sweat. Much of what had to be done to erect a building was heavy labor of exactly this back-breaking kind, and little of that labor would be likely to have been supplied by slaves when there was no clearly profitable way in which to maintain a slave force of unskilled workers for a seasonal and occasional industry. Slaves have to be fed, clothed, housed, and kept alive whether they are at

work or not, and hence they constitute an expensive commodity unless they can be kept continuously and profitably busy. Construction must always have provided a demand for unskilled manpower during the proper seasons of the year and when other circumstances so dictated. If slaves were not an economical way to meet this occasional demand, then the existence at Rome of a large number of free but poor men who had, to some degree at least, to support themselves (and their families if they had them) must have been exactly what the building contractor most wished for. Such men could be hired solely for the duration of a project or by the day, as needed, and their wages could have been controlled so that their employment was profitable to the contractor yet financially rewarding to them, too. This appears to be the way in which most unskilled labor was provided for building in Rome. The slight evidence that remains indicates that the same pattern of employing nonslave labor for building projects was in use not just in the city, with its readily available mass of urban poor, but throughout Italy and in much of the Empire. It is reasonable, then, to assert that the brute manpower for building tended to be supplied, all over the Roman world, through the hire of free but unskilled laborers for specific projects, rather than by gangs of unskilled slaves maintained by contractor, landowner, or magistrate.

A further point for consideration is implied by this analysis and is put forward by Brunt: since nonslave labor was regularly and widely employed for the construction of public works at Rome – and indeed throughout the Roman world – it can at least be thought probable that the frequent policy of pursuing public building programs and expending a great deal of public funding on public works was, though only secondary or even tertiary to the urbanistic and propagandistic motives behind such policy, intended to keep a substantial proportion of the masses of urban poor employed at least part of the year. But while this point seems eminently reasonable, it is attested nowhere in our surviving ancient sources. Indeed, the ancient sources for laborers, especially the unskilled, are paltry at best. Even the epigraphical record does not attest to anyone, slave or free, who provided the brute manpower clearly essential to the building industry or to any other labor-intensive enterprise in the Roman world. The poor Roman laborer certainly could not afford to put up an elaborate, or even a simple, epitaph for himself and his family and probably could not have read it even had he possessed the means to commission it. Beyond inscriptions, the evidence is random and open to interpretation. [. . .]

Beyond these fragments the ancient writers are silent, as are Roman inscriptions, which give a notoriously incorrect impression of Roman manual laborers, since only those able to afford them were members of *collegia* [guilds] or set up epitaphs (two of the three varieties of inscription that mention workers at all), while the manufacturer's marks encountered on bricks, pipes, and pots are largely random and contribute little to evidence outside their own narrow spheres. The lack of testimony concerning labor and laborers extends even to the legal texts, which were concerned, practically enough, with legal precedent and potentiality, that is, cases that had actually been tried in court or seemed likely to be. The truly poor do not turn to the law or the courts, since they can seldom afford lawyers or court costs. Hence the lack of anything resembling a body of labor law in the Roman jurists is only to be expected. [. . .] So, while it seems reasonable and logical to follow the brilliant analysis offered by Brunt, we have to keep in mind that the evidence is minimal. For the building industry it seems, on the whole, most likely that nonslave labor was the major source of manpower for construction projects throughout the Roman world, that slaves attached to the building industry tended to be

skilled specialists in a particular phase or element of the building process and were regularly manumitted and continued in the profession as freedmen, and that this system was used precisely because it permitted the maximum profit to be made by the contractor and the maximum use to be made of the biggest source of unskilled labor available. But it is conjecture and hypothesis, not demonstrable fact, however reasonable and appealing a reading of the fragmentary evidence. [. . .]

Brick and tile

[. . .] As baked brick took on a role in construction during the late first century BC, and [. . .] came into extensive use in the first century AD, the demand for brick led to the development of one of the more remarkable industries of Rome. While the brick industry functioned in many parts of the Empire, the evidence for it is most extensive and best known in and around Rome, where demand was greatest, and it is at Rome that we can consider the production of brick and tile as a true example of an ancient industry. The evidence for the industry, its organization, and its participants is almost entirely epigraphical.

Beginning near the end of the first century BC, just as baked brick became important, its makers began from time to time impressing stamps onto the bricks and the texts of these stamps form the raw data of our information. The brick stamps from the time of Augustus until the time of Nero tended to be simple, usually just one person's name plus occasionally the name of the brickyard (*figlinae*) from which the brick itself came. Late in the first century, probably on account of the gigantic increase in demand that the brickmakers experienced after AD 64, the stamps tended to become increasingly elaborate and correspondingly more informative, probably because of the rapid development of organized procedure that would have been needed in the brick industry. By the end of the first or beginning of the second century AD the shapes of the stamps had become quite varied as had the quantities of information they might contain. In AD 110 for the first time so far known to us the names of the consuls of Rome for that year were included in the text of a brick stamp, thus providing an absolute date for the manufacture of bricks so stamped; this occurs periodically thereafter until AD 164. Brick stamps continue to appear on bricks into the reign of Caracalla (AD 211-217). There is an inexplicable but total hiatus in the use of stamps bearing texts between the end of the reign of Caracalla and the beginning of the reign of Diocletian (AD 284-306); in later antiquity they reappear, almost always bearing only the name of an emperor, all the way into the fifth century AD (see Figure 8.1).

The stamps can provide five different sorts of information, depending on their texts; no one stamp carries all five, however, so the gathering of their evidence is a laborious and specia-lized study. The five kinds of information that may appear on brick stamps are:

1 The type of product it is, for instance, a tile (*tegula*) or, more generally, a brick/tile product (*opus doliare*);
2 The name of the clay field from which the raw material came, or of the brickyard where it was made (*figlinae*);
3 The name of the owner (*dominus, domina*) of the place, usually a private estate, on which the clay was located (*praedia*);
4 The name of whoever was in charge of producing the brick (*officinator* or once in a while *conductor*); and
5 The names of the consuls of the year it was produced (*COS(S)*).

The names are frequently abbreviated, sometimes drastically so (to a single letter), and while this habit hampers certain identification

Figure 8.1 Examples of brick stamps, first century BC to fourth century AD [Adam J.-P. (1984) *La construction romaine: matériaux et techniques*, Paris, Picard; reproduced by permission of the author]

of the parts of the name in some instances [. . .] and while slaves are generally connected with the names of their masters [. . .] there is otherwise little consistency among the different stamps as to which elements of names are presented or in what order they appear. In a few cases persons mentioned on the brick stamps are known independently of them as well, but these are almost always either estate owners or consuls, members of the aristocracy.

What the stamps do clearly tell us is that there were two people essential to the production of brick in the Roman system: the *dominus* or *domina* who owned the estate, and hence the clay from which the brick was made, and the *officinator* who must be the manufacturer. Owners tended to be aristocrats, often of senatorial rank, and quite often female; *officinatores* came usually from the lower middle stratum of Roman society and were often freedmen or slaves. The very high social status of the *domini* has been interpreted to reflect an ever increasing stranglehold on ownership of the brick industry in ever fewer and more influential hands. This led in the end to its disappearance into an Imperial monopoly, but the relative independence of the *officinatores* from the *domini* may imply that, rather than being agents tied to the estates of the *domini*, the *officinatores* often acted as independent entrepreneurs, possibly leasing the brickyards for production with the clay fields and workshops included in the arrangement. *Officinatores* do not, however, seem to have owned their *officinae*, and they do not become *domini*. It is probably overall the best suggestion to think of them rather like *institores*, or workshop managers, who were the more-or-less independent contractors responsible for the running of a number of other businesses in the Roman world. In this social context, it also makes excellent sense to think of *figlinae* and *praedia* as essentially equivalent terms that indicate the brickyards located on or near the actual fields that produced the clay from which the bricks

were made, and the *officina* as the actual production unit, presided over by the *officinator*.

It is of particular interest that such a large number of women are attested in brick stamps, both as *dominae* and as *officinatores*. A few examples will reward closer scrutiny. In many cases these were women of the very highest aristocracy who held or had inherited landed estates and hence were *dominae* to the brickyards operating on them. A remarkable example is Flavia Seia Isaurica, who was *domina* of six separate brickyards that operated more or less contemporaneously, though on different estates, between AD 115 and 141. [. . .]

There are also attested women working as *officinatores*, whether as free entrepreneurs or directly as workshop managers in brickyards, and the female *officinator* was specifically provided for in Roman law. Children, both female and male, worked in the brickyards from a fairly early age. An inscription from Pietrabbondante in Samnium, datable to the first years of the first century BC, records two female slaves, named Amica and Detfri, slaves of Herennius Sattius, who was a well-known tile manufacturer. The two women had signed the tile when they made it and impressed their footprints on it, one set of footprints and one signature on each side. Aubert estimates from the size of the footprints that Amica and Detfri would only have been about twelve years old, though admitting that such estimation is largely speculative. What is perhaps more surprising than the possible employment of slave children, which was common in the Roman world, is that the two girls were literate. Given their possession of this skill, it is reasonable to suppose that they may have been record-keepers or office staff or even, to some degree, managers in the workshop.

The record of brick stamps, most important those that have been found *in situ* in otherwise datable architectural contexts, has contributed tremendously to the chronological study of Roman Imperial architecture and at the same time permits us to trace the chronology of the individual brickyards and thence of the whole industry with unparalleled completeness in ancient history. The brick industry grew on account of the intense demand generated after the fire of AD 64 and by the immense building programs undertaken by Nero, the Flavians, and Trajan into the early decades of the second century AD. It reached its highpoint of development under Hadrian, who may have attempted to regulate or reform the industry in some way, since our evidence shows that in the year AD 123 the names of the consuls Paetinus and Apronianus were included on every stamp used in that year. This was at the same time that M. Annius Verus, himself an important *dominus* in the industry, was also city prefect of Rome and may have insisted upon meticulous use of these stamps. We do not know exactly what Hadrian was attempting to accomplish, and whatever it was it does not seem to have worked, since brick stamps are never again used so extensively.

After Hadrian's reign, public and private building slowed down and so must have the heaviest demand for bricks. [. . .] Brickyards seem to have come more and more often into the ownership of the Emperors; [. . .] by the time of Septimius Severus an industry that had grown and flourished as a private enterprise in which land owners developed and exploited the possibilities of their lands by letting skilled brickmakers establish, either by lease or by contract, brick workshops on them, appears to have become little more than an Imperially owned and controlled monopoly. [. . .]

To conclude, we have in the Roman brick and tile industry a unique quantity of documentary evidence for the organization of and participants in one of the most important supply industries that serviced the Roman building and construction trade. Clearly, the fortunes of the brick industry and its practitioners were tied to the expansions and contractions of, primarily, the immense public building programs of the emperors of the first and second centuries AD. The land and natural resource (clay fields) essential to the business were largely owned and exploited by wealthy aristocrats, up to and including the Imperial families. Interestingly, the actual brickyards themselves – that is, the means of production as well as the raw materials – were owned by the *domini*. Actual exploitation of the fields, the manufacture of bricks, was done by *officinatores*, either free entrepreneurs or skilled slaves of the landowners, who were essentially workshop managers. Whether the stamps themselves served as abbreviated forms of the legal contracts between *dominus* and *officinator*, thus permitting levels of production to be assessed between different workshops, or as records for general inventory in the brick business that would be useful to the Imperial bureaucracy, or as some form of identifying 'trademark' or 'quality guarantee' or even 'advertisement' remains an open question. Perhaps they served various of these functions either at different periods or at one and the same time. [. . .]

Notes

1 S.D. Martin, *The Roman Jurists and the Organization of Building in the Late Republic and Early Empire*, Collection Latomus, 204, Brussels, 1989, p. 46

2 Polybius 6.13.3

3 Livy 4.22.7

4 Livy 33.42.10

5 The discussion that follows is based on P.A. Brunt's essays, 'The Roman Mob', in M.I. Finley, (ed.), *Studies in Ancient Society*, London, Routledge and Kegan Paul, 1974, pp. 87–90, and 'Free Labour and Public Works at Rome', *JRS*, 70, 1980, esp. pp. 92–7

6 Frontinus, *Aq.* 96

7 Ibid., 123

8 Cato, *De Agr.* 39.2.2–3 and 5.2, in which the landowner is strongly advised not to keep excess hands, but to hire them for special needs, including building

9

THE CONSTRUCTION OF FORTIFIED TOWNS

by Vitruvius

Source: Vitruvius, *The Ten Books on Architecture*, trans. Morris Hicky Morgan, New York, Dover Publications, 1960, pp. 17–25, 27, 31–2.

Book I

Chapter IV: The site of a city

1 For fortified towns the following general principles are to be observed. First comes the choice of a very healthy site. Such a site will be high, neither misty nor frosty, and in a climate neither hot nor cold, but temperate; further, without marshes in the neighbourhood. For when the morning breezes blow toward the town at sunrise, if they bring with them mists from marshes and, mingled with the mist, the poisonous breath of the creatures of the marshes to be wafted into the bodies of the inhabitants, they will make the site unhealthy. Again, if the town is on the coast with a southern or western exposure, it will not be healthy, because in summer the southern sky grows hot at sunrise and is fiery at noon, while a western exposure grows warm after sunrise, is hot at noon, and at evening all aglow.

2 These variations in heat and the subsequent cooling off are harmful to the people living on such sites. The same conclusion may be reached in the case of inanimate things. For instance, nobody draws the light for covered wine rooms from the south or west, but rather from the north, since that quarter is never subject to change but is always constant and unshifting. So it is with granaries: grain exposed to the sun's course soon loses its good quality, and provisions and fruit, unless stored in a place unexposed to the sun's course, do not keep long.

3 For heat is a universal solvent, melting out of things their power of resistance, and sucking away and removing their natural strength with its fiery exhalations so that they grow soft, and hence weak, under its glow. We see this in the case of iron which, however hard it may naturally be, yet when heated thoroughly in a furnace fire can be easily worked into any kind of shape, and still, if cooled while it is soft and white hot, it hardens again with a mere dip into cold water and takes on its former quality. [. . .]

5 It appears, then, that in founding towns we must beware of districts from which hot winds can spread abroad over the inhabitants. For while all bodies are composed of the four elements [. . .], that is, of heat, moisture, the earthy, and air, yet there are mixtures according to natural temperament which make up the natures of all the different animals of the world, each after its kind. [. . .]

8 Therefore, if all this is as we have explained, our reason showing us that the bodies of animals are made up of the elements, and these bodies, as we believe, giving way and breaking up as a result of excess or deficiency in this or

that element, we cannot but believe that we must take great care to select a very temperate climate for the site of our city, since healthfulness is, as we have said, the first requisite.

9 I cannot too strongly insist upon the need of a return to the method of old times. Our ancestors, when about to build a town or an army post, sacrificed some of the cattle that were wont to feed on the site proposed and examined their livers. If the livers of the first victims were dark-coloured or abnormal, they sacrificed others, to see whether the fault was due to disease or their food. They never began to build defensive works in a place until after they had made many such trials and satisfied themselves that good water and food had made the liver sound and firm. If they continued to find it abnormal, they argued from this that the food and water supply found in such a place would be just as unhealthy for man, and so they moved away and changed to another neighbourhood, healthfulness being their chief object. [. . .]

Chapter V: The city walls

1 After insuring on these principles the healthfulness of the future city, and selecting a neighbourhood that can supply plenty of food stuffs to maintain the community, with good roads or else convenient rivers or seaports affording easy means of transport to the city, the next thing to do is to lay the foundations for the towers and walls. Dig down to solid bottom, if it can be found, and lay them therein, going as deep as the magnitude of the proposed work seems to require. They should be much thicker than the part of the walls that will appear above ground, and their structure should be as solid as it can possibly be laid.

2 The towers must be projected beyond the line of wall, so that an enemy wishing to approach the wall to carry it by assault may be exposed to the fire of missiles on his open flank from the towers on his right and left. Special pains should be taken that there be no easy avenue by which to storm the wall. The roads should be encompassed at steep points, and planned so as to approach the gates, not in a straight line, but from the right to the left; for as a result of this, the right hand side of the assailants, unprotected by their shields, will be next the wall. Towns should be laid out not as an exact square nor with salient angles, but in circular form, to give a view of the enemy from many points. Defence is difficult where there are salient angles, because the angle protects the enemy rather than the inhabitants.

3 The thickness of the wall should, in my opinion, be such that armed men meeting on top of it may pass one another without interference. [. . .]

4 The towers should be set at intervals of not more than a bowshot apart, so that in case of an assault upon any one of them, the enemy may be repulsed with scorpiones [military throwing engines] and other means of hurling missiles from the towers to the right and left. Opposite the inner side of every tower the wall should be interrupted for a space the width of the tower, and have only a wooden flooring across, leading to the interior of the tower but not firmly nailed. This is to be cut away by the defenders in case the enemy gets possession of any portion of the wall; and if the work is quickly done, the enemy will not be able to make his way to the other towers and the rest of the wall unless he is ready to face a fall.

5 The towers themselves must be either round or polygonal. Square towers are sooner shattered by military engines, for the battering rams pound their angles to pieces; but in the case of round towers they can do no harm, being engaged, as it were, in driving wedges to their centre. The system of fortification by wall and towers may be made safest by the addition of earthen ramparts, for neither rams, nor mining, nor other engineering devices can do them any harm. [. . .]

Figure 9.1 Construction of city walls. (From the edition of Vitruvius by Fra Giocondo, Venice, 1511)

8 With regard to the material of which the actual wall should be constructed or finished, there can be no definite prescription, because we cannot obtain in all places the supplies that we desire. Dimension stone, flint, rubble, burnt or unburnt brick, - use them as you find them. For it is not every neighbourhood or particular locality that can have a wall built of burnt brick like that at Babylon, where there was plenty of asphalt to take the place of lime and sand, and yet possibly each may be provided with materials of equal usefulness so that out of them a faultless wall may be built to last forever.

Chapter VI: The directions of the streets; with remarks on the winds
1 The town being fortified, the next step is the apportionment of house lots within the wall and the laying out of streets and alleys with regard to climatic conditions. They will be properly laid out if foresight is employed to exclude the winds from the alleys. Cold winds are disagreeable, hot winds enervating, moist winds unhealthy. We must, therefore, avoid mistakes in this matter and beware of the common experience of many communities. For example, Mytilene in the island of Lesbos is a town built with magnificence and good taste, but its position shows a lack of foresight. In that community when the wind is south, the people fall ill; when it is northwest, it sets them coughing; with a north wind they do indeed recover but cannot stand about in the alleys and streets, owing to the severe cold. [. . .]

8 [. . .] The lines of houses must therefore be directed away from the quarters from which the winds blow, so that as they come in they may strike against the angles of the blocks and their force thus be broken and dispersed. [. . .]

Chapter VII: The sites for public buildings
1 Having laid out the alleys and determined the streets, we have next to treat of the choice of building sites for temples, the forum, and all other public places, with a view to general convenience and utility. If the city is on the sea, we should choose ground close to the harbour as the place where the forum is to be built; but if inland, in the middle of the town. For the temples, the sites for those of the gods under whose particular protection the state is thought to rest and for Jupiter, Juno, and Minerva, should be on the very highest point commanding a view of the greater part of the city. Mercury should be in the forum, or, like Isis and Serapis, in the emporium; Apollo and Father Bacchus near the theatre; Hercules at the circus in communities which have no gymnasia nor amphitheatres; Mars outside the city but at the training ground, and so Venus, but at the harbour. It is moreover shown by the Etruscan diviners in treatises on their science that the fanes of Venus, Vulcan, and Mars should be situated outside the walls, in order that the young men and married women may not become habituated in the city to the temptations incident to the worship of Venus, and that buildings may be free from the terror of fires through the religious rites and sacrifices which call the power of Vulcan beyond the walls. As for Mars, when that divinity is

enshrined outside the walls, the citizens will never take up arms against each other, and he will defend the city from its enemies and save it from danger in war.

2 Ceres also should be outside the city in a place to which people need never go except for the purpose of sacrifice. That place should be under the protection of religion, purity, and good morals. Proper sites should be set apart for the precincts of the other gods according to the nature of the sacrifices offered to them. [. . .]

10

ON THE WATER SUPPLY OF THE CITY OF ROME

by Frontinus

Source: Frontinus, *De aquaeductu urbis Romae*, trans. Harry B. Evans as *Water Distribution in Ancient Rome: the evidence of Frontinus*, Ann Arbor, University of Michigan Press, 1994, pp. 20-1, 32-3, 37-40, 43-4

16 With so many indispensable structures carrying so many aqueducts you may compare the idle pyramids or the other useless, although famous, works of the Greeks! [. . .]

18 All the aqueducts arrive in the city at different elevations. As a result, certain ones serve higher places, and others cannot be raised to more lofty areas; indeed even the hills have grown up little by little from rubble on account of the great number of fires. The height of five aqueducts permits them to be raised into every part of the city, but of these, some are forced by greater pressure, others by less. [. . .] But the old aqueduct builders constructed their lines at lower elevation, either because the fine points of the leveling art had not yet been ascertained, or because they deliberately made it their practice to bury aqueducts underground to prevent them from being cut easily by enemies, since a good many wars were still being fought against the Italians. Now, however, in certain places, whenever a conduit has been worn out by age, the lines are rebuilt to be carried across valleys on substructures or arches to shorten their length and avoid underground detours around the low ground. [. . .]

74 I have no doubt that some will be surprised that a far greater supply has been found by measurements we have taken than was indicated in the imperial record books. The reason

for this is the error of those who originally made an estimate of the volume of each line with too little attention. There is no reason to believe that they fell so far short of the true volume because of fear of summer or dry spells; I myself have discovered that the volume of each line indicated above, taken from measurements made in the month of July, remained constant afterwards throughout the whole summer. Yet, whatever the earlier cause of the discrepancy, the following certainly has been revealed: 10,000 *quinariae* [the standard measurement with no equivalent in modern units] have been lost, while the emperors have limited their grants according to the volume listed in the record books.

75 A subsequent discrepancy is the result of one system of measurement being used at the intakes, another much smaller in the settling tanks, and another, the smallest of all, maintained in the delivery system. The reason for this is the dishonesty of the watermen who we discovered were diverting water from public conduits for the use of private parties. Moreover, many of the landowners, past whose land the water is carried, tap the conduit lines, with the result that public aqueducts stop their courses for private individuals . . . [1]

76 Concerning wrongs of this sort, however, nothing more or better can be said than that spoken by Caelius Rufus in his public speech

entitled *On the Waters*: 'Would that we were not proving, by provoking indignant reactions, that all these wrongs are being habitually practiced with comparable impunity: we find fields being watered, taverns, even garrets, and lastly, all the brothels equipped with constantly flowing taps.' [. . .]

87 This supply of water as calculated used to be distributed in this manner up to the time of the Emperor Nerva. Now, thanks to the foresight of our most industrious emperor, whatever was either unlawfully tapped by the watermen or lost through neglect has been added to the total supply, as if by a new discovery of sources. The total capacity indeed has been nearly doubled and subsequently delivered through so careful a distribution system that more than one aqueduct is supplied to regions previously served by only one line, such as the Caelian and Aventine, to which the Aqua Claudia alone used to be brought through the Neronian arches. As a result, whenever some repair had interrupted the supply, these densely inhabited hills lacked water. Now more than one aqueduct supplies them, and in particular the Aqua Marcia, which has been brought back on a large-scale project from Spes Vetus [the area where the main eastern aqueducts enter the city] all the way to the Aventine. Moreover, in every part of the city many basins, new as well as old, have received double taps from different lines, so that if an accident may interrupt one of them, service is not stopped, since the other continues its supply.

88 The queen and mistress of the world, who stands as a 'goddess on earth, to whom nothing is equal and nothing comes close,'[2] feels day by day this concern of her most devoted Emperor Nerva. The health and well-being of the eternal city will be affected by it even more, now that the number of *castella* [distribution tanks], conduits, fountains, and basins has been increased. No less an advantage is supplied to private consumers from the increase of the

emperor's benefits; even those who in fear used to tap water illegally now enjoy it without worry as a result of his grants. Not even waste water is lost; the causes of contagion have been removed, the appearance of the streets is clean, the air is more wholesome, and that atmosphere within our city which was always notorious in the past has been removed. [. . .]

94 It follows that we should mention the laws concerning the introduction and safeguarding of water, one of which pertains to restricting private consumers to the limits of the grant they have legally obtained, the other to the protection of the conduits themselves. In this, while I was reviewing laws passed concerning individual aqueducts, I found that certain practices were followed differently among our ancestors. In early times all water was distributed for public functions, and it was so prescribed 'that no private citizen draw any other water than that which has fallen from the basin to the ground' – these are the words of that law, that is, water that has overflowed from a basin; we call it 'overflow' water. This water in fact used to be allotted for no other purpose than use by baths and fullers' establishments, and was taxed for a set fee to be paid into the public treasury. Some amount also used to be allotted the homes of leading citizens of the state when the others permitted.

95 [. . .] From this it is obvious how much more important a concern our ancestors had for common benefits than for private pleasures, since even water that private consumers drew was related to public needs. [. . .]

103 I will now add those things that the commissioner of the water supply ought to observe and the law and decrees of the senate applicable to determining his conduct of office. Regarding the law on drawing water by private consumers, it must be observed that no one tap water without written authorization from Caesar, that is, that no one tap water not granted, and further, that he not tap more than he has legally been granted. In this way, then,

we will insure that the additional volume, which we said has been acquired, can be used for new fountains and new grants of the emperor. In each case, however, great care must be taken to counter all sorts of dishonesty. The conduits must be repeatedly checked with care outside the city to verify the authorized grants, and the same must be done in the *castella* and public fountains, so that the water may flow without interruption day and night. [. . .]

105 Anyone who wishes to draw public water for his own personal needs will have to obtain it by grant and bring written authorization from the emperor to the commissioner. The commissioner must expedite Caesar's grant and immediately send notice in writing to the freedman of Caesar, imperial deputy of the same office. [. . .]

Notes

1 The text is confused, but Frontinus seems to be stressing the contrast between private owners and public needs
2 Frontinus appears to be quoting Martial 12.8.1–2

II

A MODEL OF AGRICULTURAL CHANGE

by Neville Morley

Source: Neville Morley, *Metropolis and Hinterland: the city of Rome and the Italian economy 200 BC-AD 200*, Cambridge, Cambridge University Press, 1996, pp. 55-70.

Demand and supply

The dramatic growth of the population of Rome was impossible without an equally dramatic increase in the city's food supply. In part, this was achieved through state action. The political imperative to feed the urban *plebs* led magistrates and emperors to intervene in the grain supply, buying additional supplies in times of shortage and distributing grain collected as tax at a reduced price or without charge. The system of the *annona* [public supply of food, usually grain] became increasingly sophisticated, with the appointment of a *praefectus annonae* and incentives offered to shipowners to supply the city. [. . .]

State action alone was insufficient to keep the metropolis fed; a sizeable proportion of the city's demands for grain was met through the free market. Moreover, until the *annona* was expanded to include oil (at the turn of the second century AD) and wine and pork (in the 270s), Rome's supplies of all other foodstuffs were brought in through private channels. The elite may have fed their households from the produce of their own estates; the majority of the urban population had no option but to rely on the vagaries of the market. [. . .] In general, taking into account the times when harvest failure, war, piracy or simple logistical problems led to shortages, the food supply of the city seems to have kept pace with the growth

in population. It was of course the ability of producers as well as traders to respond to this demand that allowed the city to continue to grow.

Such an increase in the food supply could be achieved in two ways. Firstly, the area from which supplies were drawn might be expanded; prices in the city might be sufficiently high to offset the cost of transporting goods from more distant regions. The effects of these demands on the regions in question would depend on the way in which the surplus was mobilised for consumption at Rome. If it was taken as tax in kind or as rent on a share-cropping basis, the effect on local agriculture is likely to have been minimal; the region as a whole might be impoverished, if a surplus previously consumed locally was now being taken to Rome, but individual producers would have no incentive to change their production strategies. If, on the other hand, the surplus was exchanged for money, whether to pay rents and taxes in cash or to buy other goods, there might be significant changes in the local economy as a result.

This brings us to the second way in which the supply of food might be increased: by increasing the amount obtained from existing areas of supply. Taxes and rents could simply be raised; in all likelihood with deleterious consequences for farmers, which might lead in the long term to a decrease in the revenue obtained. Alternatively, farmers might be persuaded to sell more of their produce (potentially risky, if it increased their reliance on the market for their own subsistence) or to sell it to the city rather than local markets. Finally, producers might attempt to increase their marketable surplus by changing their farming strategies. Dramatic changes in agricultural practice might thereby follow: greater specialisation in certain crops, larger inputs of labour and capital, the adoption of different techniques and changes in the organisation of production. In general, we might expect a greater orientation of agriculture towards the urban market, and closer integration of the region into wider economic (and consequently social and cultural) networks.

Study of the development of Rome's food supply has tended to concentrate on the demands it made on its expanding empire. At the end of the third century BC, Sicily and Sardinia became regular suppliers of grain through tax; they were joined by the newly created province of Africa in 146 BC, and by Egypt in the reign of Augustus. As major grain-producing regions, these provinces must also have contributed a sizeable portion of the city's imports outside the *annona* scheme. The growth of the city goes hand in hand with the expansion of the empire, both through the increase in the size of the tax base and through the securing of areas of supply and supply routes. Changes in agricultural practice in the provinces can also be seen, with the development of vineyards in Gaul and Spain and oil production in Spain and Africa; a large proportion of this production was destined for the Roman market, and these changes must surely be linked to the demands of the city.

The role of Italy in supplying Rome has often been neglected. From 167 BC, Roman citizens were not subject to the *tributum* [taxation] and after the Social War this privilege was extended to all Italians; they were therefore free from that pressure to increase productiv-ity. However, much of the country was in a good position to respond to the incentives offered by the Roman market. It has been argued that the influx of provincial grain made cereal production unprofitable in Italy, just as provincial imports are said to have later ruined Italian viticulture. It is hard to believe that a region like Campania was greatly disadvantaged in comparison with Africa or Egypt, either in fertility or in its access to the Roman market. At the most, Italian grain may have become proportionally less important to the metropolis as provincial imports grew, without the actual volume of production necessarily declining.

More significant, however, was Italy's response to the city's demands for wine, oil and other foods, not to mention textiles, wood and assorted luxuries. By the time of Augustus, a large portion of the urban population was being provided with a sizeable proportion of its subsistence needs by the state; it may be assumed that recipients of the corn dole had more income to spend on other goods, and the aggregate demand this implies is considerable. Until the middle of the first century AD, Italy held a *de facto* monopoly on the supply of most food items apart from grain, due to its proximity to the market and to the undeveloped state of agriculture in many of the provinces of the empire. It is difficult to imagine how Italian agriculture could have remained unaffected by the demands of the city.

[. . .] The influence of a metropolitan city on its hinterland is pervasive, but it can take very different forms in different contexts. Madrid's influence on Castile was essentially negative, leading to the stagnation of the region's economy, while London's influence on England was in some sense 'progressive'. In both cases, costs and benefits were distributed very unevenly between different classes in society. In other words, the optimistic assessment of metropolitan influence is by no means the only model available. It is possible to construct a theory whereby the growth of Rome may be reconciled with the decline of Italy; a return, in fact, to the image of Rome as parasite.

The argument here leads towards the opposite conclusion. Rome was not London, but there are clear resemblances; the demands of the city promoted changes in agriculture and other areas of economic life that may, with due caution, be termed 'progressive'. The picture does of course change over time, in response to the rhythm of the city's growth and to changes elsewhere in the empire; the idea that Italian agriculture suffered a crisis at the end of the first century AD must be considered in this context. The scope for economic development – another loaded term – was limited in the ancient world by a number of factors; Rome's relationship with Italy shows how far things might progress, but also offers the chance to consider why they progressed no further.

Patterns of land use

In classical economic theory, agricultural producers aim for optimum land use, in terms both of the type of crop (or mixture of crops) grown and of the intensity of cultivation (that is, the ratio of inputs to land area), based on the prices that can be obtained in the market. If prices change, the optimum type of land use and level of intensity also change, and farmers should alter their production strategies accordingly. It may prove advantageous to change the type of land use, turning arable land over to pasture or specialising in particular crops. Alternatively, the level of inputs may be altered to match the new point of diminishing marginal returns. If prices rise, for example, marginal land may be brought into cultivation, additional labour may be hired (or the existing labour force made to work harder), and it may prove worthwhile to invest in new tools, machinery or animals. There might also be changes in

farm management, to make certain operations more efficient.

Farmers do not of course respond instantaneously to every fluctuation in the market; they may follow longer- or shorter-term strategies, but in general they are planning at least a year ahead, working not only on current prices but on the likely vagaries of the weather. There is therefore a tendency towards conservatism in decision-making, with the persistence of traditional strategies and methods; changes in market conditions must be significant and apparently sustainable to elicit much response. However, the growth of the city of Rome and the consequent rise in demand (and therefore prices) was sustained over at least two centuries. Such a change in market conditions would be large and reliable enough in the long term for the optimising farmer to respond to it, changing his cultivation strategy to a significant – and hence historically visible – extent.

Some responses to rising demand are likely to leave clearer traces in the literary or archaeological records than others. Thus, a decision by farmers to increase labour inputs by working more hours per week will hardly be visible to the historian, whereas a fundamental change in the organisation of labour might be more obvious. Furthermore, those changes which can be detected, in patterns of land use, agricultural technology or rural prosperity, cannot be classified automatically as responses to the growing demands of the city of Rome. It is necessary to consider alternative hypotheses, perhaps social or political rather than economic, even when alterations in the rural landscape correspond to the predictions of economic theory.

So far, discussion has focused on the possible responses of an ideal farmer at an unspecified location. It may at times be necessary, owing to the nature of some of the evidence, to talk in vague terms of changes in 'Italian agriculture', but ideally we should produce a more specific model. For this we may turn to the theory of land use patterns developed by J.H. von Thünen in *The Isolated State*, first published in 1826. This deals with the spatial distribution of agricultural activities around a market; and, while the model is presented in the idealised world of economic theory, and was derived from von Thünen's experiences with his own estate in Germany, the principles set forth in the book can easily be adapted for consideration of the agricultural activities around a particular market, the city of Rome.

Von Thünen developed a concept of economic rent independently of Ricardo, the economist who is most frequently associated with the idea. This concept, which underlies all questions of competition for the use of land, may be defined as the net value of the returns on production on a given piece of land in a given time period, taking into account not only the cost of inputs but also the opportunity costs, the value that these inputs might have if put to alternative uses. On any piece of land, the enterprise that yields the highest economic rent will be conducted.

Von Thünen's chief contribution lies in his stress on the importance of location in calculating the costs of production, whereas Ricardo was concerned with the effects of differences in fertility. The cost of transporting goods to the market must be taken into account when calculating the net return from a particular land-use type; the costs of certain inputs may rise with distance from the market. Every land-use type has a characteristic rent–distance function – that is, the way in which the economic rent it yields varies with distance from the market. Thus, the economic rent on perishable goods like soft fruit declines sharply beyond a certain distance. The rent–distance function is affected also by the bulk of the goods relative to their value, and by the level of inputs required in their cultivation.

To demonstrate the effects of distance on land-use patterns, von Thünen described *Der isolierte Staat*:

Imagine a very large town, at the centre of a fertile plain which is crossed by no navigable river or canal. Throughout the plain the soil is capable of cultivation and of the same fertility. Far from the town, the plain turns into an uncultivated wilderness which cuts off all communication between this state and the outside world . . . The problem we want to solve is this: What pattern of cultivation will take shape in these conditions?; and how will the farming system of the different districts be affected by their distance from the town? We assume throughout that farming is conducted absolutely rationally.[1]

Different crops have different rent–distance functions; thus at a certain distance from the market one crop yields the highest economic rent and hence is cultivated, while further away it is replaced by a different crop. This is shown in Figure 11.1. Figure 11.2 shows the pattern of land use that von Thünen envisaged for his Isolated State; a series of zones, concentric on the market, moving from horticulture and dairying closest to the town to forest, intensive arable, long-ley arable, three-field arable and finally ranching. Both the type of crop and the intensity of cultivation vary with distance.

The model as it stands is static, a description of the distribution of agricultural activities around the market of a given moment. It is not difficult to see how it might change over time. An increase in demand for all types of goods, based on a rise in the population of the town, would allow the zones of cultivation to spread outwards, as higher prices in the market offset higher costs of transport. As supply rises to meet demand, prices return towards their original levels, resulting in a contraction of the zones; in other words, there is a continual process of adjustment, especially at the boundaries of different zones. If demand continues to grow in the long term, the result is a permanent extension of cultivation over a wider area.

When the market in question is a metropolis like London or Rome, the zones of cultivation

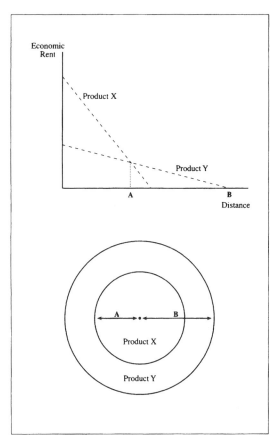

Figure 11.1 A simple model of agricultural location

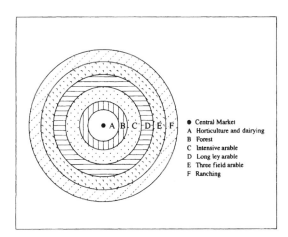

Figure 11.2 Von Thünen's *Isolated State*

extend over whole countries or beyond. [. . .] From a very early date, London promoted changes in the agriculture of East Anglia and Kent; over the seventeenth and eighteenth centuries, the whole of England became increasingly oriented towards the demands of the capital, with the spread of local specialisation (the most obvious example being the development of fruit farming and market gardening in Kent) and the adoption of new techniques.

Of course, with the exception of the band of intensive horticulture in the immediate vicinity of the city, English agriculture was not neatly arranged in concentric circles around the metropolis. Von Thünen's model is designed to isolate a particular set of economic relationships, and to this end it involves a number of clearly unrealistic assumptions; the uniform fertility of the plain and the uniform ease of access to the market are the most obvious. These assumptions, and the consequences for the model if they are relaxed, will be considered in detail below. The important point remains that, according to the model, different regions of Italy, and indeed of the rest of the empire, would respond to the demands of Rome in different ways, ease of access to the city being one vital factor in determining both the form and the extent of their response.

Distance and transport costs

Von Thünen noted that one of the ways in which his Isolated State differed from reality was the assumption of equal ease of access to the urban market from all parts of the territory; most major cities, he observed, were sited on navigable rivers. Transport costs for water travel were much lower than those for land – von Thünen thought they might be as low as one tenth of the overland rate. The existence of a navigable river therefore leads to the extension of the zones of cultivation along its length, as it alters the rent-distance function of each crop. The presence of a superior road, and above all easy access to the sea, would similarly increase the area from which the urban market could draw its supplies.

Comparative examples are not hard to find; the dependence of Paris on the network of French rivers, and the importance of the Thames and the North Sea for the development of London. Various goods, especially bulky staples, could be transported over long distances by water, allowing such cities to grow dramatically and permitting their immediate hinterlands to specialise in non-staple crops. The plight of land-locked Madrid offers a still stronger instance of the importance of water transport; since it was prohibitively expensive to transport grain for any great distance overland, it proved necessary to compel the farmers of the city's immediate hinterland to specialise in cereals. Madrid's geographical position, at least as much as social or political factors, explains the city's stagnating effect on Castilian agriculture. Other great cities in this period could rely on grain from the Netherlands or the Baltic, imported via sea and river, and so had less reason to attempt to regulate farmers; the Parisian authorities, for all their concern about the city's grain supply, concentrated their efforts on the activities of merchants, millers and bakers.

One source from early eighteenth-century England suggests that the cost ratios of different forms of transport were of the order of 1 for sea, 4.7 for river and 22.6 for land; in other words, a given good could be transported 5 miles overland, 25 miles by river and 115 miles by sea for the same price.[2] The figures for transport costs in Diocletian's Price Edict suggest that carrying grain overland would cost 55% of the load's value for every 100 miles, whereas sea transport cost only 1.3% per 100 miles; a first-century Egyptian papyrus gives a figure for river transport of 6.38%. The cost ratios in the Edict are 1 for sea transport, 4.9 for river and 42 for road (although carriage by camel was 20% cheaper).

A commonly cited example of the consequences of the cost of overland transport is the famine that struck Antioch in AD 362-3, despite the fact that grain was available fifty miles down the road. The importance of Rome's location, on a navigable river with good access to the sea, was noted by ancient writers. Strabo describes merchandise being brought down the Tiber; [. . .] Cicero observes that 'the city can not only bring in by sea but also obtain from the land, carried on its waters, whatever is most essential for its life and civilisation' – while noting also that Rome was far enough inland to escape the moral corruption common to most maritime cities.[3]

It is clear that Rome could grow so large because it had access to (and control over) the Mediterranean, and was therefore able to draw on the resources of a very wide area (above all the grain-producing regions). It is equally clear that the relative costs of different forms of transport would determine which areas of Italy were most strongly affected by the city's demands; not all parts of the peninsula enjoyed the same ease of access to the urban market. Figure 11.3 offers some indication of this differential effect; using a ratio of 1:5:25 for sea, river and land transport, it shows the areas from which supplies could be drawn for the cost of moving goods 20 and 30 miles overland. A clear pattern emerges; long stretches of the Italian coast are as close to Rome in terms of the cost of transport as parts of inland Latium and Etruria. We might expect the zones of cultivation oriented towards the urban market to follow a similar pattern. However, it is necessary to consider the question of the cost of transport in the ancient world in more detail.

The first point is that the cost of sea transport in Diocletian's Edict seems suspiciously low. This might be explained in several ways. The cost is derived from the rate quoted for carriage between Alexandria and Rome. The importance of the capital's food supply might

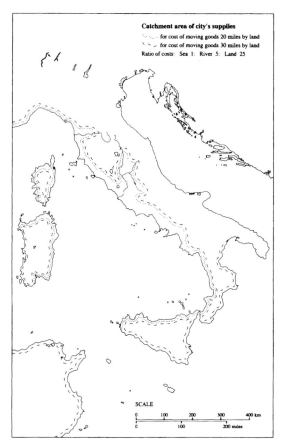

Figure 11.3 The effect of transport costs.

have persuaded the emperor or his advisers to keep this figure at an artificially low level (even at the risk of antagonising shipowners). Another possibility is that the route was so well-travelled that the price *was* relatively low. Alternatively, some but not all of the figures quoted in the Edict may have been affected by inflation; the price of wheat may be exceptionally high and quite out of line with the cost of transport.

The figure quoted takes no account of the costs of loading and unloading vessels, a task that involved the hiring of large numbers of manual labourers at either end of the journey (although during the voyage fewer workers

were required than would be necessary for overland carriage). Finally, the real costs of sea carriage must include the risk of shipwreck (and the cost of insurance) and vulnerability to the vagaries of climate – not only the restricted sailing season, but also the chances that a voyage could be delayed by contrary winds. Nevertheless, there is still no reason to doubt that sea transport was the cheapest way of transporting bulky staples over long distances.

The bulk of our evidence concerning river transport comes from Egypt, which was dominated by the Nile; it is of limited use for Italy. Most Italian rivers were too small and too irregular – either floods or trickles, depending on the season – to be much use for transporting goods. The importance of river transport on the Tiber has been noted; in the rest of Italy, only the Arno and the Po are large enough to have been used regularly for transport. There was also the canal which ran alongside the Via Appia through the Pomptine Marshes, which was regularly used by travellers and presumably also for merchandise. [. . .]

In most parts of Italy, therefore, the choice was between land and sea transport – or most likely a combination of the two. Varro describes the mule trains bringing oil, wine and grain down from the Apulian highlands to the coast, and something similar must have occurred in other regions.[4] Such an arrangement is found in Egypt, with wagons and donkeys used to carry goods to the nearest river or canal. The important question is how far goods could be carried overland to the sea before the cost became prohibitive. If land transport was as expensive as Diocletian's Edict suggests, Rome's influence, however far it extended along the coast of Italy, would be limited to a very narrow coastal strip. In many areas of the peninsula, of course, this coincides with the location of the more fertile land. However, there is reason to suspect that the Edict gives a misleading impression of the cost of overland transport, just as it exaggerates the cheapness of sea travel.

Clearly it would matter what kinds of goods were being transported. It was prohibitively expensive to cart grain overland for any great distance because the charge for transport was levied in terms of weight, and grain has a low value per unit weight. Transport charges would not absorb so much of the value of more expensive and less bulky items, even staple goods like oil and wine (especially the more expensive wines). Various types of goods are better suited to land transport; sheep and other livestock will walk themselves to market, while carrying them by sea would be expensive and liable to reduce their value. Maps based on the transport costs of individual goods will vary significantly.

The transport of livestock would not involve hire charges for vehicles and draft animals; the only costs in monetary terms would be rent for the use of pasture on the journey, and the maintenance of the drovers. Hire charges would be equally irrelevant to a farmer or estate owner using his own means of transport, whether mule or ox-drawn wagon. If the cost was calculated at all, it would be in terms of time lost from other activities. The farmer's trip into market could be multi-purpose, taking in the purchase of other items and various social activities as well as selling produce. The limit of Rome's penetration inland might therefore be set not by transport costs but by the distance farmers were prepared to travel in a day; goods could be drawn from a wider area than strict economic calculation would consider feasible. Richer estate owners might make use of their own ships as well as animals, and it is arguable how far the cost of this transport (let alone the notional 'opportunity cost' of the investment in a vessel) would be reflected in an inflated selling price. Finally, if cost is measured in time rather than money, it is worth noting that land transport could be both faster and more reliable than sea travel; when transporting perish-

able goods like fruit or lettuces, it is hardly profitable to risk having to wait for a favourable wind.

With regard to hired transport, comparative evidence suggests that a distinction may be drawn between long- and short-haul carriage. In both early modern Spain and seventeenth-century England, the former, carried out by town-based professional carriers, was far more expensive than the service offered by local peasant carriers. Roman laws dealing with land transport recognise different forms of hiring; one could hire the equipment alone, or pay the carriers a daily wage, or hire an outfit for a specific task. The risks and legal responsibility were apportioned differently in each case, and we may suppose that the costs were also. Short-haul transport was, however, highly seasonal, based on the use of peasants' animals in slack agricultural seasons.

Finally, there is the question of the economic effect of Roman roads. The agronomists consider that proximity to a road is as important in locating a farm as access to the sea or a navigable river. In the immediate vicinity of Rome, roads may have lowered transport costs by permitting the use of large wheeled wagons (over much of Italy, the terrain was more suitable for pack animals). Other beneficiaries included those involved in bringing animals, especially sheep, down from the mountains.

Even if transport costs overland were halved, this would do little to offset the massive advantages – with all the caveats noted above – of sea transport. It should be noted that Figure 11.3 was drawn using a ratio of costs of sea to land of 1: 25, rather than the 1: 40 which is suggested by the figures in Diocletian's Edict. It is clear that Rome was not surrounded by neat concentric zones of production; its economic hinterland extended along the coast of Italy much further than it reached inland, even along the Tiber. The map serves to emphasise the enormous advantages which the coastal regions of central Italy had over the rest of the peninsular, not to mention other provinces of the empire, in their access to the Roman market. This is the area in which we may expect the most significant response to urban demands. Further afield, we can expect considerable differences in the development of the rural landscape between coastal areas and inland regions along much of the Tyrrhenian seaboard.

Nevertheless, some reconsideration of the map is necessary. The fact that land transport might be cheaper than is generally accepted, or that its cost might not be taken into account at all, means that the 'narrow coastal strip' may be less narrow than strict cost-accounting would indicate. Goods with a greater value per unit weight could be carried over longer distances, and many inland regions produced goods which walked themselves to market. In particular, there is the possibility that it might be economical to carry more expensive goods overland from the Adriatic coast rather than attempting the long sea journey: from Umbria and Picenum to the navigable stretches of the Tiber, and from the Po Valley across to the Ligurian coast. [. . .]

Climate and soil fertility

The idea of equal access to the market from all parts of the hinterland is not the only unrealistic assumption in the Isolated State. Variations in soil fertility and other environmental factors are of immense importance in determining the success and profitability of agriculture; Ricardo derived his theory of economic rent from study of the effects of differences in soil fertility. The optimum land-use type and combination of inputs will differ for different types and qualities of land: at the same distance from the market, it will be profitable to grow a particular crop on some land and not on others. This may be true of adjacent farms, or even of different fields within the same farm. Thus it is unlikely that there will be complete homogeneity

within a region, either of crop types or of intensity of cultivation, and particularly favourable conditions may permit the production of some crops at a greater distance from the market than von Thünen's theory would indicate.

Rising demand encourages the cultivation of more marginal lands when the supply of the best land is inelastic. For example, the area within which horticulture is profitable is strictly limited by the speed of transport; around a large city, all but the poorest land will be turned over to intensive cultivation. The growth of the market therefore tends to increase the homogeneity of land-use types in particular zones, as well as leading to the extension of those zones. Further from the city, marginal land may be left unexploited – or, to be more exact, left for less intensive exploitation, like hunting or pig-keeping. In general, however, variations in fertility and in the availability of resources like water mean that we can expect to find a variety of farming systems in most regions of Italy.

The climate and terrain of Italy are often classified as 'Mediterranean', permitting generalisation about the types of farming systems appropriate to such an environment, both in the Mediterranean itself and in regions with a similar geography like California or Chile. In very broad terms – in comparing 'the Mediterranean' with 'north-west Europe', for example – there is some truth in such a category, but it serves to conceal a great deal of variation; not only between Italy and other Mediterranean countries (and still more southern California) but also within Italy itself. The mirage of a uniform climatic regime and farming system is particularly unhelpful for a discussion of the effects of Rome's demands on different parts of the peninsula.

One basic distinction is that between mountains, hills and plains. Less than a quarter of Italy lies below 300m or can be described as 'plain', and most of it is in the Po Valley. Plains were often prone to drainage problems and malaria, and their alluvial soils were sometimes too heavy to be cultivated with animal-drawn ploughs. Meanwhile, large areas of the peninsula lie above the limit of cultivation of olives and vines, and are inhospitable to most forms of arable cultivation. The classic Mediterranean farming system is largely confined to a narrow strip of land, consisting for the most part of low hills, between the mountains and the sea.

The picture is still more complicated; climate and geology both show considerable variation, and interact with one another to affect farming conditions. Humidity in the peninsula is strongly affected by relief, due to its origin in complicated movements of air pressure in winter and summer; thus not only is the north wetter than the south, but the west receives more rain than the east. Rainfall over the peninsula (the Po Valley and the Alps have very different climatic patterns) is very seasonal in character; in the south, most rain falls in the winter, while further north it tends to produce autumn and spring maxima. Rain generally falls in short, heavy showers, punctuating long periods of sunshine and evaporation, a pattern which tends to accelerate the erosion of certain soils.

The geology of Italy is similarly complex, with the quality of soils in different regions deriving not only from the character of the parent rock but also from the effects of cultivation, the level of forestation and the degree of erosion. On the basis of physical features and climate, Walker distinguishes six distinct regions in central Italy and five in the south,

each with its own peculiar geography.[5] However, he also stresses the degree of variation within these regions; especially, but not solely, the distinction in terms of soil, climate and forms of cultivation between hills, mountains, and valleys or plains.

It is clear that some regions of Italy were in a better position to respond to the demands of Rome than others; further, that different regions would respond in different ways. The distinction between the predominantly arable lowlands and the predominantly pastoral highlands is most obvious; the latter might grow cereals for their own consumption, but in supplying the city both location and environment would encourage specialisation in animal products. In the lowland areas, some soils seem to be particularly favourable to cereal cultivation (Spurr highlights Etruria, Campania and Apulia), others to vines (for example, on the Alban hills today they amount to a virtual monoculture).[6]

Notes

1 Hall P. (ed.) (1966) *Von Thünen's Isolated State: an English translation of 'Der isolierte Staat'*, Oxford, Pergamon, pp. 7–8

2 Duncan-Jones, R.P. (1982), *The Economy of the Roman Empire: quantitative studies*, 2nd edn, Cambridge, Cambridge University Press, p. 368

3 Strabo 5.3.7; Cicero, *Rep.* 2.3–5

4 Varro, *RR* 2.6.5

5 Walker D.S. (1967) *A Geography of Italy*, 2nd edn, London, Methuen, pp. 95–229

6 Spurr M.S. (1986) *Arable Cultivation in Roman Italy*, London, Society for the Promotion of Roman Studies, p. 8

12

THE TRANSFORMATION OF THE ROMAN SUBURBIUM

by Neville Morley

Source: Neville Morley, *Metropolis and Hinterland: the city of Rome and the Italian economy 200 BC–AD 200*, Cambridge, Cambridge University Press, 1996, pp. 83–95, 103–7

It is within the immediate hinterland of the city, the region with the best access to the urban market, that we may expect to find the most significant and visible changes in agricultural practice; a greater degree of orientation towards the market, with specialisation in a particular set of crops and more intensive cultivation. An examination of the history of this region is therefore the obvious starting point for this study.

The definition of this 'immediate hinterland' is inevitably somewhat arbitrary, and depends above all on the particular questions that are being considered. It could be defined as the area characterised by a particular set of activities – the intensive horticulture which von Thünen's model predicts for land close to the urban centre – so that its expansion or contraction over time can be charted. Alternatively, a region could be defined on the basis of territorial homogeneity. The advantages of this latter approach are that it emphasises the variety of activities that may take place within the same geographical area, interacting and competing for resources. Production is not considered in isolation from consumption or from social and demographic change.

The Roman lowlands, bounded by the Monti Sabatini, Sabini and Tiburtini, the Colli Albani [Alban Hills] and the sea, can conveniently be considered as a territorial unit (see Figure 12.1). Most of this area lies within 30 km of Rome, extending a little further up the valley of the Tiber; it can fairly be described as the immediate hinterland of the city. Not all parts of the region enjoy the same ease of access to Rome: the Tiber and the Anio are both perennial and navigable in their lower reaches. The region's geology consists overwhelmingly of rough tufa; small streams, most of them seasonal, have over the centuries formed this landscape into a confusion of ridges and valleys, making cross-country travel difficult. To the north-west and south-east are ranges of hills formed by volcanic activity, often containing lakes within former craters; to the north-east are the foothills of the Apennines.

Historically, the entire region was linked to Rome from a very early date. By the third century BC it was firmly under Roman political and military control [. . .] The economic effects of this domination, and above all of the drain on manpower resulting from Roman military demands, must be taken into account.

The fact that Rome, with its Mediterranean empire, could draw supplies from beyond its immediate hinterland, coupled with the tradition relating to the decline of peasants in Latium and Etruria, has led some historians to regard the Roman Campagna in antiquity as neglected and deserted, analogous to the malaria-ridden wasteland of the nineteenth

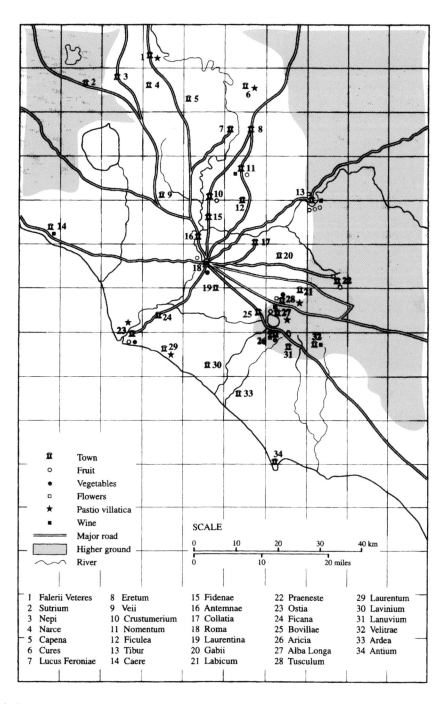

Figure 12.1 Production in the Roman *suburbium*

and early twentieth centuries. Archaeological survey, and reappraisals of the literary sources, have done much to correct this picture, reviving Dionysius of Halicarnassus' portrait of a densely occupied region:

> If anyone wishes to estimate the size of Rome by looking at these suburban regions he will necessarily be misled for want of a definite clue by which to determine up to what point it is still the city and where it ceases to be the city; so closely is the city connected with the country, giving the beholder the impression of a city stretching out indefinitely.[1]

The nature of the region's relationship with Rome must have been affected by the building of the Aurelian Wall in the third century, which decisively separated city from countryside, but it continued to be densely populated at least until the Gothic War. The Campagna was finally abandoned only in the sixteenth century, when a combination of ravaging armies, bandits and landowners brought about a flight of peasants to the city and the spread of extensive pastoralism, which in turn led to the neglect of drainage works and the spread of malarial marshes. This phenomenon was intimately related to the demands of the city for meat and of the land-owning elite for profit; the region's economy remained tied to the city, even though in this case the result was ecologically disastrous.

In earlier periods, too, the fortunes of this immediate hinterland were closely tied to the economy of Rome. Imports of foreign grain in no way reduced its economic importance for the city; instead, the *annona* freed the region from the burden of having to supply Rome with staples, permitting the development of more diverse forms of agricultural production. Of course, in the vicinity of a metropolis there might be competition for land and other resources for uses other than production, but first let us consider the evidence for a particular style of cultivation in the region.

Production

According to von Thünen's model of agricultural location, high economic rents in the region near the city, based on high prices in the urban market and relatively low costs of transport, serve to promote intensive, market-oriented, specialised production of expensive perishables; fruit, certain vegetables, dairy products. The city is not only a market for this produce but also a source of manure, tools, labour and above all the grain and other staples that permit farmers in the suburbs to specialise in non-staple crops. Meanwhile, other forms of cultivation are pushed onto marginal land, if not forced out of the city's immediate hinterland altogether.

Evidence for land prices in the vicinity of Rome, as with Italy in general, is scanty, and somewhat contradictory. That they were generally perceived as high is shown by two letters of Cicero, concerning his attempts to buy a plot of land for a grave, and by a passage in one of Seneca's letters[2] [. . .]

The Roman agronomists were aware of the importance of location in choosing an estate; this suggests that they would be prepared to pay more for land close to the city, and hence that the price of land in the *suburbium* would indeed be higher than elsewhere. Cato's advice, that the farm should be near 'a flourishing town, or the sea, or a navigable stream, or a good and much-travelled road', is echoed by Varro and Columella.[3] If the estate is close to the city, its owner may be more inclined to visit it and keep an eye on its management.[4] Varro notes that 'farms which have nearby suitable means of transporting their products to market and convenient means of transporting thence those things needed on the farm, are for that reason profitable'; Columella amplifies this by observing that access to the market increases the value of stored crops.[5] The town is a source of goods and services as well as a market; few farms can be wholly self-sufficient, and in some cases it is cheaper to rely on outside suppliers, especially in the case of specialist workers like physicians or fullers.

Any farm, therefore, is likely to be profitable if it has good access to a market. The agronomists go further, recommending particular crops if the farm is close to a town:

> It is especially desirable to have a plantation on a suburban farm, so that firewood and faggots may be sold, and also may be furnished for the master's use. On the same farm should be planted anything adapted to the soil, and several varieties of grapes . . . Plant or graft all kinds of fruit – sparrow apples, Scantian and Quirinian quinces, also other varieties for preserving, must-apples and pomegranates.[6]

> Near a town it is well to have a garden planted with all manner of vegetables, and all manner of flowers for garlands . . . The suburban farm, and especially if it is the only one, should be laid out and planted as ingeniously as possible.[7]

> It is profitable near a city to have gardens on a large scale; for instance, of violets and roses and many other products for which there is a demand in the city; while it would not be profitable to raise the same products on a distant farm where there is no market to which its products can be carried.[8]

This advice is remarkably close to the precepts of von Thünen; close to the market one should grow perishable goods, exploiting the land intensively. It is interesting to note that Cato recommends the production of wood; forest formed the second zone of cultivation in the original German model, but the adoption of mineral fuels during the nineteenth century made this activity unnecessary and unprofitable. The importance of wood in a pre-industrial economy, and the cost of transporting such a bulky commodity over long distances, would make it profitable to have land (probably land which could not otherwise be cultivated – Cato suggests that poplars should be planted in wet ground and on river banks) given over to plantations.[9]

As well as recommending the cultivation of certain crops, Columella suggests different marketing strategies for livestock and related products in the vicinity of a town. The farmer is advised to sell most young lambs, keeping only enough to replace his stock, and to profit from the mothers' milk; similarly, he should sell sucking-pigs and induce the sows to have more litters, whereas in more distant regions raising the stock is the only thing that pays.[10] Near towns, new-born chicks can be sold for high prices, or the farmer may get involved in fattening hens for market.[11] All kinds of activities, generally referred to under the heading of *pastio villatica*, such as the raising of thrushes or pigeons or the keeping of fish, were profitable only in the vicinity of Rome. Varro makes this clear; having described the huge profits to be made from the sale of fieldfares, he continues: 'But to reach such a haul as that you will need a public banquet or somebody's triumph . . . or the *collegia* dinners which are now so numerous that they make the price of provisions go soaring.'[12] This form of cultivation was very closely tied to the Roman market, and also, we may suppose, limited in geographical extent to the city's hinterland.

The recommendations of the agronomists for the cultivation of farms near the city can be supplemented from literary sources. In several poems, Martial gives an indication of what the ideal suburban estate was expected to produce. In *Epigrams* 3.47, Bassus is shown travelling out to his estate with a wagon full of cabbages, leeks, lettuces, beets, fieldfares, hare, sucking-pig and eggs – everything that his villa should have provided for him, but all bought from the market. In 7.31, Martial himself offers gifts of hens, eggs, figs, kids, olives and cabbages as if they came from his country place rather than from the market [. . .] Long lists of the products that a 'proper' farmer should be growing for himself can be compiled. The majority of the goods are perishables, most of them on the luxurious side; fruit, vegetables, poultry and assorted delicacies, as well as olives and wine. Grain is mentioned in two poems

(3.58.51 and 12.72) as something that might be produced on the farm. However, in 6.80 Martial observes that, whereas once Egypt sent winter roses to Rome, now Rome produces roses and relies on Egypt for its grain. Not all of Rome's grain needs were met from abroad, of course, but very little was drawn from the city's immediate hinterland by this date. [. . .]

Figure 12.1 is based on a list of such references, and from it a few observations can be made. The majority of places named in the literary sources fall within the Roman Compagna [. . .]; the main exceptions are Signia in southern Latium, just off the eastern side of the map, whence came apples; Tarquinii, further up the coast to the north-west, which was associated with various forms of *pastio villatica*; and a number of places producing vegetables like turnips and cabbages, which could survive being transported over long distances. Otherwise, whenever the cultivation of fruit, vegetables and flowers and the practice of *pastio villatica* are given a concrete location, it lies within the immediate hinterland of Rome. [. . .]

The map is certainly not evidence for specialisation in different parts of the region. The advice of the agronomists is meant to apply to anywhere within the city's hinterland, and, while Martial's small estate happened to be at Nomentum, his presentation of the 'ideal villa' is more widely applicable. In other words, we might expect the distinctive patterns of cultivation depicted in the sources to be found in most parts of the Roman Campagna, and the paucity of references to, for example, places within South Etruria is not necessarily evidence that the region was farmed in a significantly different manner. Of course, inter-regional variation is certainly possible, as localities differed in their soils and their access to important resources.

For the moment, however, we may continue to consider production within the Roman Campagna as a whole. A clear picture emerges from the various sources; an emphasis on the profits to be made from supplying the city with specialised, perishable goods and luxury foodstuffs. The hinterland of the city is perceived as closely associated with particular patterns of land use; Livy, for example, describes a region of Greece as 'covered with many trees and gardens, as in suburban districts', while Pliny notes the many fruit trees in the district surrounding the city.[13]

It is more difficult to offer a chronology for this development from the literary sources, except through crude comparisons of different writers – for example, the small amount of space that Cato devotes to *pastio villatica* in the second century BC contrasted with the extensive account in Varro's work in the late first century. Etruria and Latium cease to be mentioned as major suppliers of grain to Rome in the second century BC, although this could easily imply a decline in their importance relative to the grain-exporting provinces rather than any absolute decline in production, displaced by more specialised cultivation. The elder Pliny dates the arrival of various forms of *pastio villatica* (fishponds, enclosures for keeping snails, the fattening of peacocks for the table) to the first century BC; Varro considers that 'our generation' was responsible for the spread of luxurious manners of raising game and poultry.[14] Other authors date the onset of moral decline, including the development of this taste for exotic foods, to the second century [. . .]

Since the second and first centuries were the period of the great expansion of the population and wealth of Rome, these are plausible dates for the beginning of the extension of the zone supplying fruit, vegetables and luxury products to the city into the surrounding countryside. Gardens were much older than that – Pliny notes that the garden was once seen as 'a poor man's farm; the lower classes got their market supplies from a garden'[15] – but previously, we may suppose, they had formed a halo of limited extent around the city, as seen for example in maps of papal Rome. Their expansion into the countryside, and the development of increasingly luxurious and expensive products for urban consumption, were regarded as relatively novel (and somewhat dubious) phenomena.

Consumption

The expansion of this zone of specialised production into the Roman *suburbium* was not a simple process. Firstly, the establishment of market-oriented gardens and estates dedicated to *pastio villatica* involved fierce competition for resources; above all for land, but also for water and for access to the market. [. . .] Certain groups, the wealthy land-owning elite above all, had an advantage in responding to the incentives offered by the urban market, and this can indeed be seen in the development of the city's immediate hinterland. In addition, competition for resources took place not only between different groups in the countryside but also between town and country, and between different ways of exploiting those resources.

The immediate hinterland of the city was characterised by particular forms of production, but there was also a distinctive surburban form of consumption. The *suburbium* was a place of *salubritas*, *otium* and *amoenitas* [wholesomeness, leisure and delightfulness]. It was a refuge from the heat, crowds and insanitary conditions of the city; a place of leisure, for writing, reading and conversing with one's cultivated neighbours; a place of delightful beauty. From the second century BC onwards, the Roman elite built luxurious villas in the region, where they could enjoy these amenities. In succeeding centuries the presence of the imperial court, and Trajan's law forcing candidates for the senate to invest in Italian land, increased the number and splendour of these residences.

As with production, what gave suburban forms of consumption their distinctive character was the proximity of the city. The life of leisure was a complement to political life, not an alternative; the suburban villa filled the 'very urban need' for a temporary refuge in which to restore one's energies before returning to the fray. It was almost unthinkable that a member of the Roman elite should turn his back entirely on the life of the city. [. . .] The younger Pliny eulogises his villa at Laurentum because it was close enough to Rome that he could spend the night there without having to cut short his day's work.[16] When writers mention the location of properties described as suburban, they inevitably lie within easy reach of the city, especially in the east of the *suburbium* near towns like Tibur, Praeneste and Tusculum.

The suburban villa was an adjunct to the political life of the capital; it was not expected to be productive in the same way as the owner's properties elsewhere in Italy and the empire [. . .] After a particularly poor year, Pliny contrasts his unproductive estates in Tuscany and beyond the Po with his Laurentine villa, which had at least produced some writing – clearly a reversal of the usual division of productive and unproductive properties.[17]

In pursuit of the country life that they idealised, the Roman elite spent vast sums in creating a peculiarly sanitised version of the countryside. The residences that they built there were at least as opulent as their town houses, with baths, gardens and fountains; they brought with them all the furniture, books, works of art and other luxuries of town life. Large areas were enclosed as game parks, stocked with wild boar and deer that came to be fed at the sound of a horn; huge aviaries were constructed, 'larger buildings than whole villas once used to have': at villas on the coast (especially those further south in Campania), expensive salt-water fishponds were constructed, costing far more than the fish in them could ever be worth.

The countryside around the city of Rome became urbanised, not only in the density of settlement there and the lack of a clear boundary with the city, but in its economy. Like the city, it was a place in which the Roman elite spent the income derived from their productive estates elsewhere. This might support producers in the region – Pliny's villa at Laurentum was supplied from local farms – and gangs of builders and other artisans, but alongside the picture of specialised production outlined in the last section must be set another picture, of large areas of land turned over to residential use and park land.

Of course, the suburban villa was not necessarily utterly inimical to production. [. . .]

Pastio villatica is recognised as a reputable branch of farming, in which considerable profits might be made, and we may recall Varro's dictum that farming should be for profit and pleasure together.[18] [. . .]

Having divided *pastio villatica* into three classes, the aviary, the hare warren and the fishpond, Merula observes that 'each of these classes has two stages: the earlier, which the frugality of the ancients observed, and the later, which modern luxury has now added'.[19] The contrast is at times between the properly productive and the decadently extravagant (which may nevertheless yield a profit).

The *suburbium* contained a wide variety of different types of villa. Varro's treatise covers both vast game reserves at Ostia and Tusculum and the tiny estate near Falerii that specialised in honey, as well as villas that were clearly dedicated to consumption alone.[20] [. . .] Some owners chose to make their suburban properties pay, others kept them simply as places of relaxation; some took advantage of windfalls, selling birds or game if the market was good without changing the essentially consumptive nature of their estates. At times production and consumption were combined; at other times they were in fierce competition.

A further demand for land came from the city

itself. As the population grew, so did the demand for living space; this might be met in part by expansion upwards, as *insulae* increased in height to accommodate more people, but it also involved expansion outwards. Open spaces like the Campus Martius were increasingly filled up with monumental buildings, and the city expanded far beyond the limits of the old Servian Wall (and even, it would appear, beyond the line later marked by the Aurelian Wall). Another pressure came from the development of vast gardens, *horti*, on the immediate outskirts of the city; from the middle of the first century BC into the Julio-Claudian period, culminating in the building of Nero's Golden House, vast areas of the city were turned over to the landscape artists. As Pliny commented: 'Nowadays indeed under the name of gardens people possess the luxury of regular farms and country houses actually within the city itself.'[21] It was once more the elite who could afford to use large tracts of expensive land in an entirely unproductive manner.

The city also had a considerable waste disposal problem; not so much with excrement, which could be flushed into the Tiber or sold to farmers – a plentiful supply of manure is essential for horticulture – but with the large numbers of dead bodies which a population of a million produced every year. Under the Republic, the poor were probably cremated or disposed of anonymously in cemeteries or in burial pits like those found on the Esquiline. From the first century BC however, 'standards of dying' were dramatically raised; whereas once commemoration with tomb or grave-marker was the prerogative of the rich, or at least comfortably-off, the practice now became diffused among the lower orders of society. Naturally this added considerably to the demand for land in the vicinity of the city.

This land was expensive; only the very rich could afford to use it for entirely unproductive purposes like gardens or tombs. The elite could also indulge in benefaction, building the *columbaria*, collective tombs, in which freed-men, slaves and other dependants might hope to be interred. Free citizens without links to elite *familiae* had to develop other strategies. They might obtain proper burial with the help of the *collegia*, the associations of craftsmen, tradesmen and other members of the lower orders. Another solution was to use the tomb plot productively, cultivating it as a garden and making it pay for itself; the practice is well known from Alexandria, and several Roman inscriptions attest to it. The streets leading out of Rome, as well as the banks of the Tiber and the Anio, were lined with tombs, each occupying a small, regular plot which was growing vegetables for the urban market. The names on the inscriptions are, according to Purcell, indistinguishable from the ordinary population of the city, who must have travelled out regularly to their tiny properties.

As with the surburban villa, production and consumption could therefore be combined; indeed, only those who owned large tracts of land elsewhere could avoid the need to exploit suburban property as intensively as possible. The Roman hinterland was occupied by a variety of properties, each heavily dependent on the presence of the city. Different groups competed for a limited amount of land and other resources, and inevitably some were unsuccessful. [. . .]

Water

The city of Rome depended on its immediate hinterland for supplies of various foodstuffs and other goods like wood and building materials. It obtained these in a number of ways; by market exchange, as rent taken by landowners, and by owners moving supplies between their suburban and urban properties. The result was the intensive exploitation of the region, and increasing prosperity. The city sometimes added to the pressure on the supply of land

by using it for building and burials, but in general the relationship may be said to be one of mutual advantage.

In the case of another important resource, however, city and country were in direct competition. Water was an essential feature of civilised existence for the Romans, and the city daily consumed vast quantities for drinking, washing and recreation, in baths, fountains and private houses. It was also in great demand in the *suburbium*. Vegetables, fruit trees and flowers all require irrigation, and Varro recommends that aviaries, even those for non-aquatic birds, should have a regular water supply.[22] Running water also had an important place in the residential part of the suburban villa, with bath houses, fountains and ornamental gardens. Given the normal pattern of rainfall in the region, water was in short supply for much of the year; the Tiber and the Anio were perennial rivers, but most watercourses were seasonal, and there was a limited number of perennial springs.

The city's response to this situation was to take what it required and to introduce laws to protect its supplies. Perennial springs were tapped and the water transported to Rome in aqueducts, which were built on land obtained through forced purchase and subject to strict regulation – as were estates adjacent to the course of the channel.[23] In a single case, the water of a spring near Tusculum was restored to the use of local proprietors, after being tapped illegally by the *aquarii* [water carriers].[24] We may surmise that other areas were not treated so generously; for example, four of the greatest aqueducts were supplied from springs which had previously fed into the Anio, doubtless to the detriment of users at Tibur and other places along the course of the river.

Frontinus, *curator aquarum* under Nerva and Trajan, regarded his duties as concerning 'not merely the convenience, but the health and even the safety of the city', 'that the public fountains may deliver water as continuously as possible for the use of the people'.[25] His priority, as well as that of the aqueduct builders and the emperor, was the urban supply. Most of his comments about what goes on in the countryside are about abuses which threatened to disrupt this supply; illegal tapping of the conduits, damage to the channels and refusal to allow workmen onto the land. [. . .] The notion that the city was in fact stealing water that properly belonged to the countryside is not considered. Frontinus states that he intends to sort out these abuses informally, by intimating to those responsible that they should seek the emperor's pardon, but he had a considerable weight of legislation behind him if required.[26] Other laws covered 'public' rivers like the Tiber (which was important not only for water but for the transportation of supplies to Rome): farmers were forbidden to change the course of such waterways, or even to take water from them to irrigate their fields if this was likely to impair navigation.

Some aqueduct water was distributed in the countryside, about a third of the total capacity of the system. The Aqua Alsietina was notoriously muddy and was therefore used entirely outside the city; the Anio Vetus was similarly used 'for watering the gardens, and for the meaner uses of the city itself'.[27] This distribution was made possible only by Frontinus' diligence in stopping water being taken illegally; it is the disposal of a surplus once the needs of the city have been met from the purest water, like that of the Aqua Marcia. Rights to aqueduct water might be obtained either through a direct grant from the emperor [. . .] or through a more formal system: 'As soon as any water rights are vacated, this is announced and entered in the records, in order that vacant water rights may be given to applicants.'[28]

Clearly the advantage in obtaining water rights lay with those who had regular contact with Rome and links with the networks of patronage and influence; in other words, the wealthy villa-owners rather than their tenants. The only place definitely known to have been supplied with aqueduct water was the luxurious villa of Manlius Vopiscus at Tibur, fed from the Aqua Marcia by a siphon under the Anio. Archaeologists have found that the Anio Novus has two channels when it passes through the region of Tibur; this must have made it easier to clean, but it is also likely that one of these was a special branch line to supply properties in the area. The cisterns in many Tiburnian villas seem too small to supply the needs of such complexes, suggesting that they had access to aqueduct water.

In the minds of those who administered the water supply there was a clear divide between city and countryside, with the city having an overwhelming advantage in obtaining what it required. A few privileged residents of the *suburbium* also benefited from the system; the remainder were left to assert their objections through water-stealing or vandalism, and to compete amongst themselves for the water that remained after the city had drunk its fill. This competition, for both productive and leisure uses of water, must have been fierce, especially in a dry year. The Roman jurists produced a detailed set of rulings on the administration of water rights, which must reflect the likelihood of litigation in this area; the word 'rivals' derives from *rivales*, those who drew water through the same channel and who were thereby liable to come to blows.

Water rights were attached to the land rather than the owner. They could be forfeited if water was taken at the wrong time without prior arrangement, or if the right was not regularly exercised; the emphasis throughout is on continuity, that everything should be done 'as it was last summer'. The main problem with this philosophy is that it assumes a constant supply of water every year; in the event of a drought, a man taking his legitimate share of water might still be depriving a neighbour of *his* rights. The legal response was that both men should be placed under an interdict. [. . .] Another set of laws covered the problem of drainage and erosion, presumably as a response to the sorts of arguments that would arise following the heavy spring rains; at certain times of year, too much water could be as awkward as too little.

There is evidence for the existence of more organised schemes of water distribution in the areas around Rome. Frontinus writes of the Aqua Crabra near Tusculum that 'this is the water which all the estates of the district receive in turn, dealt out to them on regular days and in regular quantities',[29] and Cicero paid a rent to the *municipium* [town] of Tusculum for water rights to this stream [. . .] It may be strongly suspected that demand for water in the *suburbium* was such that sizeable rents could be charged for its use, and carefully regulated distribution schemes were necessary to keep the peace.

Summary

The demands of Rome for perishable goods like fruit and vegetables and for luxury foodstuffs supported the development of particular forms of production in the *suburbium*, resulting in intensive exploitation of the land and in increased prosperity. This led in turn to dramatic changes in the social landscape, seen above all in the fate of urban centres in the region. The fortunes of the *suburbium* became increasingly bound up with those of Rome, an orientation embodied by the roads and aqueducts which radiated out from the city and drew in people and resources from the region.

The relationship between the metropolis and its immediate hinterland was not a simple one – one need only think of the conflict generated by the urban demands for water, burial space and leisure as well as for food – but it was undoubtedly intimate and enduring.

Notes

1 Dion. Hal. 4.13.4
2 Cicero, *Att.* 12.23.3, 13.31.4; Senecca, *Ep.* 87.7
3 Cato, *Agr.* 1.3; Varro, *RR* 1.16.1-2; Col. 1.3.3
4 Col. 1.1.18-20
5 *RR* 1.16.2; Col. 1.3.3
6 *Agr.* 7.1
7 *Agr.* 8.2
8 *RR* 1.16.3
9 *Agr.* 6.2 on poplars
10 7.3.13, 7.9.4
11 8.5.9, 8.7
12 *RR* 3.2.16
13 Livy 33.6.7; Pliny, *HN* 17.1.8
14 *HN* 9.168-74, 10.45; *RR* 3.3.6-10
15 *HN* 19.50-8
16 *Ep.* 2.17.2
17 *Ep.* 4.6
18 Varro, *RR* 3.2.4-6
19 *RR* 3.3.6
20 *RR* 3.13, 3.16.10-11
21 *HN* 19.51
22 Varro, *RR* 3.5.2
23 Frontinus, *Aq.* 127-9
24 *Aq.* 9
25 *Aq.* 9, 104
26 *Aq.* 130
27 *Aq.* 11, 92
28 *Aq.* 109
29 Frontinus, *Aq.* 9

Part 2
MEDIEVAL AND EARLY MODERN CITIES

Reading 13 by Bryan Ward-Perkins makes an apt transition from the section on ancient cities to this one on cities of the medieval and early modern era, dealing as it does with the shift in the use made of baths, aqueducts and wells from the Roman emperors' tradition of munificence to the much more restrictive Christian emphasis on charity and cleanliness. The contrast between ancient and medieval Roman baths shows that urban water supply was not only a matter of technology and economics, but also of values and beliefs. Another religious contrast with antiquity – the replacement of the open agora by the mosque in Islamic cities – is dealt with by Hugh Kennedy in reading 14. But the thrust of the article is to question the attribution of the distinctive unplanned layout of the Islamic city to religious precepts; rather, it stems from the Umayyad tradition of minimalist urban government, a legal code that valued above all the rights of the individual household, and the dominance of a merchant class that looked to commercial activity and not monumental buildings, as the main feature of the city. The reading thus provides above all valuable context for the implementation of technologies; though there is a reference to the requirement in Islamic law that streets should be wide enough to permit the passage of two pack animals. Technology is centre stage in reading 15 by Lynn White, Jr. White has long been targeted as an exemplar of technological determinism to be knocked down. In this section, he explores the implications of the 'agricultural revolution of the early middle ages': the replacement of the scratch-plough by the wheeled heavy plough, then of the ox by the newly harnessed and shod horse, and finally of three-field crop rotation. Not only did these innovations enable the fertile but heavy soils of Europe to turn to agriculture, they paved the way for urbanization and a medieval industrial revolution based on water and wind. Some of the details of White's account have been challenged; but as far as technological determinism is concerned, the logic of White's argument is only that these agricultural innovations were necessary, not sufficient, for subsequent urbanization and industrialization in northern Europe.

Reading 16 returns to the von Thünen model applied by Morley to ancient Rome in the first section. This time the focus is on medieval London, and the products of the forest zone that von Thünen predicted would lie next to the area of horticulture and dairying closest to the city. The authors find that the elongated area of wood production in about 1300 broadly conforms with the transport–cost model, given the reliance of the wood trade on water transport, though in time the growth of London's economy and population would shift the economic balance to the consumption of 'sea coal'. The article also underlines the importance of wood fuel in medieval London's industries, such as baking, brewing, and brick- and tile-making. But the medieval period was also notable for developments in road transport, such that David Herlihy in reading 17 talks of a 'Tuscan road revolution'. The late thirteenth-century revolution consisted in the transfer of roads from high ridges to the swampy river valleys; this was achieved by elevated roads, also functioning as dikes, and by flint paving, and the construction of three new bridges. Again, this was no neutral feat of technology; it was facilitated by a tyrannical government's control of resources and organization.

In reading 18, Elizabeth's Ewan's paper on medieval Aberdeen returns to the theme of relations between cities and the country. Scottish medieval towns were unusual in having no surrounding wall, but she rejects the view that the physical separation of other cities from their surrounding countryside by a ring of walls marks a sharp dividing line between the medieval urban and rural worlds. Aberdeen was far from unique in having its population maintained by rural immigrants, or in the dependence of its citizens on the countryside for their food supplies, building materials, fuel (in this case peat), and raw materials for a variety of urban crafts, or in having some of its craftsmen (fullers, tanners and, perhaps, potters) located along with hospitals, gallows and religious houses on the outskirts of the city. One such edge-of-city craft activity is covered in reading 19 by Richard Goldthwaite. By the fifteenth century, thanks to its flourishing economy, Florence was much more solidly built because of the widespread substitution of brick and stone for timber, part of a development in early modern Europe that Goldthwaite describes as a 'veritable revolution in housing'.

Florence was well-endowed with building materials: sand for mortar, gravel for foundations, high-quality limestone and clay and wood fuel to make bricks. A distinctive brick architecture emerged, based on an organized brick-making industry. Some kilns were situated within the city itself, away from the denser areas of population; but because of the smoke and smell, and for greater proximity to clay and limestone, kilns were usually located beyond the city walls. A less quotidian aspect of Italian Renaissance building activity is the subject of reading 20. Nevertheless, the successful re-erection of an obelisk in the square in front of St. Peter's, undertaken by Pope Sixtus V's architect, Domenico Fontana, says much about the religious context of construction, notably the absolute power of the Popes over the built environment of Renaissance Rome. It also depended upon the well-tried methods available to execute the papal will: wooden scaffolding, capstans, levers and wedges, rollers, pulleys, an earth ramp, and the muscular effort of harnessed horses and gangs of men. The only innovation, compared with ancient building projects, was the use of the horse as a working animal.

Taken together, the readings in this section have thus far been equivocal about the contribution of technological innovations to urban developments in the medieval and early modern period. This question is tackled directly by Christopher Friedrichs in readings 21 and 22. According to Friedrichs, the physical form and fabric of early modern cities was very stable; this stability reflected persistent technological limitations, especially in power sources and communications, the particular urban effects of gunpowder and printing notwithstanding. There were innovations in military architecture, but the repertory of building types remained constant; in sum, the early modern period was marked by technological stagnation, and contrasts with the urban innovations of the medieval era and the nineteenth century. Quite a different tone is evident in reading 23; Jonathan Israel speaks of 'a dazzling array of innovations' in Dutch cities after 1648. These included an intercity transport network of horse-drawn passenger barges, unique to the Netherlands, a more exportable system of public street-lighting, and a method of fire-fighting, both devised by Jan van der Heyden, new large public clocks, and, going beyond the technological, a regulated medical service. The diffusion of street lighting and fire-fighting technology to England and France is among the topics considered in readings 24 and 25. The excerpt from Stephen Porter's book on the Great Fire of London makes it clear that fire engines of German provenance, using a bucket-chain rather than van der Heyden's pumps, were in use in the English capital from the second quarter of the seventeenth century, and were common in British towns by the 1660s. According to Leon Bernard, van der Heyden's device was probably adopted

in Paris in the last quarter of the seventeenth century; however he attributed
the city's extensive candle-lit municipal street lighting system of 1667 to the
Parisian Chief of Police La Reynie, an innovation that seems to contradict
Israel's claim for van der Heyden's priority in this area. But perhaps the most
intriguing item in the Bernard reading is the 'seventeenth-century revolution
in transportation' represented by the transition from the litter chair to the
horse-drawn carriage, and in particular, the admittedly short-lived experiment with public transport in the form of the *carrosses à cinq sous*, an
innovation attributed to Blaise Pascal.

Reading 26 is taken from *Londinium Redivivum* (London Reborn), a plan
for the rebuilding of London written by the diarist and courtier John Evelyn
within days of its destruction by fire in September 1666, though not published until after his death. His service on royal commissions, including one
on the reform of London's streets, and the publication in 1661 of *Fumifu-
gium*, a tract advocating solutions to the pollution of the city, evinced a
long-standing concern with the condition of the capital. His wide-ranging
interest in science and technology was reflected in his membership of the
fledgling Royal Society. His familiarity with other European capitals, and
their technologies, such as street paving and sleds, is evident in this excerpt.

The final reading of this section goes well beyond the geographical areas
usually encompassed by the periodization including medieval and early
modern; but this is surely legitimate, as what is at issue in Louisa Schell
Hoberman's article is the influence of Spain, a country with a Christian and
Islamic heritage, on the Mesoamerican urban culture of the Aztecs. In the
post-conquest era, Mexico City was confronted by a heightened risk of
flooding, due to the colonialists' disturbance of the delicate ecological
balance in the area of the lake city. Canals formerly assisting drainage
were filled in and converted to streets, and lakeside forestation cleared to
provide building timber, and areas for growing wheat. Soil erosion and
silting of the lake resulted, only to be made worse by the grazing of sheep
and cattle brought over from the Old World. Hoberman discusses the many
technological, economic, social and political impediments to the implementation of an ambitious drainage project, the *desagüe*, the plan of which
included a four-mile-long tunnel. Though begun in 1607, the *desagüe* was
not completed until 1789; such examples of the problems of technology
transfer demonstrate the value of a contextual approach to the history of
technology.

13

WATER SUPPLY IN EARLY MEDIEVAL ITALY

by Bryan Ward-Perkins

Source: Bryan Ward-Perkins, *From Classical Antiquity to the Middle Ages: urban public building in northern and central Italy (AD 300-850)*, Oxford, Oxford University Press, 1984, pp. 135, 138-41, 144-6, 152-4

In the very period that patronage of the secular public baths and aqueducts was disappearing, there appears in Italy considerable evidence of water and water-supply being supported by the new Christian patrons for motives of charity, hygiene, and ritual. [. . .]

Baths for the clergy are interesting because they not only represent a degree of continuity from classical times, but also accentuate the difference between the classical and the Christian traditions. Many of the clerical baths which functioned in late antique and early medieval Italy were established not to replace the old secular baths, after these had ceased to function, but while they were still working, as rival establishments. This was certainly because the secular baths were not considered morally salubrious. [. . .]

The new Christian patronage catered not only for the clergy itself, but also for the needy and sick. The bath for this purpose near St. Peter's is particularly well documented. In the area between the basilica and the river, Pope Hadrian I (772-95) restored two charitable *diaconiae* [centres for the distribution of alms], and laid down that the duties of their clergy were to include a weekly procession on Thursday 'ad balneum' ['to the bath'], (probably the bath near St. Peter's), where they would minister and give alms to the poor. The choice of Thursday for this must certainly have been in imitation of Christ's washing of the disciples on Maundy Thursday. [. . .]

Before passing on to other matters, some obvious general points about ecclesiastical public baths need to be made. Firstly, it is quite clear from the evidence that the sections of the population catered for by these baths were strictly limited – the clergy itself and those poor enough to need charity. The large area of society in between was apparently not catered for at all. This represents a radical change from classical public baths, which were intended for the enjoyment and cleanliness of the whole town, with the probable exception only of the very poor (since a token entrance-fee of a *quadrans* [coin of small value, a quarter of an *as*] was normally demanded, unless waived by a munificent gift). There is obvious continuity in some respects between the baths of classical times and those of the early Middle Ages (even the gifts of free soap echo the gifts of oil for washing of the classical period). However, there is also a great difference in the nature of the recipients of such patronage, because this had moved away from munificence, designed to benefit fellow citizens, towards charity, designed to succour Christ's poor.

It is also clear that these baths were not meant to be enjoyable. They may well in fact have been so, at least for the clergy. From the descriptions of several of them it is clear that they were heated (Pavia's clerical bath, and that of S. Sofia of Benevento) and decorated (the orthodox bath at Ravenna, and both Leo III's baths at St. Peter's). But pleasurable bathing, not surprisingly, is no longer mentioned as the intention, and classical phrases to describe baths, such as 'the delight of the people' (*gaudia populorum*), are cast aside. Instead we find Nicholas I's repair of the water-supply for the charitable bath at St. Peter's compared to Christ's descent from heaven to minister to the poor, and we also find a strong ritual element running through much of the evidence. Although a careful distinction was drawn between physical and spiritual cleanliness, the Church encouraged washing of the body before purging the soul on Sundays, and, above all, before Easter. Easter week and, at St. Peter's at least, all Thursdays were also popular periods for washing the poor, an activity intended not only to benefit them, but also to benefit the souls of the washers.

A similar ritual concern was the main motive for another area of Christian patronage of water, that of fountains set in the *atria* before churches. From two letters of Paulinus of Nola, it is clear that these were intended for the washing of the hands before entering church.[1] An inscription set up by Leo I (440–61) over his fountain in the *atrium* of S. Paolo fuori-le-mura also asked the faithful to wash their hands before entering the sanctuary, but made it clear that they should not confuse cleanliness of body with purity of soul:

> Water removes dirt from the body; but faith, purer than any spring, cleanses sins and washes souls. You, who enter as a suppliant at the shrine of St. Paul, made holy by his merits, wash your hands in the fountain.[2]

[. . .]

Another unique feature of Rome, at least as far as I know, was the public lavatory set up in the square in front of St. Peter's by Pope Sym-

machus (498–514). This may well have been flushed with the waste water from a fountain erected at the same time [. . .]

There is one area of potential Christian patronage for which no definite evidence survives from Italy – the building of aqueducts in order to supply abundant free water for private drinking and washing. This was certainly an area of possible Christian concern: in ninth-century Francia, for instance, Bishop Aldric of Le Mans (832–57) built an aqueduct for his town, which had previously relied on water brought, at a price, from the river Sarthe. In Rome, however, for which we have the best evidence for Christian patronage of water in all Italy, the private needs of the population are never clearly referred to as a motive for papal aqueduct-repair. The nearest we get to the needs of the general population are vague and ambiguous phrases, such as that used to describe Hadrian I's repair of the Aqua Virgo, whereby he 'supplied [*satiavit*] almost the whole city' [. . .]

From the *Liber Pontificalis* ['Book of the Pontiffs': collection of the lives of the bishops of Rome] it seems that the supply of domestic water played a very minor role, if any, as a motive for papal aqueduct-repair. This impression is supported by two fact. Firstly, the papal aqueducts, as far as we know, rarely reached the most densely populated areas of the city. [. . .]

Secondly, it is probably significant that, whereas we have no clear evidence of the Christian patronage of aqueducts for ordinary water-supply, we do have evidence of such patronage of wells, which were the most likely replacement of the aqueducts as a source of domestic supply. Outside S. Giovanni a Porta Latina an early medieval decorated and inscribed well-head invites the reader to drink in the words of Isaiah: 'Every one that thirsteth, come ye to the waters', and another, outside S. Marco (the gift of a priest John), repeats this phrase and adds 'and if anyone charges money for this water, may he be cursed'. Neither well is dated, but both are datable on epigraphic and, in the case of that at S. Giovanni, on artistic grounds as well, to the eighth or ninth centuries. These two well-heads obviously represent a sorry state of affairs in Rome, once so abundantly watered; but they also suggest that, although there was now room for the charitable provision of domestic water, this was now normally done through the provision of wells, not aqueducts. [. . .] Besides the baths for the clergy and those for the poor, the only other piped water-supply and bath-buildings recorded in early medieval Italy are those with strictly limited access, in palaces or in large private houses.

The bath and water-supply of this type we know most about were those in the papal Lateran palace. In contrast to the situation in the rest of Italy, where bishops seem to have shared their bath-buildings with the urban clergy, in Rome the pope maintained a personal and lavish establishment. The Lateran bath-house is first mentioned in 664 [. . .]

The limitations of early medieval water-patronage were not only dictated by the narrowing of the areas of concern; they were in part enforced by financial stringency. This is very difficult to prove, since patrons were keen to stress not the limitations but the scope of their patronage; but it is strongly suggested by restrictions both in the quantity and in the scale of the Christian amenities that replaced those of classical antiquity. [. . .]

Although archaeological research has revealed many traces of small 'very late' repairs to several of the aqueducts, on none of them has a really major early medieval campaign of rebuilding, buttressing, etc. been detected. The *Liber Pontificalis* shows what the popes would like to have done, but the material evidence is more reliable for what they were actually able to do; and this was strictly limited.

Whether restrictions on the patronage of water in the early Middle Ages had a marked effect on the hygiene of towns is, I fear, impossible to tell. The end of secular public bath-buildings certainly meant that people bathed less; but it did not necessarily mean that they ceased to wash, since they could still do this in basins and tubs at home, though one might suspect that it would at least lead them to wash rather less thoroughly and less often. Equally, the decline of the public lavatory, and of waste aqueduct-water to flush drains, must have had some effect on hygiene, but may largely have been compensated for by a probable decrease in the density and size of urban populations, leaving plenty of open spaces for cesspits.

What is certain, however, is that even if early medieval townsmen did manage to keep clean and dispose of sewage satisfactorily, this will have involved them in more effort and less enjoyment than their classical predecessors.

Notes

1 Paulinus, *Ep*. xxxii. 15 about Nola; *EP*. xiii, 13 about St. Peter's (here referring also to washing the face)

2 De Rossi, *Inscr. Christ*. ii, 80–1 no. 13

14

FROM POLIS TO MADINA: URBAN CHANGE IN LATE ANTIQUE AND EARLY ISLAMIC SYRIA

by Hugh Kennedy

Source: Hugh Kennedy, 'From *Polis* to *Madina*: urban change in late antique and early Islamic Syria' *Past and Present*, 1985, Vol. 106, pp. 15-23, 25-6

[. . .]

The coming of Islam made one important contribution to the built environment of the town. A new sort of public building appeared, the mosque. In its most obvious aspect the mosque replaced the church as the place of worship for the political and social élite of the city: in Damascus in the early eighth century the church was taken over and demolished, in Aleppo cathedral and mosque coexisted on opposite sides of a narrow street until the twelfth century, while in Emesa (Hims) mosque and church were simply two halves of the same building throughout the early Middle Ages. But the mosque also replaced the agora as the main outdoor meeting-place in the city. In Damascus, the great court of the Umayyad mosque forms the only open space of any size within the walls of the old city while in Aleppo the mosque was actually built on the old agora, its wide court occupying the area of the classical open space. The mosque also replaced the agora and the theatre in a functional sense. Plays and mimes formed no part of the life of the *madina* [town], but the theatre had also had a political function as the scene of public meetings and formal political ceremonies, and it was these functions which were inherited by the mosque. It was

here that the oath of allegiance, the *bay'a*, was taken to new rulers and the *khutba*, the weekly sermon in which the ruler's name was acknowledged, took place. It was here, too, that governors and caliphs could address the Muslims on matters of public importance. [. . .] Similarly in the confusion which followed the death of the young caliph Mu'awiya II in 684 the various contenders for power met in the mosque in Damascus. Later in the Umayyad period [the dynasty of Umayyad caliphs lasted from 661 to 750] Yazid III, who had assassinated his predecessor Walid II in 744, addressed the people in the mosque at Damascus, laying before them his plans for reform and soliciting their support. The public and political functions which would have taken place in theatre, agora or hippodrome in Byzantine times now happened in the mosque. There is an interesting contrast here with medieval Italy where urban continuity was also strong. The cathedral of an Italian town did not provide a public open space in the way the courtyard of a Syrian mosque did. When the citizens of early medieval Pavia wished to gather to make their views heard, they did so in the square (*platea*) by the cathedral. This may have been one of the reasons why open squares survived in the cities of Italy and not in those of Syria.

The mosque also took over the functions of other public buildings. It was usually in the mosque that the Muslim judge (*qadi*) held court, although there are records of early *qadis* using their own houses for this purpose. Until the appearance of the *madrasa* (theological college) in the eleventh century, the mosque also served as the centre for education in the religious and legal sciences, once again taking over the function of other forms of public architecture. The transformation of the monumental city of antiquity cannot be understood without appreciating the many different activities which took place in the mosque.

Early Muslim society did not deliberately choose to develop towns with narrow winding streets out of any conscious aesthetic or cultural preference, and the idea that there is something in the spirit of Islam which leads to the enclosed, private and secret world of the 'Islamic city' should not be entertained by serious urban historians. The most important evidence for this comes from early Islamic planned towns. When Muslim rulers laid out new cities, they adopted orthogonal plans, dividing blocks of housing by straight and sometimes wide streets. The clearest example of this comes from the early eighth-century settlement at Anjar, in the Biqa' valley just south of Heliopolis (Baalbak) [in the Lebanon]. Here the early Islamic city has four wide streets which meet at a central tetrapylon and the entire plan is ordered and regular. The same features are apparent in the contemporary settlement at Qasr al-Hayr East in the Syrian desert, where the small, planned *madina* has at its centre an open rectangular square surrounded by arcades. On a larger scale, the vast development of the ninth-century 'Abbasid capital at Samarra in Iraq shows a similar concern for order. Aerial photographs show clearly the very wide main street (much wider in fact than the main street, *cardo maximus*, of any

Roman town in Syria) and the narrower streets which lead off it at right angles and divide the city into rectangular blocks for houses and gardens. These examples suggest that where cities were planned, early Islamic surveyors (the *muhandisun*) had very similar ideas to those of their classical predecessors [. . .] Planned and unplanned cities always existed in Syria. The contrast is that in classical antiquity most cities including the largest and wealthiest were planned and ordered, in Islamic society they were not.

The picture which emerges from this study suggests that urban change in the Middle East took place over a number of centuries and that the development from the *polis* of antiquity to the Islamic *madina* was a long drawn out process of evolution. Many of the features which are often associated with the coming of Islam, the decay of the monumental buildings and the changes in the classical street plan, are in evidence long before the Muslim conquests. In other ways the evolution of the traditional Islamic town was not completed until much later; regular street plans were still laid out, if only occasionally (and it should be noted, we have no idea of the street plan of the greatest early Islamic new towns at Kufa and Basra in Iraq). The *khans* [buildings for travellers' accommodation], caravansarais, *qaysariyyas* [central textile warehouses] and *madrasas* of the traditional city seem to be developments of the eleventh and twelfth centuries. We should perhaps think in terms of a half millennium of transition.

Before considering the causes of these changes it is perhaps important to make two general points. The first is that we should avoid making inappropriate value judgements. The development of the Islamic city is often seen as a process of decay, the abandonment of the high Hippodamian ideals of classical antiquity and the descent into urban squalor. On the contrary, the changes in city planning may, in some cases, have been the result of increased urban and commercial vitality, as in early Islamic Damascus and Aleppo for example. It was rather that the built environment was adapted for different purposes, life-styles and legal customs. The changing aspect of the city was determined by long-term social, economic and cultural forces, not by administrative incompetence or aesthetic insensitivity.

The second consideration is that public, open spaces, be they narrow *suqs* [markets] or wide colonnaded streets, will always be under pressure. They will only survive if they fulfil a perceived and generally acknowledged purpose and are protected by an active and vigilant civic authority. If the usefulness of such spaces is not accepted, then inevitably they will be encroached on and built over. As far as the planning of public open spaces is concerned, the historian must seek the reasons why the constraints which had prevented such encroachment in classical times were no longer operative in late antique and early Islamic cities. [. . .]

The differing role of government is important. In the early imperial period the patronage for public building came mostly from rich local citizens who provided funds for the construction of massive monumental complexes. With the decline of civic self-government in late antiquity this patronage passed to the emperor and his local representatives [. . .]

The pattern of government patronage in the early Islamic period was very different from the classical model. In many ways the early Islamic state was a minimalist state which saw no reason to interfere in the activities of its subjects except when disturbances might result. It provided physical security for the Muslims in the shape of city walls, mosques for them to worship in and, usually, a supply of running water. Muslim authorities considered the supply of water by canals or aqueducts as an essential service, partly for ritual reasons since ablution was essential before worship. The pattern visible in Italy, where the aqueducts of the classical period are frequently replaced by wells in the early Middle Ages, is not usually found in the Islamic world. The Umayyads certainly spent money on building, too much their critics alleged, but their projects, apart from the great mosques, were palaces in both town and country, and agricultural developments with their related settlements [. . .]. They did not spend money on beautifying the streets of Damascus or on putting on public entertainments. It seems probable too that the government in Umayyad times was comparatively poor. In most areas of the Caliphate the taxes collected were distributed to the Muslims in the provinces concerned; that is, most of the revenues collected in Iraq were spent in Iraq and only a very small surplus, if any, was forwarded to the government. In modern economic terms, the government controlled a much smaller proportion of the gross imperial product than under the Roman empire. Consequently both the need and the resources for government patronage of urban building were greatly reduced, and government patronage of monumental secular building, which had become increasingly erratic during the last century of Byzantine rule, ceased altogether in the early Islamic period.

The changing legal system may have been a factor in urban development. Roman law made a sharp distinction between state and private property and it was the function of government to prevent private building on the public domain. It did not matter whether such trespassing caused problems or not, it was still illegal and the local governor was enjoined to stop it. Clearly there were many cities where such laws were not fully enforced in the sixth century but the legal provision did exist and could be called on if necessary. Roman law also concerned itself with the aesthetic aspects of the townscape, forbidding structures which degraded its appearance. Here again such laws could only be effective if the resources and will to enforce them existed but they did show that the appearance of the city was a feature with which rulers should concern themselves.

Islamic law on property starts from a quite different basis. For Muslim jurists the important unit was the family and its house. Broadly speaking, it was held that they should be allowed to do anything they chose as long as it did not harm their neighbours. Furthermore, the house was held to have some rights over the adjoining public space. This legal framework could have important repercussions for urban planning. At its most simple it meant that a man could extend his house into the street or build an overhanging balcony without needing to seek permission from anyone. If his neighbours felt that the new construction was harming them, by preventing access to their own property for example, it was up to them to take the case to the *qadi* who could, if he felt that it was necessary, order that the new structure should be demolished. But the enforcement was the result of private prosecution by those who were harmed rather than by the state authorities. Equally the *muhtasib* (market inspector) would only take action against obstructions if they caused a nuisance. When in 918 Rotgerius of Pavia, an Italian city where the old Roman concept of public streets was still very much alive, wished to build a balcony from his house over the street, he was obliged to get a licence, presumably at some expense and inconvenience, from the king; if he had lived in a Muslim city, no such permission would have been necessary and he and people like him would have been much more tempted to enlarge their properties at the expense of the public street. The jurists also held that if a man owned property on both sides of a street, he could lawfully cover it over with an arch and build rooms on it, thus converting the street into a tunnel. In the case of a small cul-de-sac, the owners of the properties could, if they all agreed, place a gate across the entrance, thus converting a previously public

road into a semi-private court, a feature typical of the medieval Islamic townscape. When it came to individual constructions, the law was equally easygoing. Aesthetic considerations played no part whatever; the fact that a building was an unsightly ruin did not mean that the owner could be compelled to tidy it up and only if it was actually dangerous could those threatened take action. While the law took no account of appearance, it was deeply concerned with privacy. If one man built his house so that it overlooked another's then the offended neighbour could go to law, since he had been harmed.

These legal changes obviously affected urban development, but too much importance should not be attached to them. Archaeological evidence shows clearly that the strict injunctions against trespassing on the public domain were unable to save the classical city layout when other social and economic factors proved too strong. Similarly, Muslim law did allow remedies for gross interference in urban functions. If someone blocked up a major traffic artery, for example, the *qadi* would decide against him and the structure concerned would have to be removed. If the Muslim community had perceived that wide colonnaded streets and spacious agoras were vital to their well-being, then they could have proceeded to law to protect them. It is clear, however, that they did not consider this to be the case, and while Muslim legists agreed that important streets must remain open, they only needed to be wide enough to allow two loaded pack animals to pass each other.

Another contributory element in the transformation of the urban pattern was the changing social structure of the cities. In general, classical cities do not seem to have been the centre of great industrial or commercial activity. Obviously there were exceptions; we have ample evidence of the trade of Tyre and Palmyra in the classical period and there was certainly some commercial activity in Tyre and

Caesarea in Palestine up to the late sixth and early seventh centuries. In the main, however, commerce and manufacture do not seem to have been the most important factor in the prosperity of towns. [. . .] The classical city seems to have depended for its existence on the fact that neighbouring landowners lived in it to take part in social and political activities; in the words of A.H.M. Jones, 'The city was a social phenomenon, the result of the predeliction of the wealthier classes for the amenities of urban life'.[1] This seems to have remained true in the Byzantine period, when the city also became the centre of the ecclesiastical administration as well [. . .]

The Islamic city was very different. Muhammad himself was a merchant from a merchant city, and many famous early Muslims had engaged in trade, including the first caliph, Abu Bakr. Early *hadiths*, the traditions of the Prophet, emerge to the effect that honest commerce was more meritorious than government service, and the prosperous trader was regarded as a pillar of society. When Muslim geographers describe a city, they mention the mosque and the markets, their extent, prosperity and the different sorts of goods for sale. For them it is the commerce of the city rather than its monumental buildings which are the chief source of interest. It was from the merchant class too that the much respected jurists and *qadis* were drawn, not from the ranks of government servants or of the military. It was natural then that the design of the city reflected the needs of this class and that the Muslim city allowed the commercial considerations to outweigh the dictates of formal planning. This is most obvious in the case of the market areas. The main consequence of the change from the open colonnaded street to the crowded *suq* was to increase the number of retail shops in the city centre as the old shops were subdivided and new structures were erected in the old roadway. Urban design now responded directly to commercial pressures and no government action was taken to counter such pressures in the name of the inviolability of public lands or of aesthetic considerations. [. . .]

Note

1 A.H.M. JONES, 'Cities of the Roman Empire: political, administrative and judicial institutions', *Recueils du Société Jean Bodin*, vi, 1954, p. 170

15

MEDIEVAL TECHNOLOGY AND SOCIAL CHANGE

by Lynn White, Jr.

Source: White, Lynn, Jr., *Medieval Technology and Social Change*, Oxford, Oxford University Press, 1962, pp. 41–4, 54, 57, 59–60, 62, 66, 69, 75–6, 78, 83–4 and 88–9

The plough and the manorial system

The plough was the first application of non-human power to agriculture. The earliest plough was essentially an enlarged digging-stick dragged by a pair of oxen. This primitive scratch-plough is still widely used around the Mediterranean and in the arid lands to the east where it is reasonably effective in terms of soil and climate. Its conical or triangular share does not normally turn over the soil, and it leaves a wedge of undisturbed earth between each furrow. Thus cross-ploughing is necessary, with the result that, in regions where the scratch-plough is used, fields tend to be squarish in shape, roughly as wide as they are long. Cross-ploughing pulverizes the soil, and this both prevents undue evaporation of moisture in dry climates and helps to keep the fields fertile by bringing subsoil minerals to the surface by capillary attraction.

But this kind of plough and cultivation was not well suited to much of northern Europe, with its wet summers and generally heavier soils. As agriculture spread into the higher latitudes, inevitably it was largely confined to well-drained uplands with light soils, which were inherently less productive than the alluvial lowlands: the scratch-plough could not cope with these richer terrains. Northern Europe had to develop a new agricultural technique and above all a new plough.

One of the obstacles was that heavy, moist soils offer so much more resistance to a plough than does light, dry earth, that two oxen often are not able to provide enough pulling power to be effective. [. . .]

The wheels on the typical heavy plough both make it more mobile in going from field to field and assist the ploughman to regulate the depth of his furrow – a matter more difficult with several yokes than with a single team. But to understand why the heavy plough eventually affected the whole of northern European life, one must understand how it attacks the soil. Unlike the scratch-plough, the share of which simply burrows through the turf, flinging it to either side, the heavy plough has three functioning parts. The first is a coulter, or heavy knife set in the plough-pole and cutting vertically into the sod. The second is a flat plough-share set at right angles to the coulter and cutting the earth horizontally at the grass-roots. The third is a mouldboard designed to turn the slice of turf either to the right or to the left, depending on how it is attached. Clearly, this is a far more formidable weapon against the soil than is the scratch-plough.

For purposes of northern European agriculture, its advantages were three.

First, the heavy plough handled the clods

with such violence that there was no need for cross-ploughing. This saved the peasant's labour and thus increased the area of land which he might cultivate. The heavy plough was an agricultural engine which substituted animal-power for human energy and time.

Second, the new plough, by eliminating cross-ploughing, tended to change the shape of fields in northern Europe from squarish to long and narrow, with a slightly rounded vertical cross-section for each strip-field which had salutary effects on drainage in that moist climate. These strips were normally ploughed clockwise, with the sod turning over and inward to the right. As a result, with the passage of the years, each strip became a long low ridge, assuring a crop on the crest even in the wettest years, and in the intervening long depression, or furrow, in the driest seasons.

The third advantage of the heavy plough derived from the first two: without such a plough it was difficult to exploit the dense, rich, alluvial bottom lands which, if properly handled, would give the peasant far better crops than he could get from the light soils of the uplands. [. . .]

The saving of peasant labour, then, together with the improvement of field drainage and the opening up of the most fertile soils, all of which were made possible by the heavy plough, combined to expand production and make possible that accumulation of surplus food which is the presupposition of population growth, specialization of function, urbanization, and the growth of leisure. [. . .]

[. . .] This plough, with its coulter, share, and mouldboard, offered much greater resistance to the soil than had the scratch-plough, and thus, at least in its earlier forms, it needed not one yoke but four—that is, eight oxen. Few peasants owned eight oxen. If they wished to use the new and more profitable plough, they would therefore pool their teams. But such a pooling would involve a revolution in the pattern of a peasant group. The old square shape

of fields was inappropriate to the new plough: to use it effectively all the lands of a village had to be reorganized into vast, fenceless 'open fields' ploughed in long narrow strips. [. . .]

While the arrival of the new plough in Scandinavia cannot yet be dated, one suspects that its effect upon population may be seen in the Viking outpouring which began *c*. 800. In any case, the Norse took the heavy plough and the method of land division most appropriate to it with them when, in the later ninth century, they settled the Danelaw in England, and then in Normandy. [. . .]

The discovery of horse-power

The wide application of the heavy plough in northern Europe was only the first major element in the agricultural revolution of the early Middle Ages. The second step was to develop a harness which, together with the nailed horseshoe, would make the horse an economic as well as a military asset. [. . .]

But even a shod horse is of little use for ploughing or hauling unless he is harnessed in such a way as to utilize his pulling power. Thanks to the studies of Richard Lefebvre des Noëttes, it is now recognized that Antiquity harnessed horses in a singularly inefficient way. The yoke harness, which was well suited to oxen,[1] was applied to horses in such a way that from each end of the yoke two flexible straps encircled the belly and the neck of the beast. The result was that as soon as the horse began to pull, the neck-strap pressed on its jugular vein and windpipe, tending to suffocate it and to cut off the flow of blood to its head. Moreover, the point of traction came at the withers, mechanically too high for maximum effect. In contrast, the modern harness consists of a rigid padded collar resting on the shoulders of the horse so as to permit free breathing and circulation of the blood. This collar is attached to the load either by lateral traces or by shafts in such a way that the horse can throw its whole weight into the task of pulling. Lefebvre des Noëttes proved experimentally that a team of horses can pull only about 1,000 pounds with the yoke-harness, whereas with collar-harness the same team can pull four or five times that weight.[2] Obviously, until the modern harness was available, peasants could not use the swifter horse in place of the plodding ox for ploughing, harrowing, or heavy hauling. [. . .]

Moreover, a horse has more endurance than an ox, and can work one or two hours longer each day. This greater speed and staying power of the horse is particularly important in the temperamental climate of northern Europe where the success of a crop may depend on ploughing and planting under favourable circumstances. Moreover, the speed of a horse greatly facilitated harrowing, which was of more importance in the north than near the Mediterranean where cross-ploughing broke up the clods fairly well. [. . .]

Not only ploughing but the speed and expense of land transport were profoundly modified in the peasants' favour by the new harness and nailed shoes. In Roman times the overland haulage of bulky goods doubled the price about every hundred miles. [. . .] In contrast, in the thirteenth century the cost of grain seems to have increased only 30 per cent for each hundred miles of overland carriage – still high, but more than three times better than the Roman situation. Now it was becoming possible for peasants not situated along navigable streams to think less in terms of subsistence and more about a surplus of cash crops. [. . .]

The three-field rotation and improved nutrition

The three-field system of crop rotation has been called 'the greatest agricultural novelty of the Middle Ages in Western Europe'.[3] It bursts upon us in the late eighth century [. . .]

[. . .] The new supply of oats made available by the three-field system increased the num-

bers and prowess of horses. But people likewise were shaped by the new food resources.

In addition to oats and barley, the spring planting was habitually composed of legumes. [. . .]

Our recently acquired knowledge of nutrition [. . .] provides us with new insight into the dynamics of the later Middle Ages. While the legumes available to medieval Europe did not in themselves supply a complete series of the biologically necessary amino-acids, by a happy coincidence the smaller quantities of proteins found in the common grains were the perfect dietary supplement to those present in legumes, and particularly in field peas. It was not merely the new quantity of food produced by improved agricultural methods, but the new type of food supply which goes far towards explaining, for northern Europe at least, the startling expansion of population, the growth and multiplication of cities, the rise in industrial production, the outreach of commerce, and the new exuberance of spirits which enlivened that age. In the full sense of the vernacular, the Middle Ages, from the tenth century onward, were full of beans.

The northward shift of Europe's focus

A more durable solution of the historical problem of the change of the gravitational centre of Europe from south to north is to be found in the agricultural revolution of the early Middle Ages. By the early ninth century all the major interlocking elements of this revolution had been developed: the heavy plough, the open fields, the modern harness, the triennial rotation – everything except the nailed horseshoe, which appears a hundred years later. [. . .]

The agricultural revolution of the early Middle Ages was limited to the northern plains where the heavy plough was appropriate to the rich soils, where the summer rains permitted a large spring planting, and where the oats of the summer crop supported the horses to pull the heavy plough. It was on those plains that the distinctive features both of the late medieval and of the modern worlds developed. The increased returns from the labour of the northern peasant raised his standard of living and consequently his ability to buy manufactured goods. It provided surplus food which, from the tenth century on, permitted rapid urbanization. In the new cities there arose a class of skilled artisans and merchants, the burghers who speedily got control of their communities and created a novel and characteristic way of life, democratic capitalism. And in this new environment germinated the dominant feature of the modern world: power technology. [. . .]

The sources of power

And in fact in the very late tenth or eleventh century we begin to get evidence that water-power was being used for processes other than grinding grain. By 983 there may have been a fulling mill – the first useful application of the cam in the Occident – on the banks of the Serchio in Tuscany. In 1008 a donation of properties to a monastery in Milan mentions not only mills for grinding grain, but, adjacent to them along the streams, *fullae* which were probably fulling mills. In 1010 the place-name Schmidmülen ['forge mills'] in the Oberpfalz indicates that water-driven trip-hammers were at work in the forges of Germany. About 1040 to 1050 at Grenoble there was a fulling mill, and *c.* 1085 one for treating hemp. By 1080 the Abbey of Saint Wandrille near Rouen was receiving the tithes of a fulling mill, and by 1086 two mills in England were paying rent with blooms of iron, indicating that water-power was used at forges. Before the end of the eleventh century, iron mills were likewise found near Bayonne in Gascony. [. . .]

In 1086 the *Domesday Book* lists 5,624 mills for some 3,000 English communities. There is no reason to believe that England was techno-

logically in advance of the continent. By the eleventh century the whole population of Europe was living so constantly in the presence of one major item of power technology that its implications were beginning to be recognized. [. . .]

During the next hundred years windmills became one of the most typical features of the landscape of the great plains of northern Europe where they offered obvious advantages in terms of the topography. Moreover in winter, unlike the water-mill, their operation could not be stopped by freezing. As a result, during the thirteenth century, for example, 120 windmills were built in the vicinity of Ypres alone. [. . .]

Particularly in southern Europe there continued to be technologically retarded pockets: Don Quixote's amazement at windmills was justified: apparently they were introduced into La Mancha only in Cervantes's time. Nevertheless, despite our dearth of fundamental studies of the process, it is clear that by the early fourteenth century Europe had made extraordinary progress towards substituting water- and wind-power for human labour in the basic industries. For example, in England during the thirteenth century mechanical fulling of cloth, in place of the older method of fulling by hand or foot, was decisive in shifting the centre of textile manufacturing from the south-eastern to the north-western region where water-power was more easily available. Nor was England especially advanced: the guild regulations of Speyer in 1298 show that fulling mills had completely displaced earlier techniques in that area as well. Similarly, mills for tanning or laundering, mills for sawing, for crushing anything from olives to ore, mills for operating the bellows of blast furnaces, the hammers of the forge, or the grindstones to finish and polish weapons and armour, mills for reducing pigments for paint or pulp for paper or the mash for beer, were increasingly to be found all over Europe. This medieval industrial revolution based on water and wind would seem to have reached its final sophistication when, in 1534, the Italian Matteo dal Nassaro set up a mill on the Seine at Paris for the polishing of precious stones, only to have it taken over in 1552 by the royal mint for the production of the first 'milled' coins.

[. . .]

Notes

1 R. Lefebvre des Noëttes, *L'Attelage et le cheval de selle à travers les âges* (Paris, 1931), fig.448

2 Ibid., p. 159, shows that a team of horses or mules which would be expected to pull 1,980 to 2,480 kgm. today could pull only about 492 kgm. in ancient harness. A.P. Usher, *History of Mechanical Inventions*, 2nd edn (Cambridge, Mass., Harvard University Press, 1954), p. 157, concludes, on the basis of late nineteenth-century tables of work normally expected from horses, that 'the net effectiveness of ancient draft animals in harness was not more than one-third of the modern expectation'. He adds, however, that 'the figures in the modern table are distinctly low', and that the assertion 'that animals in antiquity achieved only one-third of the modern expectation is really a moderate statement, under- rather than overstated'. We may therefore accept Lefebvre des Noëttes's estimates as being close to the truth. A. Burford, 'Heavy transport in classical Antiquity', *Economic History Review*, 2nd series, xiii (1960), pp. 1–18, emphasizes the inadequacy of ancient horse harness but properly underscores the fact that, despite this relative inefficiency, the ancients achieved great results using oxen. [Burford's essay is reprinted as Reading 5 in the present volume on pp. 29–36].

3 C. Parain in *Cambridge Economic History*, Cambridge, Cambridge University Press, Vol. 1, 1941, p. 127.

16

FUELLING THE CITY: PRODUCTION AND DISTRIBUTION OF FIREWOOD AND FUEL IN LONDON'S REGION, 1290–1400

by James A. Galloway, Derek Keene and Margaret Murphy

Source: James A. Galloway, Derek Keene and Margaret Murphy, 'Fuelling the City: production and distribution of firewood and fuel in London's region, 1290–1400', *Economic History Review*, 49, 1996, pp. 447-72[1]

Of all the constraints upon the growth of pre-industrial cities, perhaps none was more critical than the fuel supply. Domestic heating and cooking, commercial food preparation, and industries all contributed to a concentrated demand for fuel in towns. The cities of northern Europe, faced with colder winters and home to fuel-hungry brewing industries, almost certainly required more per caput than their counterparts in the warmer, wine-drinking south. The basic needs of the 80,000 or so inhabitants of London are thus likely to have generated one of Europe's largest aggregate demands for fuel in the decades around 1300. Although the later fourteenth-century capital had a much reduced population, the general rise in living standards almost certainly brought with it an increase in the consumption of fuel per head. London was also the focus of a large industrial demand for fuel, from the metal-working and textile-finishing crafts within its boundaries, and from the pottery, metal-working, glass-making, and baking industries which were a notable feature of its hinterland. The demands of the city's building industry were also substantial: for burning lime, for firing the tiles commonly used for roofing by 1300, and for producing increasing quantities of bricks from the later fourteenth century onwards.

Moreover, London was not the only major city to draw its fuel supply from the woods of southern England, which also served the needs of highly urbanized areas overseas.

Throughout the middle ages, wood grown in the accessible hinterland was the main direct source of heat in London, but before 1200 a trade in other fuels had grown up to meet specialized industrial requirements. By the 1160s there was a charcoal market in a central neighbourhood favoured by the iron-working crafts. The rapid expansion of building in stone had, by 1180, promoted the shipping of 'sea coal', presumably from Newcastle upon Tyne, to a site in the city's western suburb for use by limeburners, whose activities came to be especially associated with the fuel.[2] By the mid-thirteenth century London smiths regularly used coal, which was unloaded at wharves just downstream of London Bridge and distributed widely within the city. In 1300 coal appears to have been used in London predominantly as a source of concentrated and persistent heat for industrial purposes, and consumption was sufficiently great for its noxious by-products to be a matter of concern. At that time London was the most important site in a coal distribution network extending from the Tyne to the south-western ports, and it seems to have occupied a comparable position in the late fourteenth century.

The influence of fuel upon the growth, and potential for growth, of major cities was mediated by its bulk and relatively low value. A high proportion of its final price was represented by the cost of transport, especially in the case of firewood. Oven-dried wood, with zero moisture content, can attain energy values as good as, or better than, low-grade coal. However, coal from north-eastern England was of a generally high quality, and, as the relative prices suggest, had a significantly higher energy value per unit of weight than the firewood available to fourteenth-century Londoners. Its disadvantages were its perceived noxiousness and, above all, the distance over which it had to be shipped to London. The great cost of shipping coal to London is reflected in the fact that the London price in 1259 was at least four times the Durham price in 1300. These factors mean that, if reserves of woodland were adequate, we would not expect to find sea coal used in the capital other than for the industrial processes where it offered advantages. Domestic consumption of coal in London does indeed seem to have been very small *c.*1300. In this respect conditions resembled those in 1550, when the city was probably about the same size. The rapid growth of the city after 1550, however, was accompanied by a great increase in coal consumption for domestic as well as industrial purposes, suggesting that at both dates the city's domestic fuel needs may have been approaching the limit of the capacity of economically accessible woodland to supply it.

The old notion that the later middle ages and the early modern period witnessed a progressive shortage of wood, associated with industrial and military expansion and with the clearance of woodland, has been widely criticized. Historians of the countryside have shown that the woods of eastern and southern England were a carefully managed, renewable

resource which, far from becoming exhausted, demonstrate remarkable continuity as sources of both fuel and building timber. Scepticism regarding wood scarcity may, however, have gone too far. The scale of the industrial use of sea coal in London by the fourteenth century perhaps suggests that firewood was in increasingly short supply. The small quantities of peat which were imported from overseas point in the same direction.

More striking is the sharp rise in firewood prices during the later thirteenth century. In the 1270s and 1280s the demesne [estate] at Hampstead, 5 miles from London, sold the standard firewood unit known as the faggot at around 20d. per hundred. The price rose dramatically in the late 1280s and 1290s reaching a peak of 38d. in 1291–2. Prices appear to have dipped slightly in the opening years of the fourteenth century, but by the 1320s were creeping up again. Similarly, firewood prices at three manors in Surrey up to 19 miles from London rose by some 50 per cent between the 1280s and 1330s. These increases reveal the particular strain on London's fuel supply at that time, for other key agrarian indicators, such as arable land values and London wheat prices, display relative stability or a tendency towards decline during this period, while in districts beyond the immediate influence of the city fuel prices did not rise. Over the same period in London the price of wood fuel appears to have increased more rapidly than that of coal. In 1259 sea coal was purchased in London for 12d. and 14d. a quarter, while in about 1270 the London price for what may have been the equivalent weight of firewood (27.75 faggots) was 7.8d. By 1300 the London price for coal was probably less than the 16d. per quarter for which it sold in northern Hertfordshire, while that for the equivalent weight of faggots was 13.3d. This represents a price increase of less than 23 per cent for coal by comparison with 70 per cent for faggots, and presumably a corresponding increase in the propensity to consume coal.

The London region considered in this article contained some of the most wooded parts of medieval England. In 1086 counties close to London and the lower Thames were conspicuously more wooded than those further away. The region contained two major tracts of woodland: the Chiltern Plateau and a part of the Weald. More important than the extent of woodland resources, however, was their accessibility to centres of consumption. In the context of a pre-industrial economy, von Thünen placed wood production close to the central city in his model of land-use zoning. Because of its bulk in relation to its sale value, wood could profitably be produced only in places close to the market [. . .] The best return on the use of a particular piece of land will be determined by the price of the commodity produced and the cost or time involved in carrying it to the market, as well as by inherent qualities such as soil fertility. That return, independent of other factors of production, is known as the economic rent, and in a region where a central market exercises a dominant influence, will generate a concentric pattern of specialization, in which a product such as wheat can most profitably be grown at a greater distance than wood. In von Thünen's model, only the perishable products of horticulture and dairying occupied land closer to the city than commercial wood production. Wood itself is differentiated within the model, with charcoal and building timber able to sustain greater transport costs than firewood and thus being produced at a further remove. The model has been shown to explain certain patterns of agrarian production in the London region c.1300.[3] Although wood production has sometimes been considered the least plausible aspect of the model, empirical confirmation of the existence of a 'forest zone' around one pre-industrial city has been found.[4]

This article aims to test the transport–cost model, on the lines of von Thünen's theory, in relation to the supply of firewood to London in the fourteenth century. It puts forward estimates of the likely scale of the demand generated by the city and its region, and considers its probable structure. It attempts to assess the ability of the region to meet those needs by using two sets of measures focusing on the period around 1300 and on the later fourteenth century. These derive from databases comprising regional samples of manorial accounts, which detail production on the demesnes of ecclesiastical, royal and lay estates, and inquisitions *post mortem* (hereafter IPMs), which describe in simple terms the resources of manors in lay ownership. By drawing on a variety of other sources, including manorial accounts not in the databases, the article examines the productive and distributive systems employed. It also assesses the degree to which the provision of an adequate fuel supply for London placed a strain upon its hinterland, and the ability of the region to participate in wider networks of trade in fuel.

I

Firewood (*boscus, busca*) was a recognized commodity in London from an early date, although less standard, and more dangerous, fuels were also used in the city's hearths. Thus, soon after 1212 London bakers and brewers were ordered to burn firewood rather than reeds, straw, or stubble. Most wood fuel entered the city in a prepared condition, as firewood cut into equal lengths and of various thicknesses, or as charcoal. Mid-fourteenth-century attempts by the city authorities to regulate the trade in firewood provide a good indication of the types then most commonly used. These were faggots and talwood, the latter made of beech and oak and both measured in units of 100; charcoal, sold by the quarter; and wood called billets. Accounts for lay and religious households in or close to the city, and accounts for manors close to

London, confirm the prominence of these wood products as urban fuel. Manorial accounts for the London region rarely identify the species sold as fuel, but in their references to the sale of whole trees (which would have provided both timber and fuel) most commonly name oak, beech, and ash.

Faggots were bundles of brushwood, rods and sticks of roughly equal size tied together. Kindling quickly to produce a short hot blaze, they appear to have been the principal fuel of the city's bakeries and brew-houses and were also used in a variety of industrial processes such as brick- and tile-making.[5] They were probably also used as kindling on the domestic hearth. A specialized type of small faggot used in kilns, and in baking and brewing, was known as the bavin. In the fourteenth century Westminster Abbey used thousands of bavins each year.

Evidence for their price and transport indicates that talwood and billets were heavier, thicker pieces of wood than the individual pieces which were tied together to make a faggot, although they too were sometimes sold in bundles. Their principal use was for domestic heating and for some food processing and industrial purposes. In Westminster Abbey talwood appears to have been the principal heating fuel: in 1402–3 the infirmarer used 1,275 talwood in his fireplaces. Billets were smaller than talwood and may have had specific uses. In 1305 a London dyer purchased 1,000 billets from a city woodmonger for 9s., suggesting that billets were used in furnaces for heating water as well as in domestic fireplaces. Billets and talwood were perhaps the main components of the heaps of logs (*trunci*) which obstructed the street outside many houses in the thirteenth-century city. Charcoal had a number of industrial uses, but was also used in braziers as a domestic heating fuel, occurring, for example, in fuel allowances to Westminster Abbey corrodians.

In order to satisfy London's demand there had emerged in the city before 1200 an identifiable group of specialized traders in wood. These men are most often described as *buscarii*, although the vernacular descriptions or bynames *buscher*, *wodemongere* and *tymbermongere* were also used. In the period from *c*.1275 to *c*.1375, 90 such individuals involved in London's wood market have been identified. [. . .]

Castle Baynard Ward, and especially Woodwharf in the parish of St Peter the Less, upstream of London Bridge, was the centre of the London firewood trade, probably from well before 1200. Later, significant numbers of woodmongers owned houses, wharves, and other property there. [. . .] Queenhithe, a little further downstream, was another destination for shouts (river boats) carrying firewood, probably by the early thirteenth century, while the existence by *c*.1200 of the nearby wharf known as Timberhithe underlines the significance of traffic down the Thames for supplying the city with wood of all sorts. At least as early as 1276 faggots were also brought upriver to wharves just below London Bridge, and the name Wood Street suggests that before 1150 there was a market for wood close to the inland commercial heart of the city. Water transport was crucial for the wood trade. Many woodmongers owned shouts and boats, from which they customarily sold firewood directly to citizens who carried it away in carts. Woodmongers also stored stocks of wood on their wharves. [. . .] In time of shortage they were required to sell off such stocks on pain of confiscation, as in 1379, when 6,000 billets stored near Billingsgate by two woodmongers were seized by the city authorities.

Processing wood before bringing it to the city added value and reduced transport costs. The significance of processing is likely to have been greatest where the market was most concentrated. Manorial accounts indicate that within the region London exerted a major influence on woodland management. Manors close to the centre of demand employed their own workforces to cut underwood (*subboscus*) and to make faggots, bavins, talwood, billets, and charcoal. By contrast, those located away from major markets tended not to use these specialized terms for the firewood they produced, often selling their underwood by the acre for others to fell and prepare for sale. [. . .]

Manorial accounts for the London region show that throughout the fourteenth century faggot production was a specialism of the counties of Middlesex, Surrey, and Essex, and of parts of Buckinghamshire, Hertfordshire, and Kent (see Figure 16.4). Bavins appear to have been even more locally specialized: the term occurs only on the Middlesex manors of Hampstead, Hendon, Colham, and Hyde, and at Iver in the extreme south-east of Buckinghamshire, all within 19 miles of London. Later fourteenth-century accounts for those manors specializing in making faggots and bavins provide detailed information on quantities produced and disposed of. Increasingly, woodland was perceived in the same way as arable land, with an annual 'crop' which formed a readily marketable commodity. [. . .]

The larger types of firewood are less well documented. Production and sale of talwood is recorded for manors in both periods and shows a clear geographical pattern, with northern Kent and the vicinity of the Thames between Henley and Datchet prominent. Manors within about 12 miles of London rarely sold talwood in the fourteenth century, perhaps because their woods were on such short cropping cycles that the wood rarely achieved the required thickness. The manor of Iver, about 19 miles from London, sold 3,700 talwood in 1378 at 4s. per 100. While this is the highest price recorded for talwood in the 1375–1400 database, it falls considerably short of the price of 7s. per 100 ordained in London for talwood in 1362–3. This price differential may reflect differences in the size or species of talwood,

but also accords well with cost–distance from London. [. . .]

Rendering wood into charcoal, which compares well in calorific terms with coal, greatly increased its energy and monetary value per unit of weight, and therefore extended the distance from which it could profitably be brought to London for sale. In the city, charcoal was sold in sacks nominally containing one quarter and costing between 10d. and 12d. each. Records of attempts to prevent sale of charcoal in smaller sacks from 1368 onwards show that it was regularly brought to the city by pack-horse and cart from such places as Croydon (Surrey), Cheshunt, and Hatfield (Hertfordshire), and from as far away as Hatfield Broadoak in Essex, some 26 miles from London. [. . .]

Specialized production of fuel formed part of a wider strategy of tree management, in which a balance was struck between different uses and products, including building timber from mature trees, firewood and poles from smaller trees and branches, and grazing and pannage [pasturage for swine] for cattle and swine. It was possible to combine all of these uses, but normally one assumed pre-eminence. In particular, if there was an accessible market for wood products then they took precedence over less intensive uses. The frequency of underwood cropping was varied in order to produce different grades and sizes of wood. Management of underwood in the London region, as revealed by a small number of IPM extents, was characterized by short cropping cycles, with a mean and mode of seven years c.1300, and as such seems to have been directed towards faggot production. By the later fourteenth century cycles had lengthened to a mean of 11 years and a mode of eight years, perhaps influenced by an increased market for logs and a shift towards less intensive systems of wood-pasture management as labour costs and livestock numbers increased. However, the change, if real, might also reflect lower woodland fertility. Given the small sample, and the fact that none of the documented cycles relate to London's immediate vicinity, it is not possible to be more specific concerning metropolitan influence upon this apparent change.

II

Some quantitative estimates can be made of metropolitan and regional demands for firewood in the fourteenth century, and of the capacity of the region's woodland (i.e. land primarily occupied by trees) to meet them. There are no contemporary records of total quantities consumed or produced, but it is possible to estimate the fuel required to process an average per caput grain requirement into bread and ale, while some household accounts enable annual fuel consumption to be calculated. In quantifying the domestic needs of the general population, and the fuel consumed by industry, however, it is necessary to draw upon evidence from a later period.

It has been estimated that in 1600, by which date coal had just replaced wood as the 'general fuel' in London, per caput consumption stood at 0.75 tons per annum. Although metropolitan living standards were perhaps a little lower in 1600 than in 1300, and certainly lower than in 1400, industrial consumption was higher, particularly in brewing and brick-making. The figure of 0.75 tons of coal per caput can therefore be used as a starting point in assessing the fuel needs of the medieval city.

The dry wood equivalent of 0.75 tons of coal may be estimated as 1.76 tons.[6] On that basis, London would have consumed 141,000 tons of wood per annum in 1300, with a postulated population of 80,000, and 88,000 tons per annum in 1400, when the population may have been 50,000. The per caput consumption of fuel was probably lower outside London, and so the total consumption of the 10 counties of the London region may have been equivalent to some 1,035,000 tons of wood in 1300, falling to some 649,000 tons in 1400. Some of that requirement would have been met by sea coal and so, for several reasons, these figures should probably be regarded as maxima.

Baking and brewing in London may have required between 29,100 and 31,800 tons of firewood per annum in 1300, 21 to 23 per cent of the city's total estimated consumption. Even allowing for significantly increased intakes of ale in the later fourteenth century, baking and brewing are unlikely to have represented more than 23 to 27 per cent of total consumption at that time, and the actual quantities used for these purposes may have been similar to those in 1300. Changes in diet may thus not have had a major impact upon the overall levels of fuel consumption.

These estimates imply that London's total fuel needs required the entire underwood output of some 70,000 acres of intensively managed woodland in 1300, falling to 44,000 acres in 1400, assuming an output of 2 tons per acre. These areas of woodland are equivalent to 11 per cent and 7 per cent, respectively, of the area of Middlesex and Surrey combined. In 1300 the region as a whole may have required the output of some 518,000 acres and in 1400 of 325,000 acres, equivalent to 10 per cent and 6 per cent, respectively, of its land area. Rather more than these percentages might have been needed to produce the fuel required, since the underwood also supplied wood used for fencing and some building purposes. On the other hand, both coal and loppings from trees felled for timber contributed to the fuel supply. The estimated output of 2 tons per acre may be too high. If it was 1.33 tons, then the areas of woodland required for London's supply would be equivalent to 16 and 10 per cent of the area of Middlesex and Surrey, in 1300 and 1400 respectively; and the region as a whole would have required 14 and 9 per cent of its area.

How do these proportions compare with likely patterns of land use in the region? The sources permit estimation of the area under woodland, but rarely indicate the presence of trees in hedgerows, orchards, or along watercourses. A minimal figure is provided by IPMs for lay estates, which show that in the period 1270–1339 the mean woodland acreage on demesnes in the London region was 13.5 acres, representing approximately 6 per cent of the total acreage. The much smaller sample of IPMs from the period 1375–1400 indicates a figure as low as 4.2 per cent. These figures, however, almost certainly underestimate the proportion occupied by woodland, since many of the most extensive woods of the region were in ecclesiastical or royal ownership. [. . .] Furthermore, IPMs neither cover woodland within areas of commons and 'waste', nor quantify the significant amount of woodland lying within private parks. Some extents record only the acreage and value of cropped underwood, ignoring those parts of a wood which yielded no income in the current year.

Higher figures for the extent of woodland in

Table 16.1 Distribution of woodland in the London region

| County | Wood as a percentage of | | Percentage of IPM[c] |
	land in lay hands c.1300[a]	land area c.1350[b]	wood entries specifying acreages c.1300[a]
Beds.	11	10	88
Berks.	5	14	64
Bucks.	6	16	66
Essex	4	14	85
Herts.	10	20	90
Kent	10	17	90
Middx.	9	20	100
Northants.	2	6	38
Oxon.	4	11	46
Surrey	8	31	79
All 10 counties	6	16	78

Notes and sources: [a] *derived from IPMs*
[b] *derived from Rackham's estimate (see note 7)*
[c] *IPM: inquisitions post mortem*

the London region are obtained by adapting Rackham's estimates for 1086, an exercise which suggests that in 1350, 16 per cent of the London region lay under woodland (Table 16.1), compared to 10 per cent for England as a whole.[7] This proportion seems high, certainly in comparison with the estimated 9 per cent of the region occupied by woodland in the early nineteenth century, but if the estimate is correct it appears that in aggregate the London region during the fourteenth century contained sufficient woodland for its fuel needs and probably could have generated a surplus for export.

Woodland, however, was unevenly distributed within the region, and demand did not fall evenly upon it. In particular, the concentrated demands of the capital weighed more heavily upon its immediate hinterland. Given the high cost of transporting wood, it seems likely that some extensive tracts of woodland within the region (large parts of the Weald and the Chiltern Hills, for example) could not economically supply the London market. Using information on the costs of production and transport and on sale prices it is possible to define the area from within which, in normal years, faggots could be supplied to the London market without making a loss. In the early fourteenth century, when the sale price of 100 faggots in London was approximately 48d., the cost of carrying them overland in Middlesex was around 2.5d. per mile. At that rate, allowing for production costs, it would have been possible to supply faggots to London from up to 17.4 miles distant overland, the outer limit of the zone being represented by such places as Watford, Brentwood, and Westerham. The calculation, however, makes no allowance for transaction costs and profit. If they represented 20 per cent of the sale price in London, the radius would reduce to 13.7 miles. There is insufficient information for comparable calculations concerning talwood or billets, which may have been able to withstand a greater distance, or charcoal, which certainly could.

The cost of transporting wood in the vicinity of London effectively doubled between the second and fourth quarters of the fourteenth century, with the dramatic effect of reducing London's theoretical faggot supply radius to about 11 miles overland, or to 9 miles allowing 20 per cent for profit and transaction costs. The larger radius would place Enfield and Romford at the edge of the supply zone, while the smaller would just include Acton to the west of the city and Eltham to the south-east.

When water transport was available, more distant places could be drawn into the zone. If the ratio between the costs of carrying wood by land and by water was the same as for grain, then London's theoretical faggot supply zone assumes an elongated shape, extending up the Thames valley beyond Henley in the early four-

teenth century, but only just as far as that market town in 1400 (see Figure 16.1). Virtually the whole of Middlesex would have lain within the zone in 1300, together with about half of Surrey, north-east Berkshire, the southern Chiltern district in Oxfordshire and Buckinghamshire, and those parts of Essex, Kent, and southern Hertfordshire within reach of navigable water. The zone also included a limited coastal fringe of eastern and south-eastern Kent. By the later fourteenth century there would appear to have been a marked contraction towards the river routes.

While the discovery of further information on transport might revise this picture, the conclusion that the brunt of the city's demand for firewood fell upon a circumscribed area is unlikely to be reversed. How did the distribution of

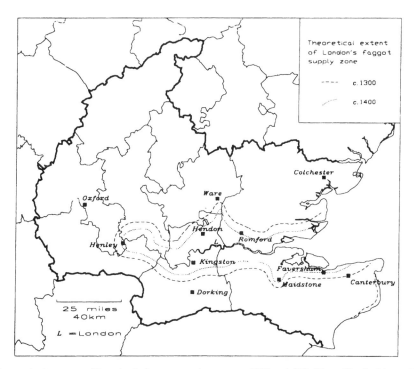

Figure 16.1 Theoretical extent of London's faggot supply zone, c.1300–c.1400. *Note:* The bold continuous line denotes the boundary of the London region study area. Historic (pre-1974) county boundaries are indicated by lighter continuous lines.

woodland within the region match up to this likely spatial impact of demand? The geographical distribution of woodland resources within the region, estimated as a proportion of the land area by two methods (Table 16.1), shows some striking variations between counties, although the methods produce significantly different absolute values. Both methods identify Middlesex, Kent, Surrey, and Hertfordshire as among the most wooded of the counties. Buckinghamshire emerges as moderately wooded, but the overall county figure obscures a contrast between the virtually woodless north and the well-provided Chiltern district of the south. Similarly, the percentage of woodland in Surrey is certainly under-represented by the IPMs, which are scarce for the Wealden south of the county.

The most notable discrepancy between the two measures is in the ranking of Bedfordshire, the most wooded county of the region according to one method, but among the least wooded by the other. IPM evidence as to annual woodland values, which cannot be used systematically because of a lack of consistency in the way in which these are recorded, suggests that the Bedfordshire woods were relatively valuable, along with those in parts of the Thames and Lea valleys, in parts of Hertfordshire, and in eastern Kent. The highest underwood values for the period 1270–1339, over 36d. per acre per year, occur at Enfield (Middlesex), Folkestone (Kent), at several locations in central Essex, and in the vicinity of Saffron Walden.

Another, and perhaps more significant, insight into geographical contrasts in the relative importance of woodland as a commercial

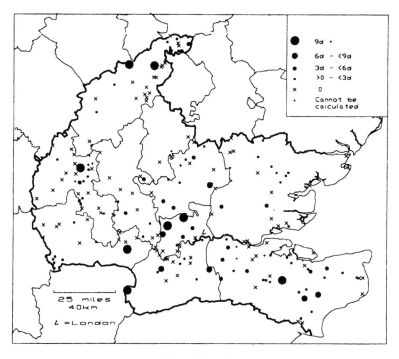

Figure 16.2 Income generated from wood sales 1290–1315, adjusted according to size of demesne (i.e. relative to acres sown with grain)

resource is revealed by variations in the detail with which it is recorded in IPM extents. While 97 per cent of the 823 IPM extents from the period 1270–1339 which mention wood specify an aggregate value, only 78 per cent provide further details concerning its acreage, with marked variation between counties (see Table 16.1). On this basis, woodland resources appear to have been most significant in Middlesex, which may have lain wholly within London's firewood supply zone *c*.1300, followed by Kent, Hertfordshire, and Bedfordshire. Slightly less prized overall were the woodland resources of Essex and Surrey, with Buckinghamshire and Berkshire decidedly less so. Woodland appears to have been least valued in Oxfordshire and Northamptonshire, although the picture was probably different near Peterborough in the northernmost part of the latter (see Figure 16.2). There appears to have been a tendency, to which Bedfordshire seems an exception, for woodland resources to be less valued in the west and north of the region than in the south and east.

London, therefore, lay within a region which still contained quite extensive areas of woodland in the early fourteenth century. Some of these reserves, however, were in parts of the region from which transport to London would have been prohibitively expensive, such as inland parts of the Kentish or Surrey Weald. Nevertheless, accessible parts of the region, such as Middlesex, also contained quite high proportions of woodland, and the evidence suggests that there the woods were most prized and most intensively managed.

III

Variations in the extent and accessibility of woodland within the London region were reflected in the systems of commercial production and the distribution networks which emerged to supply the market for wood fuel. One way of assessing the importance of the commercial production of wood fuel in the fourteenth century is to measure the degree to which it contributed to manorial income. There are problems with this approach, however, arising from variations in management and accounting practice and from the irregularity of income associated with occasional heavy felling. Manorial practice is only fully revealed in those few cases where long series of consecutive accounts are available. Furthermore, the farming-out of demesne arable while retaining woodland in direct management is a feature of many manors in the later fourteenth century, particularly in Middlesex, and this necessitates their exclusion from calculations involving total income.

About half of the 204 demesnes in the database for 1288–1315 for which the relevant evidence is complete sold some type of wood or wood product. In the period 1375–1400 a greater proportion of the 140 manors (66 per cent) record sales of wood. In both periods recorded wood sales contributed only around 5 per cent of total sales income and the average value of wood sales was between £1 and £2 per annum. Where it was possible to distinguish firewood income from general wood income it was found that the mean value of identifiable firewood sales more than doubled between 1300 and 1400, and as a proportion of total wood income, firewood sales increased from 38 to 53 per cent.

Comparison of wood income across the region is distorted by variations in demesne size. This can be partially compensated for by representing the value of wood sales relative to acres on the demesne which were sown with grain. In the period 1290–1315 the mean wood income per grain acre was 1.4d. The highest income, of 26d. per acre at Middleton Stoney (Oxfordshire), may reflect a one-off felling of timber. On 14 demesnes the income was 6d. or more. There was a clear concentration of these demesnes in the area defined in terms of costs as London's faggot supply zone (see Figure 16.2). Accounts for these demesnes virtually always identify the wood sold as firewood, and testify to direct links with the London fuel market, as also do accounts for nearby demesnes not in the database. In 1375–1400 the mean wood income per grain acre was 2.8d., with the highest income, 61.6d. at Betchworth (Surrey), reflecting regular charcoal sales. On 18 demesnes the income was 6d. or more. There had been some changes in the distribution of those manors with the highest incomes per grain acre (see Figure 16.3). Neither the increase in firewood prices between the two periods nor the decline in demesne grain acreages (averaging 25 per cent across the region as a whole) was sufficiently skewed to any part of the region to be the cause of these apparent changes in the distribution of income from wood sales. In the late fourteenth century London was still an important influence on the pattern. The suggestion in Figure 16.3 that it was a less significant force than earlier may reflect the different composition of the databases.

It is clear that the database manors alone (chosen for the completeness of their record across the whole range of agrarian activity within two strictly defined sample periods) cannot provide a fully representative picture of woodland management within the London region. In an attempt to define more precisely those areas where the production of fuel was most commercialized, all manors in the region (whether in the database or not) known to have sold more than 1,000 faggots a year in the periods 1290–1330 and 1375–1400 have

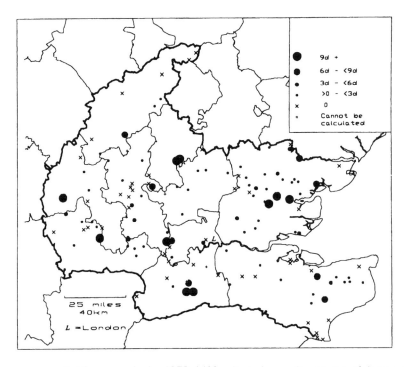

Figure 16.3 Income generated from wood sales 1375–1400, adjusted according to size of demesne (i.e. relative to acres sown with grain)

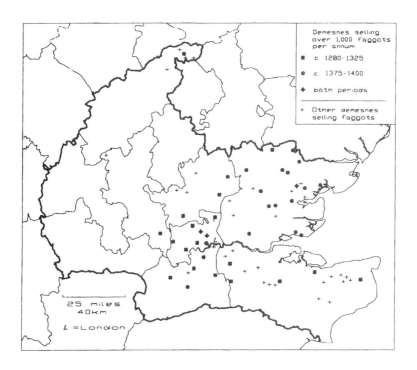

Figure 16.4 Number of faggots sold by demesnes in the London region, *c.*1280–*c.*1400

been mapped (see Figure 16.4). While few manors equalled Hampstead and Hendon in their sales, Edgware, Acton (Middlesex), Byfleet, and Farleigh (Surrey) sold between 2,000 and 4,000 faggots a year in the earlier period, representing incomes of between 60 and 100 shillings, while in the late fourteenth century Colham (Middlesex), Bocking, Bures, Kelvedon, Little Maldon, Moulsham, and South-church (all Essex) sold between 2,000 and 6,000, representing annual incomes of between 60 and 188 shillings. The distribution suggests that the chief impact of London was confined to its immediate vicinity, in Middlesex and Surrey and perhaps to the margins of the Thames estuary; that there was a growing market in north-central Essex which was separate from that of London; and that in the earlier period there was a significant market in the vicinity of Peterborough.

The activities of London woodmongers and the evidence concerning the supply of great households in the capital conform to the concentration of significant wood sales in the London area (see Figure 16.5). Londoners purchased wood directly from manors such as Dorking (Surrey) and Hampstead, Hendon, and Isleworth (all Middlesex). [. . .] On the archbishop of Canterbury's manor of Bexley, 15.5 miles from London, wood was regularly sold to Londoners, especially brewers, in the late fourteenth and fifteenth centuries. That wood was transported overland to Woolwich or Erith and stockpiled there before being sent by water to London.

Households in or close to the capital frequently drew their fuel supplies from within the same restricted territory. Westminster Abbey drew heavily on its manors of Hendon and Hampstead, and received smaller supplies

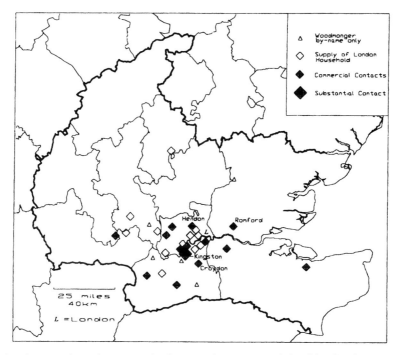

Figure 16.5 The London wood-supply zone in the fourteenth century as defined by the direct evidence of woodmongers' activities and household supply

from its demesnes at Battersea, Wandsworth (both Surrey), and Hyde (Middlesex). The abbey also purchased firewood in London, at Ham near Kingston, and at *Brokwod* (probably Brookwood in Surrey, near Woking). [. . .] The abbatial manor of Denham (Buckinghamshire) supplied talwood and large faggots, while timber was carted from Greenford (Middlesex, 11 miles by road from Westminster) to the Thames and then carried in the abbot's boat. In 1296, 2,000 *busche* came down river from Hurley (Berkshire). [. . .]

Many of these examples underline the significance of the Thames in linking the producers and consumers of wood. The nature of the firewood trade, with its heavy costs for overland transport, may have tended to restrict the emergence of entrepots such as those which dominated the grain trade. The need to bring

wood to the water as soon as possible, so as to contain costs, is likely to have encouraged the use of many smaller landing places, particularly on the middle Thames. [. . .]

Building timber, of higher value relative to its weight than firewood,[8] was brought to London from many places in the thirteenth and fourteenth centuries, but Surrey again seems to have been by far the most important source, with Kingston assuming a pivotal role. Building works at Westminster regularly involved the purchase of timber at Kingston, and its carriage to the site by water. In the later fourteenth and fifteenth centuries timber for London Bridge was purchased in numerous places in Surrey (including Kingston, Croydon, Dorking, and Beddington) and was carried to London by land and water. Timbers destined for buildings in the city were often shaped in specialist framing yards some distance away.

London was the primary focus for the wood trade in the region, but the demands of other significant urban centres had also to be met. The geography of wood sales (see Figures 16.2–16.4) reflects to some extent the relative fortunes of provincial towns as revealed by other sources. The presence of Canterbury may have stimulated sales of faggots in eastern Kent, although the evidence for the city's vicinity concerns the estates of the Cathedral Priory, which met its own substantial needs by direct transfers from its woods within 3 to 9 miles of the city. There is some indication that wood sales continued to be significant in the Canterbury district throughout the fourteenth century. Similarly, Oxford seems to have stimulated sales of wood (not, however, in the form of faggots) from manors to the north of the town *c*.1300, although its influence had perhaps lessened by 1400. [. . .]

The most striking change over the fourteenth century concerned Essex, where manors in the northern and north-central part of the county emerged as distinctive for their sales of wood and firewood (see Figures 16.2–16.4). The small towns and villages of this industrializing region created a relatively concentrated demand, both through the requirements of such crafts as dyeing and through the domestic requirements of their landless populations. Moreover, Colchester is notable for its prosperity during this period, when it was larger than both Canterbury and Oxford. [. . .]

Considerations of supply of fuel, as well as other raw materials, probably determined the location of many industries for which the metropolis was a major market. This was certainly the case with bakers at High Wycombe and in the villages around London. During the thirteenth and fourteenth centuries there was a pottery industry in south Hertfordshire (where faggots were produced and sold on a large

scale: Figure 16.4), a substantial one around Kingston, and another in south-west Essex, all of which, together with kilns closer to the city, served London. Eastern Kent had a notable tile-making industry and manors in this area used demesne woodland to fuel their kilns. In coastal areas the production of salt in pans required large amounts of fuel. Iron- and glass-making were to be found pre-eminently in the Wealden areas of Kent and Surrey, and some of the manorial sales recorded in the databases, including the charcoal sold by Dorking and Betchworth manors, may have gone to supply these industries. The bynames of some purchasers of firewood from demesnes suggest links with the baking, brewing, and pottery and tile-making industries. It is possible that the apparent commercial significance of woodland in parts of Bedfordshire is to be explained by a specialized local malting industry.

Overseas demand for fuel had a significant impact on wood production in parts of the region, although this is not immediately apparent from the evidence concerning manors in the databases. In both physical and cost–distance terms parts of coastal Kent and Sussex, and perhaps of Essex too, were closer to Flanders, with its substantial urban demand for fuel, than to London. The impact of overseas demand, apparent in a variety of evidence, was perhaps especially great where it combined with that of London. East Kent may have come under this dual influence (see Figures 16.3 and 16.4), and in the early fourteenth century the highly valued woods at Reinden above Folkestone probably took part in the trade. The Wealden woods on or near the River Rother certainly supplied overseas markets by 1300. [. . .]

IV

If the estimates of firewood consumption and of the extent of woodland presented in this article are realistic, it seems likely that the overall productive capacity of the London region was fully adequate to meet the demands placed upon it by the population of the region; indeed, it seems also to have been capable of meeting some extra-regional needs through the cross-Channel export trade. Those demands, however, fell very unevenly upon the region's woodland resources because of the high cost of transport. This meant that the market for wood fuel was far from fully integrated, and that some areas of woodland were largely inaccessible to the main centres of consumption. London's share of regional demand for firewood fell upon a limited area, largely restricted to Middlesex and Surrey, and to accessible parts of Kent, Essex, Buckinghamshire, and Hertfordshire. The shape of the city's supply zone was strongly influenced by the availability of water transport, as it seems to have been unusual to carry firewood more than 12 to 18 miles overland in the early fourteenth century, and less after the Black Death as costs of production and transport soared.

These factors led to the emergence of a zone of specialized firewood production, characterized by relatively high income from wood sales, and by the preparation and sale of particular types of wood fuel for domestic and industrial consumers. The regional associations of London's woodmongers, and the patterns of firewood production and sales on manorial demesnes, broadly confirm the definition of that area produced by a simple transport-cost model. To some extent, therefore, this study provides a further confirmation of the relevance of the von Thünen model to the English medieval economy. Specialized and relatively intensive production zones emerged in response to concentrated market demand, particularly in the case of bulky but essential products and in the vicinity of England's largest city. This zoning was more perfectly realized on the plane of production and marketing than of land-use. There was no 'forest zone' encircling London to the exclusion of other uses of

the soil. It is the case, however, that woodland occupied above average proportions of the counties surrounding the fourteenth-century capital, in conformity with the model. Here woodland was a resource which was highly prized in relation to other land uses, notably pasture and arable, and here it was profitable to divert labour into the production of fuel for specialized metropolitan needs.

The balance of economic factors which tended to produce zoning was highly dynamic. Changes in population and in the relative costs of labour and agrarian products altered the balance of the spatial economy, as did the availability of new products which could substitute for old ones. Sea coal was such a product, already an important factor in London's fuel supply by 1300. The availability of large quantities of cheap coal facilitated the growth of London's population in the early modern period to levels far above its medieval peak. Flexible and productive though the wood fuel supply system was, it is hard to believe that it could have sustained such expansion without placing a strain upon other forms of agrarian production. By contrast Paris, which did not have London's ready access to supplies of mineral fuel, did not use coal in any quantity until the nineteenth century, despite its great growth from 1550 onwards. Its continued reliance on wood production in its hinterland may have contributed to difficulties in the food supply system which were not apparent in England, even in the vicinity of London.

By 1300, within the circumscribed district upon which London's demand for wood fuel had most impact, the intensive production of fuel may have been reaching the maximum possible intensity, but did not draw into the city's supply system those more distant areas where supplies of wood fuel were abundant. When, after 1550, the capital's demand greatly exceeded former levels, it was presumably easier to develop the supply of coal than to expand the wood supply system into those less accessible districts. At that very period many of the extensive woods of Middlesex succumbed to clearance for grazing, tillage, or horticulture. Those woods were not 'used up', but rather became unnecessary and of reduced value in the new spatial economy which then emerged.

Notes

1 The article arises from the Feeding the City Project based at the Centre for Metropolitan History, Institute of Historical Research, University of London and organized in collaboration with the Queen's University Belfast

2 Sea Coal Lane, by the Fleet outside Newgate, is recorded by 1180, and Limeburners' Lane was nearby

3 Galloway and Murphy, 'Feeding the City'; Campbell, Galloway, Keene, and Murphy, *Medieval capital*

4 Horvath, 'Addis Ababa', p. 313

5 Faggots were probably made from a range of types of wood. Accounts for two manors in Kent record sales of faggots of thorns; thorns were used in Kentish tile kilns. Manors close to London produced faggots of two different sizes, 'long', which had two ties or bands, and 'short', with one tie

6 In the seventeenth century three or four cart loads of wood were thought to be equivalent, in calorific terms, to a London chaldron or 1.4 tons of coal. The weight of a cart load varied, but capacities of 1 ton and 0.865 tons appear to have been common, and may be compared with the cart load of 0.852 tons which seems to have been common in the London region, apart from Essex

and Hertfordshire, in 1300 (calculated from the capacity to carry wheat: Campbell, Galloway, Keene, and Murphy, *Medieval capital*, pp. 193-4, 196). Assuming median values for the seventeenth century (3.5 loads of wood; cart loads of 0.9375 ton), the wood equivalent of 0.75 tons of coal would weigh 1.76 tons. [. . .]

7 Calculated on the assumption that between 1086 and 1350 the extent of woodland was reduced by one-third: Rackham: *Ancient woodland*, pp. 133-4 and tables 9.1, 9.3, and 9.4

8 It has been estimated that an oak trunk could be carried for up to 50 miles overland before carriage costs exceeded an average sale price: Rackham, 'The growing and transport of timber and underwood', p. 214

References

CAMPBELL, B.M.S., GALLOWAY, J.A., KEENE, D., and MURPHY, M. (1993) *A Medieval Capital and its Grain Supply: Agrarian Production and Distribution in the London Region, c.1300*, Hist. Geog. Res. Ser., no. 30

GALLOWAY, J.A. and MURPHY, M. (1991) 'Feeding the City: Medieval London and its Agrarian Hinterland', *London Journal*, 16, pp. 3-14

HORVATH, R.J. (1969) 'Von Thünen's Isolated State and the Area around Addis Ababa, Ethiopia', *Ann. Assoc. Amer. Geog.*, 59, pp. 308-23

RACKHAM, O. (1980) *Ancient Woodland: its history, vegetation and uses in England*, London, Edward Arnold

RACKHAM, O. (1982) 'The growing and transport of timber and underwood' in S. McGrail, ed., *Woodworking Techniques before 1500*, Brit. Arch. Reps. Internat. Ser., 129, pp. 199-218

17

ROAD IMPROVEMENT IN THIRTEENTH-CENTURY PISA

by David Herlihy

Source: David Herlihy, *Pisa in the Early Renaissance: a study of urban growth*, New Haven, Conn., Yale University Press, 1958, pp. 90–105

The great changes in inland communications taking place in Tuscany in the late 13th century Johannes Plesner called a 'road revolution.'[1] In the early Middle Ages, the roads, wherever possible retracing the Roman, had like the Roman, followed mountain ridges. Less now for military reasons and fear of ambush, more because the river valleys were ill-drained, spotted with impassable swamps, and unhealthy, this favoring of ridges resulted in lengthy detours, slowness, and practical impassability for carts. The road revolution consisted in the directing of the chief roads through the valleys, in paving them, in permitting traffic in carts upon them, and in lending to them an unprecedented economic importance.

In inland communications, Pisa needed ties both with the sea and with the four inland areas she bordered – Versilia, Lucchesia, the Valdarno, and the Maremma. In the early Middle Ages the most popular routes were the rivers. Pisa in antiquity had stood at the confluence of two rivers, the Arno and the Serchio, the latter flowing down from the north, from Garfagnana and Lucca. However, in 575, to help his flood-plagued flock, the Irish bishop of Lucca, Frediano, diverted (through a miracle, our source, Gregory the Great, relates) the Serchio to its present course, through Migliarino to the sea. Pisa, however, could not be deprived of so convenient a highway. Through-

out the Middle Ages and into the Medici period, a canal called the Auser continued to trace the Serchio's old route [. . .]

The Auser provided Pisa's busiest road with the Monti Pisani [Pisan Hills]. [. . .] On it, too, was taken marble from the Monti for Pisa's cathedral. By 1304 the canal was equipped with locks placed at the Porta Leonis near the cathedral, which could be closed, raising the water level so even marble-carrying barges could float upon it. [. . .]

Pisan boats penetrated deep into the river system of the Arno's plain. [. . .] At Pisa herself, wharves stood within the city [. . .] where was unloaded the bulk of Pisa's overseas commerce. Likewise within the city the noble residences which lined the banks boasted their private wharves. [. . .] North of the city facing Florence, embracing both sides of the river, was the Piaggia. [. . .] There boats were constructed and loaded; there, too, stood public buildings, probably a customs house [. . .]

Traffic moved slowly on these inland waters; the trip to Florence took six days, less than ten miles per day. It moved surely, however. [. . .] The barges, however, could be ample. At once large enough to carry bulky grain, at times even to risk the open sea on ventures of its own, small enough to master the Arno's shallows, was the *placta marinara*, the sea-going barge. One built in 1299 was thirty feet long and eight and one-fourth feet wide. Another could carry one hundred thirty-five barrels of wine from Castiglione della Pescaia in the Maremma to the city, enough to satisfy the thirst of twenty-seven Pisans for one year.

As frequently as barges, small boats, *barche*, are met on these waters. One constructed in 1314 was twenty-four feet long and four and one-half feet wide; its slim shape assured easy control in the river's shallows. [. . .]

So far as inland communications are concerned, however, in the late 13th century Pisa's chief energies, and the energies too of other Tuscan cities, were directed to road building. We may note in passing that even the building of roads seems not to have gone unaffected by the aristocratic–democratic feud. Probably because Pisa's Ghibelline [faction supporting the Holy Roman emperors] popolo was more immediately concerned with warring against the Guelf [faction supporting the papacy] hinterland cities than with trading with them, democratic regimes seem relatively negligent in improving the inland roads. Conversely, aristocratic governments, and especially the tyranny of Ugolino della Gherardesca, were especially energetic in constructing roads and building bridges.

Pisa's flat and ill-drained contado [countryside around the city] made land travel difficult, especially during winter. No matter where the destination, on leaving Pisa the traveler did best by taking the shortest route to the nearest ridge. Further, dependence on fords meant that during the flood season of autumn and winter, travel had all but to cease. Let us begin a trip to Florence, sometime before the 1260s, passing through only the section within the Pisan contado. We are leaving from Kinsica, Pisa's most populous quarter on the Arno's left bank. We first must ford the Arno at the river's great bend above the city where the water is shallowest, at 'Guathalungo,' the 'long wade.' We then strike out for the Monti Pisani, not because this is the direction to Flor-

ence but because by hugging the base of the Monti Pisani we can best hope to find the terrain passable. Then, not twenty miles up the Arno, we must ford the river again, this time to avoid the great swamps of Bientina, at Ricavo or Calcinaia. Actually, these were fords only for low waters and only for horsemen. [. . .]

In the latter half of the 13th century, however, Pisa had embarked on an ambitious and extensive plan to improve inland communications, her share in the Tuscan road revolution. The fourth book of Pisa's statutes (1286) presents an imposing list of projects [. . .] Behind the plan and the execution stands the figure of Pisa's tyrant Ugolino Gherardesca, who, if tyrants are always the best builders, seems likewise to have appreciated the importance of roads for Pisa if she was to retain her stature as Tuscany's favorite mouth. The same year the Florentines ordered paved the road to Pisa [. . .]

To liberate roads from mountain ridges and build them on the plains, drainage was the chief technical difficulty. Over swampy land the road had to be elevated. [. . .] Pisa's cart roads, *carrarie*, thus often doubled as dikes. The dike on the Arno's left bank east of the city provided a base for the new Florentine road made of gravel, and the road west of the city was planted every four feet with trees to give strength to the dike and permanence to the road. To secure good drainage and overcome mud, pavement was essential, so that the roads, as the statutes say, would be passable even in winter. [. . .] The technique that used flint for pavement is as old as the Romans, perhaps the Etruscans; its wide application, however, seems peculiarly an accomplishment of the late 13th century. Common both in the chartularies [collections of charters] and statutes after 1285, it is apparently unknown before. Such roads might be twelve to fifteen feet wide, enough for carts to pass. And good roads meant large carts. Pisa's earliest carts, as seen from the contracts of 1263, were two-wheeled affairs, and of the growth in load size we are better informed than of the growth in wheels. In 1299 one cart could carry from Pisa to the Porto Pisano [Pisa's main port] the great chain to be placed across the harbor's mouth. The weight of the chain can be appreciated from a piece of it (or one similar) contained today in the Camposanto; records show the entire chain weighed 2,500 pounds.

Good roads, however, were little serviceable without good bridges. Up to the late 12th century Pisa had one bridge across the Arno which had apparently survived since Roman times. Its exact location has been a point of dispute among scholars. At any rate, it was apparently destroyed in 1179, and shortly afterwards two bridges were being built to link the two sides of the Arno. [. . .] Like its famous counterpart at Florence, [the Ponte Vecchio] supported shops and stores to help finance it, and its central location made those shops and stores important possessions. By the early 14th century Pisa's official weigher of florins, an important figure in the city's economic life, was stationed upon it.

In the second half of the 13th century, Pisa constructed a third bridge over the Arno. The Ponte della Spina [. . .] forms part of an ambitious project to redirect the road to Florence. Work upon it started in 1262; it was, however, neglected by the popular regimes of the late 1260s and the 1270s, and only the energy of Ugolino's tyranny carried it to completion (1286). [. . .]

In antiquity, Pisa had stood on Rome's chief road to Gaul, the Via Aurelia, built about 241 BC between Rome and Pisa, and its northern extension to Tortona and Liguria, built by the consul Emilius Scaurus in 109 BC and called the Via Emilia. In the early Middle Ages, however, when roads had to be self-maintaining, when the coastal plain was more swampy, unhealthy, and exposed to pirate forays, the road north, [. . .] retreated inland to the mountain ridges.

In the 13th century Pisa set about building another road to the north, retracing the old coastal route of the Roman Via Emilia. [. . .] The road assumes considerable importance in the statutes of 1286, as part of Pisa's effort to colonize her maritime plain to the north. In 1273 Guido of Corvaria, along this road, reached Carrara in three days, a speed of about twenty miles per day. By the 14th century, a letter of exchange dated in Pisa May 24 could call for payment in Cremona by the middle of June. Although travel was possible even at night upon it, it seems not to have been passable for carts.

To Lucca two roads ran. One climbed directly over the Monti Pisani, and while impossible for carts and difficult even for the old or weak, was the shortest and oldest road to Lucca. In the late 12th century, Pisa built a second road, skirting the Monti Pisani, through the half-mile wide gap of Ripafratta. It was longer but level, and by 1184 we hear of considerable traffic in carts upon it; it was, however, hardly passable during the rainy season. [. . .]

To the Valdarno as well two roads ran. The oldest [. . .] set out from the Porta Calcisana on the Arno's right bank and provided Pisa's link with Calci, a town important for its water-power, its mills, and its sheep, and with Buti, important, among other reasons, for its shale diggings. Most important, the road on the Arno's right bank, passing through Ghezzano and Mezzana, was the oldest road to Florence. [. . .] Near where this road entered the city, Florentine merchants had their house. By 1286, however, it had been paved only its first six miles. [. . .] The apparent neglect is explained by the fact that Pisa was building a better road to Florence.

And the Florentine road is an excellent illustration of the conquest of the river valleys, the great accomplishment of the 13th-century road revolution. The building of a new road on the Arno's left bank, the logical route to follow, faced a twofold handicap. The land outside the Gate of Guathalungo was especially swampy. [. . .] In 1262 Pisa started its third bridge, Ponte della Spina, to replace the ford. Two years later, a contract tells us that Pisa was already 'sending' a road on the Arno's left bank, that was to surmount a gravel mound to ensure protection from floods. Initiated simultaneously as parts of a single plan, both bridge and road lagged in the democratic period. The Ponte della Spina was still uncompleted by 1286, and of the road the statutes refer to the 'bad way which is there' and call for its repair. [. . .] By the early 14th century it assured Pisa contact with the Valdarno that was independent of weather. By 1322 across it was carried not only Florence's industrial products but even spices from Venice and furs from the north, and it seems to have been Pisa's most important route to the northern areas. We hear of ambassadors making a round trip to Venice, presumably on this road, in one month.

Leading south from Pisa, through Collesalvetti, crossing the Livorno hills to emerge into the Maremma at Rosignano, continuing south to the limit of Pisa's contado, Castiglione della Pescaia, was Pisa's own pilgrim highway, the Via Romea. [. . .] An entry in a chartulary of 1308 shows that Pisa was attempting to pave the Via Romea from Rosignano to Castiglione della Pescaia, a distance of some seventy miles. That the section of the road north of Rosignano to Pisa is not mentioned suggests that it had been successfully paved the years before, and indeed the section figures as a model for a good road in the statutes of 1313. This ambitious project seems an effort to replace Pisa's sea lanes, harassed by pirates who dared strip ships in the Port Pisano itself, with safer bounds of inland communications.

Leaving Pisa at the Arno's left bank near San Paolo a Ripa d'Arno, following the Arno's left bank to San Piero a Grado, and swinging south across the plain ran the road to the Porto Pisano. In the late 12th century Pisa had bridged the swamps it had to cross and provided a hospital, San Lorenzo, at the Stagno inlet for the relief of weary merchants. Of three stone bridges, the largest across the Stagno was appropriately called the *Pons usione*, the bridge of departure. The busiest of Pisa's roads, it was lined with hermits who lived from charities meted out by the rich merchants who passed; it was lined too with trees, every four feet. The road and its hermits figure most frequently in notarial wills. It was a good road too. When about 1260 a Tunis-bound merchant fell from his horse on the Stagno bridge, broke his leg, and had to be carried back to Pisa in ignominy, a cart could take him. Over this road, too, was taken the one and one-fourth ton chain for the Porto Pisano.

And at the Porto Pisano converged the lines of Pisa's maritime communications.

Pisa's river harbor was miserable. When Federigo Visconti made his Sardinian visitation of 1263, to board his galley he descended the Arno to San Rossore. Upon his return in attempting to bring him to the city, the ship stuck in the mud immediately below its destination, San Pietro a Vincoli, 'as if [the ship] were saying,' relates the archbishop, 'you'd better get off here.' To be sure, the ship had in fact reached the city; it was summer too, when the Arno was lowest. But for a galley, drawing little water, built for speed and maneuverability, to have trouble is little compliment to the Arno's navigability. It measures how far at Pisa the geological sequence characteristic of many river ports had progressed: silting of the harbor, growth of the delta, alienation of the sea, erection of and mounting dependence on a subsidiary harbor more accessible to sea-going vessels. To be sure, smaller ocean-going ships could still reach Pisa, but this honors more the mariners' courage in taking petty boats to the open sea than the Arno's success in floating large ships to Pisa. Her subsidiary harbor Pisa found in the Porto Pisano, standing south of the Arno's mouth very close to Livorno. There in springtime Pisa's fleets gathered to await the merchants who, after a Sunday farewell feast with family and friends, on Monday mounted their horses to ride to port, ship, and adventures.

Note

1 'Una Rivoluzione stradale nel dugento', ['The Road Revolution of the Thirteenth Century'] *Acta Jutlandica*, *I* (1938)

18

TOWN AND HINTERLAND IN MEDIEVAL SCOTLAND

by Elizabeth Ewan

Source: Elizabeth Ewan, 'Town and hinterland in medieval Scotland' in *Medieval Europe 1992* (pre-printed papers for conference on medieval archaeology in Europe, University of York, 1992), Vol. 1: *Urbanism*, pp. 113-18

In the last decade or so the walls which surrounded medieval towns and cut them off from the countryside have begun to be breached by urban historians. Rather than studying towns in isolation from their hinterland, recent work has tended to stress the inter-connectedness of town and country. The relationship between the two is a fascinating and complex one, operating on many levels and in both directions. This chapter will examine how one Scottish town, Aberdeen, interacted with its hinterland in the Middle Ages and suggest that the bounds between town and country were by no means as clear and distinct as either medieval administrators or later historians perceived them to be.

Aberdeen is a useful case study for a number of reasons. From at least the twelfth century, it acted as regional centre for the fertile countryside of northeast Scotland. It has the best medieval town records in Scotland although regular runs of these do not start until 1398. It has also been the subject of extensive archaeological investigation. It can thus show how documentary and archaeological evidence can work together to throw new light on the town-hinterland relationship.

Early charters gave Aberdeen a favoured position in the economy of the northeast. Commerce was a prime motivating factor behind the king's promotion of his burghs and most of the privileges granted to the townspeople were intended to give them economic power over the surrounding region. Many early Scottish burghs were given an extensive 'liberty', an area over which they had commercial privileges such as a monopoly on the trade in wool, fleeces and hides, Scotland's most important exports. Aberdeen's liberty was co-extensive with the sheriffdom of Aberdeen, running south from Banff on the north coast and west along the entire valleys of the Rivers Dee and Don. The town was the export centre for this whole area, shipping wool to the cloth towns of Flanders, hides and skins to the rest of Europe, and fish, especially salmon, to England and the Continent. The town was also a distribution centre, sending imports to other towns and estates in the northeast.

The produce of the countryside was also used in the town itself. Medieval towns, unable to feed themselves, were dependent on their hinterlands for supplies to sustain life. Scottish towns enacted various measures to ensure the provision of adequate food supplies for the inhabitants. But the hinterland provided much more than food for the townspeople. It supplied materials for other necessities of everyday life, resources for the town's industries, and people to sustain the town's growth. Many goods were brought to market by country people but the townsfolk themselves were also active in exploiting the resources of the countryside.

Much of the material for everyday necessities came from the lands around the town. Excavation, carried out largely in the backlands where the poorer part of the population probably lived, has found the most common type of house to be constructed of wattle and daub. It has been estimated that for an average-size house of the type found in excavation, about 1,000 wattles would be needed. The necessary wood was probably available in coppiced woodlands and along the riversides. Among the woods available around the town were oak, ash, hazel, elm, aspen, alder and willow. The construction of these dwellings was simple enough that it could be carried out by the inhabitants themselves, using the resources in the immediate vicinity of the town.

Such resources were used for a variety of purposes. Wattle provided boundary and live-stock fences, the linings of drainage ditches, footpaths, and reinforcing walls for pits. Other materials were also used in house construction, heather, straw and rushes to thatch roofs, clay to provide an insulating layer over the wattle, rushes on the floor and for beds.

Heating was an important concern in the cold damp climate. Peat was a major source of heat, both for warmth and cooking. Several early charters to Scottish towns include grants of peatlands. The fuel was gathered by the townspeople themselves. Wood was also used for fuel as well as for building and other industrial purposes. In 1319 the king granted the forest of Stocket to Aberdeen to help meet this need. The town showed great care in husbanding the resources of the forest, appointing officers to supervise it and prosecuting those who did damage to it through overcutting.

The staples of the urban diet, ale and bread, were produced mainly from supplies purchased from countrydwellers, but the diet could be supplemented by gleanings from the countryside around the burgh. For example, wild berries could be gathered locally – seeds of several varieties have been found in excavation. Fish were also used to supplement the diet. The town had its own fishings on the River Dee which were leased out to individual townspeople, but it seems likely that many inhabitants fished in the other rivers and streams which flowed around and through the town. Fishbones were a very common find in excavations.

The countryside was the main source for the raw materials for Aberdeen's industrial activity. Aberdeen's crafts were similar to those found in most rural villages in England and the Continent, although the occasional documentary reference and archaeological evidence such as clay moulds probably used for metal-working, suggest that there may have been a few more specialised craftworkers such as goldsmiths catering for an aristocratic market. On the whole, however, the craftworkers of Aberdeen concentrated on the working of raw materials from the countryside, especially wool and hides.

Customs accounts show wool and hides coming to the town on a regular basis for export overseas. Wool came to the burgh ready to export but hides were often removed from the cattle only when they reached the town. In 1487 countrypeople bringing cattle to the burgh had to pay a fee to pasture them on the towns links. The presence of cattle is attested to by several court cases which were brought over them. Cattle provided skins to be worked by skinners, barkers, tanners, and leatherworkers, hooves for neat'sfoot oil, horns for hornworkers, bone for bone-working, tallow and grease for candles and soaps. Wool was used to make cloth. Other locally-available materials which are rarely mentioned in documentary sources but are revealed through archaeological evidence include wood, flax for linen, and dye-plants.

The hinterland affected urban industry not only in its provision of the raw materials to be worked but also in the yearly rhythms of the activity. The raw materials needed by the crafts were only available at certain times of the year. Cattle were generally slaughtered in the autumn, so that trades involving leatherworking may have been largely a wintertime activity. It has been suggested that many craftsworkers turned their hands to more than one activity, working on different materials such as bone or antler at different times of the year. Labour might also be available in different quantities throughout the year. People might come in from the countryside at times of underemployment there or migrate out from the town at times of harvest.

Labour was another point of contact between town and hinterland. It was common for towns to draw their population from the hinterland. Medieval towns were unable to maintain their population without immigration. A common pattern was for young people to move to the town and enter service, usually domestic for women and industrial for men, in order to gain enough capital to set up their own households. Many found employment in the houses of their kinsfolk who might only have moved to the town in the previous generation. Aberdeen surnames suggest that much of the town's population was drawn from the hinterland. It was common for such families to stay in touch. In a number of court cases townspeople and countrypeople were cited as accomplices.

Town and countrypeople interacted most commonly in the twice-weekly market where the produce of the countryside was brought for sale. The privileges granted to the burgesses ensured them of the position of middleman between the rural producers and foreign merchants interested in their wares. The market was closely regulated by the town council to try and ensure regular supplies of essential foodstuffs at set prices. [. . .]

The records of forestallers include those convicted of practising outside the town a trade over which the burgh council claimed jurisdiction, for example dyeing cloth or tanning hides. Despite royal legislation, Aberdeen had no effective monopoly on many of its industrial activities. Other 'urban' activities also extended into the countryside, helping to break down the barriers between town and country. Hospitals, especially those caring for lepers, were often sited outside the town. An Aberdeen leper's hospital was situated from an early date to the north of the town, its existence preserved in the placenames Spital and Leper's Croft. The gallows of the town was situated outside the town itself. Religious houses were situated on the outskirts of the town and cared for the poor and sick of both town and country.

The relationship between town and country was one which went both ways. Many townspeople actively participated in the rural life of the hinterland. The grant of agricultural lands to the burghs implies that the townspeople were expected to continue to follow agricultural pursuits at least on a part-time basis. The earliest inhabitant of Aberdeen to appear in the records was a miller, a rural occupation in itself but also one which implies agricultural activity by his fellows.

Within the town itself were areas which could be used for agricultural pursuits. [. . .]

The town also interacted with its hinterland by expanding physically into the countryside. The crown made grants of the forest of Stocket and the lands of Rubislaw to the town in the fourteenth century. The growth of suburbs further blurred divisions between town and country. In Aberdeen the suburban areas of the town varied in character, from the fishing communities of Futy and Torrie to east and south to the three areas known as the croftlands which spread along the roads to the north [. . .]

These areas were well-suited to a number of the craft activities of the burgh. Potters often located their works outside medieval towns, partly due to the fire risk, partly due to the location of sources of clay. This may explain why little trace of pottery works has been found in excavations in Scottish medieval towns, despite the common occurrence of locally-made pottery. If the opportunity arises to excavate more areas of suburban growth, more evidence may be found. In Aberdeen, there was a source of clay to the east along the Futygate. A track leading to the clay may have been used by potters to remove it to their workshops. Clayhills to the west of the town may have been the site of medieval potteries.

Fullers and tanners might also settle on the outskirts of the town. Suburban sites offered water supplies and open space to dry their cloth and leather. Where evidence of such activities has been found in Aberdeen, it has tended to be towards the northern extent of the town, up the Gallowgate which led to Old Aberdeen. One major tanning site was uncovered near the Loch which marked the northwest boundary of the town. [. . .]

There is evidence that some wealthy burgesses were investing in croftlands, not so much for their agricultural value but for their worth as real estate which could be bought and sold. John Crab, a prominent burgess, owned a string of properties along the main road running west from the town.

These lands could have been exploited directly for producing goods for sale at the market, but the property transactions of the fourteenth and fifteenth centuries suggest that their main value to the purchasers was as investment in real estate which was then rented out to tenants. As Aberdeen grew there would be an increasing demand for land, if not inside the town proper as close to it as possible. Such properties offered a number of advantages. Those which fronted on to the major access routes provided opportunities to offer services such as smithying or accommodation to travellers. Craft activity could be combined with farming. Through bringing the crofts into the urban property market, the townspeople helped to spread commercialisation from the town to the countryside. [. . .]

Aberdeen was a small town by European standards, with a population perhaps approaching 3,000 in the early fifteenth century. However, it was one of the largest Scottish towns. It seems likely that smaller towns had equally close links with their countryside. These contacts were facilitated in Scotland by the fact that most towns were unwalled until the sixteenth century or later, relying only on natural defences. There was thus no real physical boundary between town and hinterland. It would be worth asking how the town walls elsewhere in medieval Europe were regarded by those who lived on either side of them. Perhaps they were not so impenetrable after all.

19

BUILDING RENAISSANCE FLORENCE: MATERIALS, TECHNIQUES, ORGANIZATION

by Richard A. Goldthwaite

Source: Richard A. Goldthwaite, *The Building of Renaissance Florence: an economic and social history*, Baltimore, Johns Hopkins University Press, 1981, pp. 19–21, 171, 173–81, 183, 186–8, 216, 221–2, 237

Tourists from Montaigne to the present cannot be taken to task too severely for not finding the much-vaunted beauty of Florence immediately apparent on their first walk through the city's streets.

If not a better planned city for all this building, Florence was at least a better built city and a safer place to live. Manetti (in his life of Brunelleschi) commented on the crude method of building evidenced in the older structures and still seen in his day, and their fragility is noted in the annals that record the ravages wrought on them by disasters of all kinds. Devastating fires broke out in 1293, in 1301, and again in 1304, when (according to Villani) over 1,700 buildings – palaces, towers, and houses – were destroyed. The danger of fire was so great that Paolo da Certaldo advised keeping rope and sacks around the house so that when the alarm came a person could get his goods and himself out as quickly as possible. The great flood of 1333 swept away many of the buildings all along the Arno [. . .] And then there was the mob, ever ready to turn itself loose, in the faction-ridden society of this earlier period, to wreak its vengeance on the enemy of the moment. Salutati's despair about the situation was not all rhetoric: 'How many and magnificent houses of citizens and how many palaces have been destroyed by the internal discord of our citizens! How many have been annihilated by fires sometimes set deliberately, sometimes caused by chance.'[1]

By the fifteenth century there was little talk of this kind of destruction. Buildings were more substantial, since construction materials were almost entirely brick, stone, rubble, and tile. Walls were massive enough that fireplaces were normally set in interior walls and kitchens moved from the top to the ground floor. Timber was reduced by the widespread vaulting of ground floors and the substitution of stone for the outside supports (*sporti*) of overhanging floors. Looking at these structures, seemingly built for an eternity and having already endured half a millennium, it is difficult to see how fire, flood, or even the willful action of the most violent mob could do much damage to them. In fact, hardly any evidence of such destruction can be turned up anywhere in the annals of the Renaissance city.

Florence was also a remarkably clean city by European standards of the time. At the beginning of the fifteenth century Goro Dati commented on how well the streets were paved 'with flat stones of equal size so that they were always clean and neat, more so than in any other place,' a feature that for the next two centuries or so invariably impressed visitors from across the Alps.[2] Although apparently the pavement of streets was largely a private expense, they were kept reasonably clean by an extensive drainage system of sewers dumping into the Arno that was maintained by the city. Moreover, palaces commonly had their own private source of water. Dati and Varchi both make the point that most had their own wells, making it possible (as Dati says) to get fresh water, even to the top floors. When in 1476 the sons of Messer Giannozzo Pandolfini undertook remodeling of their father's house to divide it between them, they saw to it that another well was dug to assure each residence an independent water supply. Buildings were also fitted out with internal latrines, probably very like the closets that emptied into cesspools below that can still be seen in the Davanzati palace (and that were illustrated by Francesco di Giorgio). These were the responsibility of the building's owner, and the records of the Parte Guelfa for the sixteenth century are full of litigation over problems relating to cesspools shared by several households. One clothmender made a note in his book of memoranda of the cost for the annual emptying of a cesspool (from 1520 to 1524), and a scene in the *Decameron* (VIII, 9) is set at the ditches near Santa Maria Novella, which was one of the city's dumping-grounds for this refuse.

Plumbing of all kinds was a major feature of palaces put up in the Renaissance. Building accounts are full of expenses for wells, cesspools, cisterns, sinks, latrines [. . .] and the tile piping that constituted the plumbing system; and contracts for foundations commonly included stipulations for leaving holes and openings to accommodate the system. Occasionally events about the plumbing were considered important enough to be recorded among the memoranda Florentines were ever compiling about the notable events in their lives - like the bursting of two latrines and the flooding of a room in the house of Agnolo di Niccolò Benintendi in 1473, and the discovery by Bartolomeo di Lorenzo Banderaio in 1539 that the cesspool for his latrine was actually under the house of his neighbor. When the doctor Antonio di Ser Paolo Benivieni made improvements in the plumbing of his house in 1487, he described the complete system. On investigating the drainage from a downstairs latrine, he found that it emptied into a cesspool along the street, and he rerouted the drainage from a sink so that it connected up with another latrine and thereby helped flush it. Two years later he had three new cesspools dug for the emptying of latrines and sinks above and for the drainage of rainwater from the courtyard and roof. Benivieni tells us he recorded these plumbing improvements so the system would be understood by whoever might take up residence there in the future. It may also have been his natural instinct as a medical man to take a particular interest in the sanitary system of his house. The system Benivieni installed is exactly that described in the architectural treatises of two near-contemporaries, Alberti and Filarete. [. . .]

Bricks and lime

Florence by the fourteenth century was a city largely built of bricks [. . .] In Italy, where building in brick has a tradition that goes back as far as the eleventh century (if indeed it was ever interrupted after the demise of Rome), the brick industry has a notable if also largely unremarked history. Major brick structures from the eleventh and twelfth centuries are to be found in the churches of Bologna, Modena, Parma, Pavia, and other towns of the Po Valley; in the following centuries much civic and military as well as ecclesiastic construction was executed in brick. It was amidst this flourishing of brick construction in Italy that masons refined their handling of the material and created a distinct brick architecture. They made much of the polychromatic effects of different shades of bricks; they cut and molded them into different shapes and devised special molds to prefabricate decorative details in terracotta; they built them into a variety of architectural details like moldings, arches, pilasters, columns, and capitals; they varied courses and applied decorative bands against fields of plain brick. Of more economic importance than this imaginative use of brick in the occasional architectural monument is its widespread diffusion as a building material in more ordinary construction as the awareness of the fire hazards of high-density living in the booming Italian towns of the twelfth and thirteenth centuries led to an investment in more substantial buildings.

The architectural evidence for the rise of the brick industry in northern Italian towns is complemented by the increased mention of kilnmen and kiln products in early guild and communal documents. The first guild of kilnmen on record dates from the early thirteenth century in Venice, and there was one in Verona by 1319. A list of guilds at Orvieto in 1300 includes both limeburners and tile-makers. [. . .]

Paralleling the emergence of a corporate consciousness among brickmakers was the growing concern of communal governments with the regulation of an industry so vital to the public interest. The documentary material here, too, attests the growth of the industry in Italy. In the thirteenth century towns everywhere began to pursue a vigorous policy to control prices and set measurement standards for kiln products of all kinds. Legislation was also directed to assure the public's supply of bricks; Venice (and perhaps Padua) had communal kilns to supply public works. Parma required each town in its territory with a baptismal church to build a kiln at the expense of the local inhabitants and to fire it twice a year; at Parma itself five firings were required of kilns. At Venice attempts were made to stimulate production by setting the number of firings

at each kiln, extending loans to kilnmen, and forbidding export of bricks and lime. In Florence a 1325 statute shows the government's concern that there be a sufficient number of kilns to supply that city's needs.

The evidence of guild and communal documents, not to mention the archaeological record, points to considerable activity in the brick industry in the medieval Italian towns, probably more so than for any other place in Europe. [. . .]

Everywhere in Europe the history of the industry closely follows developments in vernacular architecture related to improvements in the general standard of housing. The evidence for brickmaking so early in the Italian towns, in other words, is probably a comment on the better living conditions that generally prevailed there. In any case, the veritable revolution in housing that took place in early modern Europe finally brought bricks to the fore in many places as the essential material for all kinds of buildings, from the simplest structures to great monuments of architectural quality. The economic expansion of those centuries meant urbanization, greater wealth, and the disposition of more men to invest in substantially better housing, with the result that brick production began to pick up over much of the Continent. Whereas in medieval France, for instance, there was little vernacular building in brick, by the eighteenth century the production of bricks was important enough to merit major publications on kilns. In England brick was a vital ingredient of the so-called Great Rebuilding, and cheaper transport resulting from the canal system built up in the eighteenth century further stimulated the industry. When the industrial revolution got underway, brick-making may have been the building-material industry with the largest market in Europe. Measurement of the commerce in bricks from that time on, in fact, has been one way of getting at a general index of construction.

In this early modern chapter of the story of

bricks Holland has a unique place. Good clay and peat were among the few natural resources of the country, and they were the raw material and fuel essential to the development of a brick industry. The area was one of the few with a tradition of brick-making going back into the Middle Ages. As just noted, Flemish bricks were being exported to England as early as the thirteenth century. Dutch cities had municipal brickyards and were regulating the industry by the fourteenth century. The prosperity that commercial expansion brought to Holland in the sixteenth century had a significant impact on demand in the construction market, and since brick was the only building material that could be produced in plentiful supply, kilns became a major industry. Industrial growth was facilitated by the expansion of an inland waterway system that made possible the economic transport of clays, fuel, and the finished product – all bulky, cheap materials. It has been claimed that already in the seventeenth century the per capita production of the Dutch industry was twice that of England at the end of the eighteenth century, when the industrial revolution was well underway. Moreover, since bricks came to be used as ballast for outgoing vessels in the country's international shipping empire, the industry supplied bricks literally to the world – to England, the Baltic, and further afield to Asia, Africa, and the Americas. Dutch bricks turned up everywhere, and in some places their importation left its mark in the history of local architecture – for instance in the Baltic towns and in colonial America. Holland in these early modern centuries was the first place where an industry grew up that produced a basic building material that was bulky, heavy, and relatively cheap for more than a local market. Indeed, it is likely that no building-material industry has ever had a more widespread distribution of its products. In any case, the highly developed Dutch industry provides a useful comparison [. . .] for a study of the industry in Florence, another center of early capitalism.

Kilns

Whereas the building-material industries of woodworking and quarrying merely prepare raw materials for use, a kiln operation is a full-scale industrial process that transforms raw material into manufactured products – stone into quicklime, clay into bricks and tiles. Moreover, it is an industry whose development is subject to many variables – types of fuel, qualities of clay, the level of kiln technology, climate and seasonal changes, facilities for transport, the availability of alternative building materials, and, of course, taste. Each of these is a factor in the formula that determines the nature of brickmaking, and since each combination of them depends on physical environment and historical development, the industry varies in different parts of Europe at different times. It is not surprising, therefore, that the history of brick and lime production has yet to be written, for it requires a synoptic vision of many local operations, few of which, in fact, have ever been studied.

Roman kilns were of the pit-type, with fire chambers dug into the ground and the ovens above; they were partly embedded and partly walled, but were fully exposed to the elements at the top. In medieval Europe bricks were commonly burnt in a clamp (or heap), a largely temporary structure that was built of the bricks to be fired and dismantled after each firing. With ashes mixed into the clay so that the bricks created their own heat and with fuel packed between the stacks of bricks, a clamp simply burned itself out. Although kilns were not unknown, clamps were used extensively throughout the early modern period, the technique being the normal procedure for making bricks even in England and Germany well into the nineteenth century. Clamps were especially suitable where coal was in plentiful supply. They were elastic in their size, which

Figure 19.1 Kilns in the environs of Florence, c.1600

could be changed to meet the specific demand of the moment, and capacity could be increased up to half a million bricks. Moreover, clamps could be made on the spot at a building site, hence reducing carriage costs. As long as the demand for bricks was sporadic it made no sense to invest in permanent industrial structures. [. . .]

Florentine kilns were permanent installations, and they appear to have been fully developed as industrial structures, perhaps among the first in Europe. That they were substantial establishments is indicated by assessments of them for tax purposes of 100 to 200 florins and sometimes as high as 300 florins. These values represented the capitalization of rents, and the higher rents were as much as had to be paid for the most expensive premises in the city – for instance, a prosperous cloth shop. A kiln property, however, most likely included not only the industrial establishment itself (the kiln) but also the raw material in the form of clay-pits; furthermore, the kilnman might live on the premises and may have been able to farm some of the land. [. . .]

The making of bricks was largely, though not completely, a rural industry. The firing of bricks took place outside the city walls not only because of practical considerations of accessibility to clays (or limestone) and firewood but also because, for obvious reasons, communal legislation restricted their location in the city. Urban kilns were not altogether unknown, for unbaked bricks brought into the city (presumably to be baked) were explicitly exempted from gabelle charges. Streets at one time or another known as Via delle Fornaci were located toward the walls and away from the populated center [. . .] Later building accounts, however, show that brick and lime almost always came from outside the city. [. . .] There were, of course, kilns just outside the city gates [. . .]

Kilns were the most prominent industrial structures drawn in on the maps made for an early seventeenth-century survey of the Florentine countryside (see Figure 19.1). Some kilns survive that have great vaulted 'porticos' reaching the full height of the building in front. A few at least three centuries old are still in operation, albeit with some modernization of the chambers, and others, now converted to different uses, abound. [. . .]

The one obvious advantage of a permanent installation is its much more economic use of fuel. [. . .]

In the history of technology lime and brick kilns are treated as two different installations since each process required a different kind of oven. In Florence, however, the two processes could be brought together at one kiln site and even in one structure. Although on some building accounts suppliers of lime and suppliers of brick fall into two distinct groups, and on tax records and property documents the specific nature of a kiln is often indicated, it is nevertheless clear that many kilns produced both items. [. . .]

Kilns were usually located in the countryside for easy access to raw materials (clay and limestone), fuel, and temporary labor, not to mention the desirability of isolating a polluting and potentially dangerous industrial process. [. . .]

Stone

No European stone has enjoyed more fame and a more extensive international market than the white marble of Carrara. So much of this stone was quarried in the period of the Roman Empire that long after the shutdown of imperial quarry operations Italians continued to live off that production simply by plundering Roman monuments. Around 1300 the minor building boom of Tuscan cathedrals at Pisa, Florence, Siena, and Orvieto resulted in occasional expeditions to Carrara to obtain marble, but transportation was such a major problem that these quarries did not replace more local sources. In this early phase of the revival of the quarries at Carrara production was sporadic and in the hands of contractors or agents of clients (for the most part cathedral building committees) who organized the entire enterprise, even sending stonecutters to the quarries for months, and sometimes for several years.

Toward the end of the fourteenth century rising demand, above all from Florence for the completion of the cathedral, with the cupola and the elaborate sculptural program then underway, gave rise to efforts by workers at the quarry zone itself to organize production. This development of the local economy was then sustained by the subsequent diffusion of Renaissance taste for marble throughout Italy and even northern Europe. [. . .]

Few cities in Europe have easier access to stone than Florence. Good building stone – an arenaceous [sand-like] limestone known locally as *pietra forte* – is located no farther away than the hills around the city immediately outside the city gates. [. . .] The immense public projects undertaken in the fourteenth century – above all, the cathedral complex – created a sustained demand for more stone, including vast quantities of marble, and for elaborate sculptural decoration. Finally, at the beginning of the fifteenth century the introduction of the classical forms of moldings, columns, cornices, and a variety of other ornamental details used both inside and outside new buildings vastly enlarged the range and quality of stone production and generated new demand from the private sector for use of this kind of decoration in homes and chapels. The refinement of this taste for stonework through the fourteenth

and fifteenth centuries implies a considerable development of the stone industry. More and higher-skilled stoneworkers were needed, and stylistic innovations opened up new possibilities for the development of their skills that obviously had consequences of inestimable value to the craft traditions of the city. There was probably not another city in all of Europe with such a large number of highly skilled stoneworkers as were found in Florence by the fifteenth century, and in fact the emigration of them throughout Italy was a phenomenon of considerable importance for the diffusion of Renaissance taste.

The supply of other materials

[. . .]

Sand and gravel were the two most common natural materials needed in construction – sand for mixing with lime to make mortar, and gravel for fill in foundations. The Arno had plentiful supplies of both, and legislation established builders' rights of access to this natural resource regardless of the ownership of property along the banks. [. . .]

Notes

1 Quoted in Hans Baron, *The Crisis of the Early Italian Renaissance* (Princeton, Princeton University Press 1966), p. 109
2 Creighton Gilbert, 'The Earliest Guide to Florentine Architecture, 1423,' *Mitteilungen des Kunsthistorischen Insitutes in Florenz*, 14 (1969), p. 46

20

REPOSITIONING THE VATICAN OBELISK

by Domenico Fontana

Source: D. Fontana, 'Della transportatione dell'obelisco Vaticano, Rome, 1590', in Friedrich Klemm, *A History of Western Technology*, trans. D. Waley Singer, London, George Allen and Unwin, 1959, pp. 198–205 and Plate 11

We, Sixtus V, hereby confer on Domenico Fontana, architect to the Holy Apostolic Palace, in order that he may the more easily and quickly achieve the removal of the Vatican Obelisk to St. Peter's Square, full power and authority during this removal to make use of any and every craftsman and labourer as well as their tools, and if necessary to force them to lend or sell any of them to him, for which he will duly satisfy them with a suitable reward. Moreover he may use all planks, beams and timbers of any sort whatever that are found in suitable places for these purposes to whomsoever they may belong; but he shall nevertheless pay the owners a suitable price valued by two arbiters who shall be chosen by the parties. Moreover he may fell and lop any trees which do in any manner belong to the church of St. Peter, to the Chapter or to its Canons, especially also such as are possessed by the Cemetery, the Hospital of San Spirito in Sassia or the Apostolic Chamber, without any compensation. He can transport these through any place and can pasture therein the animals that he needs for this work without suffering any molestation thereby, but he must pay compensation which will be estimated by experts who will be chosen for this purpose. He may purchase and carry away the afore-mentioned and any necessary objects from anybody without paying duty or tax. Without license or other document, he may take all sorts of nourishment for his own use and for his servants and animals from Rome or from other towns and neighbouring districts. He may take and carry off capstans, ropes and pulleys wherever he finds them, even though they may become broken; nevertheless he must promise to repair them and to bring them back intact and he must pay a just hire for them. Similarly he may use all instruments and objects belonging to the building of St. Peter's and may give orders to the servants and officials thereof that within an appropriate space of time they shall make a clear space on the Square around the obelisk and shall prepare everything necessary for this purpose. In case of need, he may demolish the houses next to the obelisk though the form of compensation to be paid must be firmly settled beforehand.

In short, we give to the here-named Domenico Fontana full authority to do, arrange and demand everything else that may be required for the afore-mentioned purpose; furthermore, he, his agents, servants and household staff may everywhere and at every time bear every sort of arms except those which are forbidden. And we command all magistrates and officials of all the Papal States that in all the aforesaid matters they shall afford help and support to the said Domenico Fontana. All others, however who are in any respect subjects of the Apostolic See, whatever be their rank and station, we command, under pain of our displeasure and a fine of 500 Ducats or more as we may determine, that they shall not dare to obstruct this work or in any wise to molest the aforesaid Domenico, his Agents or his workers, but without delay or any excuse, shall assist, obey, support and aid him.

Given at Rome, at St. Mark's on the
5th October, 1585. . . .

Determination of the weight of the obelisk

Before I prepared for the undertaking of the removal, I wished to ascertain the weight of this obelisk that is nearly 75 feet 6 inches high.

I therefore caused to be hewn out a palmo [8·7 inches] in a cubical block of the same stone; I found that this block weighed 86 pounds . . . I deduced that the obelisk weighed $963,537\frac{35}{48}$ pounds . . . .

I reflected accordingly that a capstan with good ropes and pulleys will raise about 20,000 lb. and that therefore 40 capstans would raise 800,000 lb. For the remainder (of 163,537 lb.) I proposed to use five levers of strong timber, each 42·65 feet long; so that I should have not only a sufficiency but an excess of power. Moreover, according to my dispositions, more light machines could always be added, if the first should prove inadequate.

When my invention was published, it transpired that nearly all the experts doubted whether so many capstans could be brought into co-operation so that they could work with combined force in order to raise so great a weight. They said that the capstans could not all pull evenly, that the one with greatest load would break, thus causing confusion that would put the whole machinery out of gear. I, however, though I have never combined so many different sources of power, nor seen anything similar, nor could be certain by means of any comparison with another, nevertheless felt certain that I could do it; because I knew that four horses, harnessed to one of those ropes as I had arranged, however hard they might strain, they would never be able to break it; but that if one capstan had too great a proportion of the load to bear, it would no longer be able to haul, nor would it be able, as alleged, to break the rope; the other capstans would meanwhile be turned, until each had again taken its rightful share of the burden. Then the first, which had been too heavily loaded, would also begin again and the power of all would be combined. Furthermore in addition I had arranged that after every three or four revolutions, the capstans should be halted and that if the men then felt the ropes and found that

one was too greatly strained, they should relax it. . . .

All these arrangements were not new to me and I avoided thereby all dangers and ensured that no rope would break. As it was necessary to build a wooden scaffolding and to create space to set up the aforesaid 40 capstans, it was clear that the Square in question was somewhat too narrow, and that it was necessary to demolish a few houses and to level the ground.

The scaffolding to raise the obelisk

In order to set up the scaffolding, eight wooden pillars or posts were erected, four on one side and four on the opposite side of the obelisk, each about [1·08 m] 3·55 feet distant from the next. Each pillar consisted in section of four timbers, each of which was [490 cm] 19·3 inches thick, so that each pillar was nearly one metre thick. The timbers were attached in such a manner that the mortise-joints should not coincide. They were clamped together at several points by iron bolts and straps, so that they could easily be taken apart again. . . . Around these eight pillars were installed 48 struts. . . . This scaffolding was so firm that the largest building could have been erected on it. But at its summit it was further held by four guy-ropes that rose obliquely from the ground to which they were anchored, and which were tightened by tackles. On the bearers (above on the pillars) were laid five strong timbers, each 21·3 feet long and more than $25\frac{1}{2}$ inches thick, in each direction. Upon which 40 tackles were hung between bearers. These were operated by 40 capstans. . . .

Then the obelisk was covered by double rush-mats, in order that it should not be injured. Above these were laid planks, 2 inches thick. Above these, on either side of the obelisk three iron bars $4\frac{1}{4}$ inches wide and half that amount in thickness whose lower ends were bent under the obelisk, since it stood on bronze blocks. These iron rods were over two-thirds as high as the obelisk, and were constructed of several pieces attached to one another by hinged joints. They were encircled by nine hoops of the same iron about equally spaced along the length. . . .

[. . .] While the armature was being set up, the Square was levelled, the capstans installed and the pulleys attached to them. And in order that those who were entrusted with the supervision of the scaffolding might detect which of the capstans lagged behind or was too far ahead, I had a number marked on each capstan and likewise on its guide-rollers and tackles, so that when necessary a sign could be given from the top of the scaffolding as to which capstan should be slackened or tightened, so that the supervisors of the capstans could obey these orders at any moment without the least confusion. . . .

After all the capstans had been marked, each was worked in turn by three or four horses in order to balance the pulling power of the different horses; and after every three or four turns was revised, until they were all hauling equally. This object was achieved on 28 April 1586.

As very many people crowded to watch so remarkable an enterprise, and to avoid disorder, the streets leading across the Square were barred, and an announcement was made that on the day fixed for the Obelisk to be raised, no-one except the workmen might pass the barriers. Any other person who forced his way in would be punished by death. Furthermore, under pain of severe penalty, no-one might delay the workers, nor speak, nor dispute, nor make any noise in order that the prompt execution of the orders of the officials should not be hindered. To ensure the immediate fulfilment of this decree, the Chief of the Sheriff's Officers should be stationed with his Corps within the enclosure, so that the utmost quiet prevailed among the crowd, partly on account of the novelty of the work and partly on account of the threatened punishments. . . .

On 30th April, two hours before daybreak, two Masses were celebrated in the Church of the Holy Ghost; in order that God, for whose glory and that of the Holy Cross this remarkable undertaking was to be carried out, should grant His grace and should permit it to succeed. And, that He might grant the prayers of all, the workers, foremen and carters engaged on this great work, having by my command been to Confession on the previous day, went all together to partake of the Communion. Also our Lord [the Pope] had on the previous day given me his blessing and advised what I should do. After all had received Communion and had heard the appropriate sermons, he stepped out of the church into the enclosure, and all the workers were ordered to their places. Each capstan had two overseers whose instructions stated that each time the signal of a trumpeter was heard – whom I placed in a raised position, visible to all, the capstans would start work, and it was their duty to keep a sharp eye open that the work was rightly done; but when the sound was heard of a bell which was hung high up on the scaffolding, a halt had immediately to be called. Within the enclosure, at the end of the Square, stood the Chief of the carters with 20 strong horses in reserve and 20 men at their service. Moreover I had eight to ten reliable men scattered in the Square who walked around and took care that no disorder arose during the work. I had also instructed a detachment of 12 men to carry hither and thither as necessary reserve ropes, capstans and pulleys. These were kept in an elevated position in front of the store-house, whence they could at every signal or command carry out the orders given them, so that no capstan overseer needed to leave his place. But at each capstan I placed both men and horses to work it, that the men might the more intelligently follow the orders of the overseer, since horses alone sometimes either remain still or move too quickly. Under the scaffolding were placed twelve carpenters, who had continuously to drive wooden and iron wedges under the obelisk on the one hand to help to raise it and on the other hand continuously to support it, so that it should never hang free.

These carpenters wore iron helmets on their heads, to protect them in case anything fell from the scaffolding. I assigned 30 men to keep under observation the scaffolding, the capstans and the ropes. On the three levers to the west . . ., I set 35 men ready for service and at the two opposite levers 18 men with a little hand-worked capstan.

After a Paternoster and Ave Maria had been recited by all, I gave the sign to the trumpeter; and as soon as his signal sounded, the five levers and the 40 capstans with 907 men and 75 horses began their work. At the first movement, it seemed as though the earth was shaking, and the scaffolding cracked loudly, because all the wooden members were crushed together under the weight; and the obelisk which had leaned more than 17 inches towards the Choir of St. Peter's assumed a vertical position. . . .

The obelisk was then raised, $23\frac{3}{4}$ inches in twelve movements, which was sufficient to push in the skids and to take away the metal blocks on which the obelisk had stood. At this height it was held and strong wooden bearers and wooden and iron wedges were driven under the four corners of the obelisk. This happened at 10 p.m. of the same day and the signal was then given with a few mortars on the scaffolding, and the whole artillery expressed its joy with loud thunder. And according to the order, dinner was carried round in baskets to every capstan, that no-one should leave his post. . . .

The obelisk while being raised was, as already described, continuously underpinned by the carpenters with wedges, as though it stood on a pedestal. When this was completed, they proceeded to remove the blocks, of which only two were laid on the surface of the ped-

estal. Each weighed 440 pounds. One of them was immediately taken to His Holiness the Pope, who manifested great joy thereat. . . .

While the metal blocks were removed from the pedestal the skid was placed on the rollers. The skid was narrower than the foot of the obelisk so that it could be pushed between the supporting timbers under its corners.

The obelisk had now to be laid down which owing to the amount of the movement and the length of the stone was a harder task than the first. For this purpose the pulleys and ropes were differently arranged, so that the western side on which the obelisk was to repose on the skid should remain free. . . .

The accomplishment of these preliminary works required eight days; and on Wednesday the 7th of May 1586 in good time in the morning the whole preparation was accomplished. At the foot of the obelisk the four pulleys were fastened, and the capstans to serve them stood behind the Sacristy on the West. These began to haul at an early hour in the morning and to draw toward them the foot of the obelisk which rested on a skid which ran on rollers, while the other firmly attached capstans slackened their ropes. . . .

When it was half way down it began of its own accord to slide backwards on the rollers; it was therefore no longer necessary to haul it in this direction, but on the contrary to attach a pulley in the opposite direction to the foot, in order to regulate it according to the desire of the foreman. At 10.0 p.m. it lay firmly held on the skid which had been drawn under it without anyone having been hurt. His Holiness heard of this with the greatest satisfaction and the whole people were so happy about it that the architect was led home with drums and flourish of trumpets. . . .

As the obelisk had to be transported from this point 300 ells to its new position and as it was found on levelling that the new position was 28 feet 6 inches lower than the square where it had formerly stood (that is to say at the same height as the surface of the old base) a level embankment was made (i.e. an embankment with horizontal summit) from the old to the new position; the earth for the purpose was taken from Monte Vaticano behind the buildings of St. Peter's. It was made 71 feet wide at the base, 35 feet 6 inches at the summit and 26 feet in height. Around the site of the scaffolding it was made 67 feet wider at the top, and at the foot about 89 feet wider. It was covered with timbers which were supported by posts and struts; and in many places timbers were laid across it, in order that it should at no point yield to the great pressure. . . .

While all this was being carried out, a layer of worked limestone was placed on the foundation which had already been prepared on the approach to St. Peter's and was to support the obelisk. . . . The pedestal, of bonded white marble blocks was then again placed in the centre. . . .

Raising and setting of the obelisk[1]

On the 10th September 1586, when all was in place, before daybreak, two Masses were celebrated in the Church of the Priory Palace, and as at the laying down of the obelisk everyone

Figure 20.1 Erection of the Vatican obelisk, Rome, 1586. Employing 40 capstans, 140 horses, and 800 workmen. Engraving from N. Zabaglia, *Castelli e ponti*. Rome, 1743

who took part in the work went to Communion and prayed to God for a successful outcome. Then every man was placed in position. By daybreak all was in order, and the work began of the 40 capstans, the 140 horses and the 800 men with the same trumpet and bell signals for work and for standstill as before. While the top of the obelisk rose up, its foot was pulled by four capstans, placed on the opposite side so that the ropes which raised the top remained always vertical. The weight to be raised became progressively less, the higher the top rose and the further under it the base was drawn. When the obelisk was half raised, work was stopped and the obelisk was shored up in order that the workers might have their midday meal. After the meal each one devoted himself again zealously to the work. The obelisk was set up in 52 stages, and in many respects it was a beautiful spectacle. Countless people had assembled, and many, in order not to lose their places for the show, remained without a meal until evening. Others made platforms for the people who were streaming in, and earned much money thereby. By sunset, the obelisk was upright; but the skid which had been drawn under it while it was raised, remained beneath it. Immediately the signal was given by mortars on the scaffolding and was answered by guns, and the whole town was filled with joy. All the Roman drummers and trumpeters again hastened to the architect's house and echoed their applause. When the happy news was proclaimed from the scaffolding, His Holiness

was giving audience as he had come from Monte Cavallo to St. Peter's in order to receive the French ambassador in public Consistory [a full meeting of the Cardinals]. Here the news that the obelisk had been raised was brought to His Holiness, and gave him great joy.

The seven following days were occupied in resetting the capstans and fastening the pulleys to the four sides of the obelisk in order to be able to adjust it . . . on the day fixed for the removal of the skid, a beginning was made to work the capstans and pull down the levers so that the obelisk rose slightly; and immediately, its pedestal being broader than the skid, it was underpinned by the carpenters with wedges, so that the skid could be drawn away. Then the bronze blocks were set in their places, and those with tenons were set in lead. Then on the same day the capstans were again operated and the levers pulled downwards, and one wedge after another was struck away, letting down the obelisk gradually so that by the same evening it had come to rest on the blocks; but it was then too late to adjust it. On the following day it was set vertically. . . . Thereupon they proceeded to remove the scaffolding from the obelisk. On the 27th September it was free, and His Holiness commanded that a procession should be arranged to bless the obelisk and to dedicate the golden cross upon it.

Note

1 See Figure 20.1

21

URBANIZATION IN EARLY MODERN EUROPE: CHANGE OR CONTINUITY?

by Christopher R. Friedrichs

Source: Christopher R. Friedrichs, *The Early Modern City, 1450-1750*, London, Longman, 1995, pp. 8-12

For well over a generation the central idea around which most people constructed their understanding of European social history – and of the place of the city within it – has had to do with the theory of modernization. This is the notion that the broad currents of European social history must best be understood as a transition from a traditional, largely agrarian and 'underdeveloped' society to the modern industrial and 'developed' society with which we are familiar today. Traditional society, in this view, was static, deferential and strongly religious; powerful local elites dominated a vast, largely illiterate population engaged in unchanging and inelastic economic activity. By contrast a modern society is seen as dynamic, mobile, heavily secular and at least potentially democratic; a fully literate population relates to political life on the national level and engages in economic activity geared to constant expansion and growth. The transition from traditional to modern society is described as an interactive process in which advances in any one sphere of life contributed to changes in every other; the transition may have been irregular and even stalled at times, but eventually enough changes took place to make the onset of modernization irreversible. These transforming changes, we have been taught, first took place in northwestern Europe and then spread gradually to the rest of Europe, to

North America and eventually to other parts of the world.

The concept of modernization owes much to even earlier concepts of long-term social change. It owes something, for example, to the view of Max Weber, the great social theorist who emphasised the advent of 'rationality' as a key step in the evolution towards modern societies. It owes even more, however, to Marxism [. . .] The movement from 'traditional' to 'modern' society echoes the classic Marxist concept of a transition from feudal to capitalist society. As originally formulated by Karl Marx and Friedrich Engels, Marxism attributed a central role to the European city of late medieval and early modern times – for it was in the city that a new class and a new system of economic relations emerged. The class was the bourgeoisie; the system it created was capitalism. In time, according to the classic view, bourgeois capitalism overthrew the old feudal order and came to dominate every aspect of economic, political and social life. This view has always had its critics, of course – including neo-Marxists who give more emphasis than Marx and Engels themselves did to the capitalist transformation of agricultural relations. But the basic conceptual framework has remained remarkably durable. So, more recently, has modernization theory.

What all this implies for the history of the

city is a deep-seated, often unconscious tendency to search the records of urban history for evidence of the onset of modern conditions. It is only natural, perhaps, in studying the rich variety of social and economic changes in any particular city, to focus on those events which seem to betoken a significant shift towards modernity. The literature about European cities in the early modern era is in fact full of descriptions of landmark events when some crucial shift towards the modern allegedly took place.

Yet this tendency can in fact be highly misleading. Certainly the early modern city experienced countless changes [. . .] But these changes were by no means all unilinear, leading systematically from pre-modern to modern conditions. Some, in fact, can only be described as regressive. [. . .]

In fact, the entire emphasis on change in the early modern city can be misleading. For in many of its most important aspects, the early modern city remained remarkably unchanged during [the period 1450–1750]. This is a point that needs emphasis, for it runs strongly against the grain of most recent studies of the early modern town. To be sure, there are historians who emphasize the immobility and lack of innovation in early modern Europe – but by and large their emphasis is on rural society. The city, by contrast, is normally seen as a centre of innovation and creativity. And so it is. Yet the truly creative and transforming epochs in the history of the European city took place not during the early modern era, but before and after it. It was in fact during the last few centuries of the middle ages that the physical form of the European city and its institutional framework were developed. The basic relationship of the city to the state, the internal organization of urban life, the role of the guilds and the church, the very pattern of streets and buildings – all these things were firmly in place by 1450. Of course the early modern era was an age of dramatic movements

in European history: the Reformation and Counter-Reformation, the growth of European contact with other continents, the developments of new techniques of capitalism, the theory and practice of absolutism, the revolutions in physics and astronomy. There were specific changes in cities as well, particularly in those cities closely linked to the princely and royal courts of western Europe and central Europe. Many towns grew substantially in size, and some new towns were founded. But the impact of all these changes on the everyday life and routine of most city-dwellers was [. . .] relatively small [. . .] It was the nineteenth century, not the early modern era, that radically transformed the European city in its appearance and, more importantly, in its social and economic life. [. . .]

Of course the social norms of urban life in early modern times differed massively from those of today. Patterns of family life, work, leisure, politics, religious expression, and of course the very look and feel of the city itself have little in common with today's world. But this book does not accept the widespread argument that these obvious differences are accounted for by some deeper difference between human nature in early modern times and human nature today. To say that people were more religious, or less worldly, or more deferential, or less able to plan ahead, or more violent [. . .] [implies] that there was some fundamental variation in mental structure between people then and people now. This is not true. Men and women of the sixteenth and seventeenth centuries were motivated by much the same mixture of values and emotions as motivate people today: they were ambitious and altruistic, optimistic and pessimistic, devout and impious, assertive and passive, spontaneous and deliberate. But they operated in a different framework. What made this framework different from ours was, above all, one thing: limited technology in certain key areas of activity. And this in turn was due largely to the lack of information about certain specific topics – particularly a number of physical, chemical and biological processes whose workings are taken for granted today.

This does not mean that people in early modern times lacked scientific information or technological skills. Far from it. Indeed, many of the products of the medieval and early modern cities – from cathedrals to cloth, from cannons to clocks – were the result of enormously complex technical processes, developed over centuries of trial and error. But early modern life was burdened by technical and scientific limitations of enormous consequences. Power sources were meagre. Almost all power had to be generated by human or animal muscles, for the use of inanimate power was extremely limited. This in turn had implications for the way in which products were manufactured, the way in which buildings were constructed and the way in which people and goods were transported. Communications were slow. Information could travel no faster than the wind at sea or the fastest horse by land. It was hard to make things very hot. This restricted what could be done with metals. It was hard to make things very cold. This prevented people from preserving most foods. Control of pain

was extremely limited. Surgical intervention beyond a few outer millimetres of the human body was virtually impossible. The role of bacteria in human biology was not understood. Antibiotic medications were not available. All these factors contributed to creating an entirely different age-distribution of illness and mortality than the one we are used to today. This in turn had significant implications for relations within families and attitudes about life, death and the hereafter.

Of course there were important changes in scientific knowledge during the early modern era. But most of these changes were in the areas of physics and astronomy, and their impact on everyday life was quite limited before the later eighteenth century. There were striking technological developments as well. The early modern era opened with the inventions of gunpowder and printing: both changed the way in which cities interacted with the surrounding world and the way in which city-dwellers interacted with each other. But neither fundamentally altered the institutional structure or value system of European urban life. Only the massive scientific and technological transformations of the nineteenth century would change all that.

22

TECHNOLOGY AND THE BUILT ENVIRONMENT OF THE EARLY MODERN CITY

by Christopher R. Friedrichs

Source: Christopher R. Friedrichs, *The Early Modern City, 1450-1750*, London, Longman, 1995, pp. 21-9, 32-5

The most distinctively urban feature of the early modern city was the outer wall [. . .] To be sure, not every city had a wall, and not every walled community was a city. But the correlation was remarkably high. It is hard for people today to capture any sense of the size and ubiquity of city walls in early modern Europe. Only a few European cities [. . .] still have their walls intact [. . .] Yet in early modern times, the wall was almost a city's dominant architectonic feature.

Indeed, the term 'wall' is hardly sufficient to suggest the size and solidity of urban fortifications. The basic wall system for almost any city was a product of the middle ages, but throughout the early modern era urban fortifications were lavishly expanded and strengthened. Many cities, already girdled by one massive wall, received an additional perimeter of bastions linked by new walls in the sixteenth or seventeenth century. Access to the city was possible only through perpetually guarded gates, many of which were surmounted by high towers from which watchmen could survey the flow of traffic for miles in any direction.

City walls obviously had a military function [. . .] But the wall had other functions as well, for the system of fortifications had come into being largely to give city officials some control over the flow of goods and people in and out of their community. Access to the city was deliberately made difficult [. . .]

Not every city had a wall, nor did every city that had one maintain it carefully. In England, which had not experienced a foreign invasion for centuries, city walls were often neglected – though their importance was quickly rediscovered during the civil wars of the mid-seventeenth century [. . .]

Some cities lost their walls for political reasons. The English town of Gloucester, for example, was forced to tear down its wall in the 1660s as a result of having supported the wrong side in the recent civil war. Similarly, when Louis XIV seized control of the Alsatian town of Colmar in 1673, he ordered that the city's fortifications be destroyed [. . .] [However] throughout the seventeenth century the conquest of a city was as likely to lead to a strengthening of its fortifications as it was to lead to the opposite. After conquering the Flemish town of Lille in 1667, for example, Louis XIV immediately ordered the construction of a massive citadel connected to the city's bastion ring - partly to protect the city from reconquest, and partly, it seems, to intimidate the inhabitants themselves.

The city wall normally enclosed the built-up

areas of the community. But the city's legal boundary customarily extended for some distance beyond the actual walls. A few buildings - mills, chapels, convents, hospices - could always be found outside the walls. Some cities also had built up suburbs or *faubourgs* outside the walls. In the middle ages it had been customary from time to time to build new walls to accommodate urban growth, and in many cities this continued to be done in the early modern era. In cities which experienced rapid urbanization, however, this was not possible. London, which underwent the most spectacular growth of any major city in early modern Europe, is the paradigmatic case. In 1500 London was still contained within its medieval walls; the administrative centre at Westminster a few miles upriver was a completely separate community. By the end of the sixteenth century, however, there were extensive settlements beyond the walls of London. Growth continued in all directions, and by 1750 London had become a vast urban agglomeration. [. . .]

But London and the other great metropolitan centres were exceptions. Most cities retained their walls and gates until the end of the eighteenth century. In many places, however, there was increasing negligence in maintaining the fortifications. The outer ring of bastion walls built in the seventeenth century proved of little tactical value and their upkeep was often slipshod [. . .] In the late eighteenth century, city walls suddenly came to be seen as impediments to urban growth, aesthetic form and the flow of the traffic. Within two generations in city after city the walls were razed and parks and promenades were laid out in their place. But the idea that traffic should move in and out of a city without any impediment was still too much for the average European to swallow. Frankfurt am Main well illustrates the point. In 1804 the demolitions began and within a matter of years the once mighty walls of Frankfurt were totally gone. But on the major roads, where imposing city gateways had once served to channel

movement in and out of the city, there now stood little guardhouses and metal gates. Not until 1846 did these last faint echoes of the walled city of Frankfurt finally disappear.

Everyone has some mental image of the topography of the medieval or early modern city: a densely congested network of narrow, winding streets lined with high house fronts, a pattern relieved only occasionally by open squares and marketplaces. The image is not inaccurate, but it can be misleading. Most cities had considerable amounts of green space, even within the walls: there was no shortage of pleasure gardens, market gardens and orchards, not to mention other open areas like cemeteries and workyards. But the gardens were private spaces, behind houses or enclosed by walls, generally inaccessible to anyone but their owners or those who worked there. Public activity took place in the streets and marketplaces, or outside the city gates [. . .]

The basic street plan of almost every European city was a product of the middle ages, and once it was laid down the plan normally proved remarkably resistant to change. Only entirely new cities could ignore the constraints imposed by past construction. Early modern Europeans were quite alert to the potential attractiveness of straight streets and broad vistas: the 'ideal cities' of Renaissance theorists like Antonio Filarete or Francesco di Giorgio Martini almost always involved a rectangular grid or a pattern of identical straight avenues radiating from a central square. A few cities were even constructed along such lines [. . .]

Such designs, however, could hardly ever be imposed on existing built-up cities. Property owners [. . .] were likely to resist any proposal to realign streets if it meant impinging on their land [. . .] Nor did the sudden destruction of a city centre by fire routinely clear the way to significant changes in urban topography. It could happen, of course: when much of Stockholm burned down in 1625, the Swedish king insisted on redesigning the inner city to eliminate a cluster of small dark alleys and provide for a major new thoroughfare. But this was hardly the norm. When the heart of London was destroyed by fire in 1666, numerous proposals were offered to reconstruct the city on elegant new lines. But any such plans would have required massive schemes for compensating landlords or redistributing property. In the end some alignments were made, a few streets were widened, and less flammable building materials were mandated – but major changes in the layout of the city proved impossible to realize [. . .]

None of this means, of course, that city centres remained completely unaltered [. . .] New public buildings were constructed and old ones were extended. Occasionally a whole group of dwellings would be razed and new structures erected. [. . .] In 1605, for example, King Henri IV of France decreed that a neglected royal tournament-ground in northeastern Paris [. . .] should be transformed into an architecturally integrated residential square [. . .] The resulting Place Royale, which was completed by 1612 and soon became a highly fashionable address, has long been praised by historians as a masterpiece of urban planning – and rightly so. But it came into being only because the ground concerned was a vacant tract held by a single owner, in this case the crown. Without this advantage, even the most energetic king would have trouble putting such plans in place. For by and large, in Paris as elsewhere, once the city core was built up it became highly resistant to topographical change.

It was in the undeveloped districts outside the city wall that more dramatic changes could take place. Fields and gardens outside the walls were occasionally transformed into formal parks and promenades where well-dressed members of the public were permitted to stroll. Entire new neighbourhoods were established as well. Where no previous pattern of houses existed, it was easier to lay out a rectangular grid of wide streets which approxi-

mated the Renaissance ideals of city planning. Many a European city thus acquired elegant new districts adjacent to the old crowded centre [. . .]

Many new districts, in fact, developed in a much more haphazard way. In certain parts of London, for example, growth was almost completely uncontrolled – especially in areas east of the city and south of the Thames where the old ribbon development along major roads and down the riverfront increasingly gave way to a vast agglomeration of suburban tenements [. . .]

Some cities experienced no physical expansion at all. In many cities, the area enclosed by the city walls proved ample for a stable or even declining population. In other cases, the population grew but there was nobody with sufficient interest, incentive or influence to orchestrate the systematic development of a new district [. . .]

Walls and streets were the two main elements of urban topography imposed by human design on the physical environment. But a third element was provided by waterways and canals. Almost every city, of course, had grown up alongside a river or near a body of water. Often river water had been diverted to supply a network of waterways around and through the city. Where the terrain permitted, a moat might encircle the walls to improve the city's defensive posture. And often marshy areas had been drained and streams regulated so as to form a series of canals. These waterways served a variety of purposes. Many urban industries, such as the manufacture of textiles, leather or paper, required ready access to water for their technical processes. And a canal often served as a source of drinking water or a convenient place to flush away wastes. In some cities – as in parts of the Netherlands, where settlements had emerged behind massive dikes on marshy land below sea-level – there might be a vast and intricate network of canals, traversed in turn by bridges which linked an equally complex system of streets and roads. [. . .]

Nature provided the setting: some cities were situated on flat terrain, some developed on sloping ground above a river or a shore, a few nestled beneath an imposing hill topped by a fortress. Walls, streets and canals were the key elements of the city's infrastructure. But above all, a city was a collection of buildings, public and private [. . .]

Among public buildings the most prominent ones were generally ecclesiastical in nature. With only rare exceptions, churches were the largest buildings in any city at the beginning of the early modern era [. . .] Of course there was no shortage of major secular buildings as well. Some cities had a castle which had once been or continued to be the seat of some prince or ruler [. . .] Only cities of great strategic or administrative importance were likely to have castles. But every city by the end of the middle ages had a number of public buildings specifically designed to meet the political and economic needs of the city-dwellers themselves.

The largest of these were often devoted to the administration of city government. These included such grandiose establishments as the doge's palace in Venice, the Palazzo Vecchio in Florence, or the great town halls of Brussels and Bruges with their soaring belfries. But many other cities had more modest town halls at the end of the fifteenth century. The town hall could in fact be overshadowed by other municipal buildings. The regulation of commerce had emerged in the middle ages as one of the chief functions of urban government, and this was reflected in the construction of huge facilities for the sale, transfer, inspection or storage of commercial goods [. . .]

Another critical obligation of urban governments – the guarantee of a sufficient supply of food for the population – was manifested by the existence of municipal mills in most cities and granaries in many [. . .]

All these public or semi-public buildings were of immense importance to the life of a pre-modern city [. . .] [Nonetheless] the great majority of buildings in a late medieval city were in fact not public buildings at all, but houses [. . .]

House design varied across Europe. Local custom powerfully influenced the appearance and layout of buildings. A few characteristics, however, were common to houses in most of Europe. Houses were generally built in rows, with little or no space between them and fronting directly on to the street. Houses were hardly ever more than four or five storeys high, but differences in width and depth were common. Large houses might have an interior courtyard or a yard or garden to the rear. The ground floor would most likely include a workshop and probably a kitchen. The higher floors would have rooms in which families lived and slept. Cellars and attics were used for storage or for accommodating servants or tenants. Glass windows were still a luxury; wooden shutters were the norm.

This, then, was the basic repertory of urban building forms inherited from the middle ages. Of course, the building process is always dynamic, and throughout the early modern era new buildings went up and old ones were extended, remodelled or converted to new purposes. New architectural forms and, to a lesser extent, new methods of construction were developed. More and more major buildings were designed in accordance with foreign architectural styles: Italian models were increasingly copied north of the Alps. Some building complexes, such as shipyards, grew larger than they had ever been in medieval times. Significant innovations emerged in military architecture, as bastions and citadels came to ring the periphery of some major cities. Yet the basic categories of urban building types remained relatively constant during the early modern era. The traditionalism of economic and administrative systems and the absence of sweeping technological change diminished the likelihood of fundamental innovations within the cities. In fact it was not until the nineteenth century that radically new types of urban structures - factories, railways, department stores and the like - were introduced in European cities on a massive scale.

23

A GOLDEN AGE: INNOVATIONS IN DUTCH CITIES, 1648–1720

by Jonathan Israel

Source: Jonathan Israel, 'A Golden Age: innovations in Dutch cities, 1648-1720', *History Today*, Vol. 45, 1995, pp. 14-20

Between April and June 1648 the most elaborate and impressive celebrations which had thus far ever been held in the northern Netherlands [. . .] were held in Amsterdam and most other Dutch cities. The reason for this unprecedented outlay, disruption of normal activity and quest to impress and involve the general public was the final ratification of the Peace of Münster (April 1648). This not only ended the Eighty Years' War in the Low Countries, one of the greatest struggles of Europe of early modern times, but marked the successful conclusion of decades of effort to establish and consolidate the Dutch Republic as a free and independent state on territory formerly ruled by the king of Spain. [. . .]

At the same time, these celebrations were the Dutch contribution to a wider set of festivities held all across northern and central Europe to mark the end of the unprecedentedly destructive Thirty Years' War. As such, the festivities of 1648, both in their Dutch and wider European context, were a psychological turning-point between a dreadfully bleak era of struggle and dislocation, and the deep pessimism and gloom which had resulted, and a more hopeful era; one of building and reconstruction [. . .]

With the Thirty and Eighty Years' Wars simultaneously out of the way, city governments could now think about reconstructing their war-torn cities and, in the case of the Scandinavian capitals and flourishing Hamburg, embark on those ambitious projects and city extensions which it had seemed prudent to postpone while fighting and disruption continued. Furthermore, since at that time the Dutch Republic was economically and culturally the most dynamic and flourishing country in Europe, it was entirely natural that many of these cities whether or not they had been devastated, especially those in the Protestant north, should look to the Dutch Republic for most of the ideas, designs, methods and technology which was to shape their general renovation and reconstruction during the second half of the seventeenth century (particularly in the case of St Petersburg, Russia's window on the west) and the beginning of the eighteenth.

However, if we are to grasp how it was possible for such a small country as the Dutch provinces to have exerted such an immense influence over urban development in northern Europe, an influence which was, in most respects, far greater than that of Britain or France down to around 1720, it is by no means sufficient just to point to the general readiness for renovation and refurbishment, or to the special dynamism of the Dutch economy at that time. The phenomenon is more complex than that. For the Dutch cities were themselves then entering a major new phase of expansion and renewal, shaped by a dazzling array of innovations and new techniques and, more than anything else, it is this which gave them their special relevance and immense influence over such a considerable period [. . .]

By 1648, the impact of the Dutch cities on the European urban scene was already very considerable and had been growing, especially since the 1590s. The seemingly miraculous expansion of Dutch commerce and shipping which had begun to take over the 'rich trades' of the world in the 1590s, elbowing all rivals aside, had reached such a point that it had aroused intense envy and resentment in almost every part of Europe and not least in England. Moreover, by 1648 those parts of Europe – especially Scandinavia, northern Germany and the Baltic – which were particularly susceptible to Dutch cultural influences were already so steeped in Dutch methods, styles and ways of doing things that everything else had been pushed into the background. When Hamburg and Copenhagen rebuilt their city fortifications during the second decade of the century they did so using Dutch engineers and Dutch designs [. . .] Yet, notwithstanding this vast impact of Dutch commerce and shipping, and of Holland's art, architecture and engineering (particularly drainage, harbours and fortifications), those aspects of Dutch culture which were to have the greatest impact on urban development, refurbishment and planning after 1648, were only just beginning to be noticed.

The chief reason why the Dutch had not yet begun even potentially to make their real impact in the sphere of urban improvements, health care, town planning and public services is that the Dutch cities too, like those of Germany and Scandinavia, had since around 1620 been systematically postponing major new investment in building and city extensions. Just as Amsterdam needed a new and larger city hall long before 1648 but work on the new edifice began only in that year, and Leiden put up with old and dilapidated gate-houses,

only replacing them with magnificent new structures after 1648, so all big projects were put off. But once the Eighty and Thirty Years' Wars were finally over, the accumulation of grandiose and ambitious schemes led to a frenetic burst of building and refurbishment throughout the length and breadth of Holland. Not only were numerous large public buildings erected in the 1650s and 1660s [. . .] but those cities which achieved an impressive measure of growth between 1648 and 1672, especially Amsterdam, Leiden, Rotterdam, The Hague and also Haarlem, [. . .] also laid out whole new urban quarters, constructed new canals and roads, and planned new housing as part of integrated urban development schemes [. . .]

However, the integrated reality of Dutch city planning and improvements after 1648 could not be emulated elsewhere in its entirety because many features of the Dutch urban scene were highly specific to [the provinces of] Holland and Zeeland. Thus numerous foreign travellers of the period remarked that Amsterdam was much cleaner and less cluttered than London or Hamburg. But one of the main reasons for this was that the city government banned the use of horse-drawn coaches and wagons in the city, insisting that goods, supplies and furniture be moved by water and digging new canals and improving old ones to facilitate such traffic. This was perfectly feasible also in other Holland and Zeeland towns, but hardly practicable elsewhere.

Another feature which could not be imitated elsewhere were the regular passenger services between towns by means of horse-drawn passenger barges, with departures several times daily between the main towns, working according to a published schedule [. . .] Furthermore, not only these but also many other Dutch urban improvements of the period could only be effectively implemented because of the almost absolute power of the city governments within their cities and jurisdictions

[. . .] If a city government wished to implement an ambitious and costly urban plan there was no question of this being opposed by any authority or body outside. In this respect, Swiss and some German Imperial Free Cities enjoyed a comparable freedom of action but cities under monarchs, such as London, Paris, Copenhagen or Stockholm, did not. Monarchs had their own agendas and priorities and, in most cases, a considerable sway over resources.

But what other European cities, including London, could and did do, some sooner, some later, was to adopt in full, or in part, such individual urban improvements and innovations introduced by the Dutch cities as did not need specifically Dutch conditions for their implementation. A classic instance of such successful borrowing was the adoption of the Dutch system of public street lighting. Europe's first proper system of public street-lighting was planned, in conjunction with members of the Amsterdam city government, by the artist–inventor Jan van der Heyden (1637–1712). Van der Heyden designed a street-lamp manufactured of metal and glass with shielded air-holes able to let out the smoke without letting in the wind. The lamps burnt through the night on a mixture of plant oils with wicks of twisted Cypriot cotton.

Besides the considerable cost, the plans to light up the whole of Amsterdam at night presented appreciable problems. But the burgomasters and regents decided to go ahead, motivated by a desire further to improve orderliness in the city, and reduce crime, as well as the incidence of drunkards falling into the canals at night and drowning. The plans were finalised and approved in 1669. By January 1670, the entire city was lit up after dark – what an amazing sight it must have been – by 1,800 public lamps (increased to 2,000 before long) affixed to posts or the walls of public buildings. The lamp-posts were placed 125 to 150 feet apart, Van der Heyden having calculated that maximum lighting efficiency with

minimum wastage of oil and equipment was achieved with that spacing. One hundred public lamplighters were recruited who each re-filled, lit and, in the morning, extinguished, twenty of the lanterns. It took about fifteen minutes to light up the whole city. By 1681, another 600 public lamps had been added to the original 2,000. The advantages of lighting up the city at night were so obvious that first other Dutch cities, and then cities outside the United Provinces began to follow Amsterdam's example. Dordrecht installed the new system in 1674. Having ordered the equipment from Amsterdam, both Berlin and Cologne installed hundreds of Dutch lampposts and lit up their cities in 1682.

Another urban improvement which was widely imitated in late seventeenth-century Europe was Amsterdam's remarkable new fire-fighting service. Here again the technology involved was devised by Van der Heyden. But the key to success was the ability of the city government, having seen the potential of his innovations, to back them with funds and a sophisticated civic organisation. Although Van der Heyden was not the first Dutch expert to contrive a pump able to throw up a continuous jet of water from the canals, he produced improved pumps and was the first to join them to (leather) hoses. After being put in charge of the city's fire-fighting service, in 1672, he created an organisation the key element in which was the distribution of quantities of pumps and hoses around the city and their storage in special depots; and the assigning of able-bodied guild members in each quarter to take charge of the equipment and to ensure that it was promptly rushed to the scene and used in the event of fire. The sight of the new water pumps and hoses in action greatly impressed contemporaries, the effect being heightened by Van der Heyden's dramatic and well-publicised illustrations of fire-fighting scenes. One astonished English visitor described the hoses as being 'as big as a man's thigh which by the assistance of pumps, at which they labour continually for three or four hours, throw up water to the tops of houses and force it three hundred paces over the tiling'.

Naturally, other northern cities were quick to see the value of the Dutch pumps and hoses and, albeit with a few years' time lag, proceeded to adopt similar systems based on the Dutch equipment [. . .]

One of the most crucial of the Dutch urban improvements of the mid-seventeenth century and one which was widely imitated, especially in Germany, Switzerland and Scandinavia, was the setting up in Amsterdam of a civic medical board called the *collegium medicum* consisting of three university-trained physicians and two prominent apothecaries to inspect, supervise, license and register medical practice in the city [. . .]

When the system of control was first introduced by the Amsterdam city government, by city edict of March 23rd, 1639, the chief concern seems to have been to curb the abuses going on in apothecaries' shops, especially the selling of impure or bogus medicines and wide discrepancies in charges. The edict laid down that the physicians of the collegium were to visit and inspect all the apothecaries' shops in the city 'two or three times per year', without prior notice, and verify what was being sold. At the same time, the city published a list of authorised prices for medicines to which apothecaries were expected to adhere and it was laid down that apothecaries and their assistants would only be permitted to practice in the city if they satisfied the collegium that they had the adequate knowledge and expertise. Other Dutch cities followed and, soon, so did various German and Swiss cities [. . .]

While the original emphasis was mainly on the supervision of apothecaries, the Dutch system of regulating civic health care gradually evolved during the middle and late decades of

the seventeenth century becoming more comprehensive as well as more sophisticated [. . .]

The Dutch city governments of this period, eager as they were to attract immigrants and increase the population of their cities, made a serious and sustained effort, and with some success, to improve living conditions and health care. At the same time, they vied with each other in erecting imposing public buildings – hospitals and orphanages, as well as town-halls, gate-houses and churches – beautification and splendour being essential aspects of the urban development schemes which they so intensively devised and debated [. . .]

With both practical and aesthetic considerations firmly in mind, nothing appealed to the Dutch city governments of this period more than opportunities to combine public utility with beautification. A development which gave them precisely such an opportunity was the arrival in the middle decades of the century of new types of very large public clocks which (especially after the 1650s) also kept time more accurately than the clocks of the past. In their drive to embellish Leiden, the city government there developed a veritable mania for affixing such clocks to the city's public buildings, including one [. . .] placed at the top of the imposing new White Gate (built in 1650), near where the passenger barges loaded and unloaded the travelling public so as to facilitate punctuality in barge departures. [. . .]

The expansion of the Dutch economy, and of the Dutch cities, ended abruptly with the Anglo-French attack on the United Provinces in 1672. In that year Louis XIV invaded the Republic and occupied its eastern provinces while the French army, combined with the English and French fleets, delivered a blow to Dutch commerce and industry from which they were never fully to recover. After 1672, it was unquestionably England which was the most dynamic and fastest-growing commercial economy in the western world. Nevertheless, it is important to note that at that time there were in Britain no large cities other than London (albeit as large as Amsterdam and the six next largest Dutch cities combined) which was by all accounts a somewhat disorderly and chaotic place compared with the Dutch cities.

Moreover, despite their stagnation after 1672, the Dutch cities were at that time sufficiently far ahead of England in technological innovations, health care and urban planning to retain something of an edge not only down to the end of the seventeenth century but even for a decade or two into the eighteenth. It was not until after around 1720 that Britain can be said to have overtaken the Dutch Republic in terms of technological sophistication.

Consequently, despite the emergence of England as the world's most dynamic economic and colonial power after 1672, it was still the Dutch cities, rather than the British, which were the main model for urban planning and improvements in northern Europe for another half century.

24

FIRE-FIGHTING TECHNOLOGY IN EARLY MODERN ENGLAND

by Stephen Porter

Source: Stephen Porter, *The Great Fire of London*, Stroud, Sutton, 1996, pp. 21–7

The fire-fighting equipment that was available consisted chiefly of leather buckets, fire-hooks and ladders. The fire-hooks were used rather like grappling irons to unroof, or even to pull down, buildings in the path of the fire. Pick-axes, shovels, crowbars and chisels were also useful, for digging up water pipes and then cutting them open, and brooms and long-handled swabs were kept for beating out the flames. Hand-held water squirts were developed in the sixteenth century, but the major innovation during the period was the fire engine, which was introduced in the second quarter of the seventeenth century and widely adopted in London and many provincial towns.

A ready supply of equipment was essential and was achieved by requiring the parishes and livery companies to keep a specified number of buckets, ladders and fire-hooks [. . .] Leather fire-buckets were obtained from the founders in Lothbury, who were also the principal manufacturers of fire-engines, which had been developed in Germany. John Jones, a merchant, was responsible for acquiring the first engines to be used in England. Acting with the help of his brother Roger and at the instance of the Court of Aldermen [the ruling body of the city of London] he brought two engines to London early in 1625, apparently from Nuremburg. In February that year, he was granted a patent awarding him a fourteen-year monopoly for making and selling engines 'for the casting of water' and claimed that with a crew of ten a fire could be put out using an engine 'with more ease and speed' than 500 men equipped with buckets and ladders [. . .] In October [1625] Alderman Hammersley acquired two more engines, from Hamburg, and by June 1626 he had brought over eleven other 'Engines or water spouts', which were distributed around the city.

John Bate's *The Mysteries of Nature and Art*, published in 1634, included a description and sketch of an engine 'for spouting water', which he described as 'very useful for to quench fire amongst buildings'[1] This rather understates the reaction to the effectiveness of these early engines, which created a favourable impression, not just in demonstrations, but actually in use at fires. In 1638, following a blaze at Arundel House in the Strand, the Privy Council wrote to the lord mayor and aldermen that it had been informed of 'the excellent use to bee made in Accidents of fier, of the new Engines for spowting water'. The problem at Arundel House had been that the engines had arrived too late because none were kept in the nearby parishes, demonstrating that it was advisable that they should be 'neare & ready at hand' wherever a similar emergency might occur. The council therefore recommended that the City parishes should acquire their own fire-engines, the smaller ones joining together for the purpose and the larger ones providing their own, and it also instructed the Middlesex parishes to obtain engines [. . .]

The first towns to acquire an engine did so just before the Civil War, with Norwich, Worcester and Devizes all buying their first fire-engine in 1641. After a hiatus during the war years the process resumed in the late 1640s, when such towns as Bristol, Gloucester and Marlborough bought their first engines [. . .] Fire-engines were, therefore, common items of equipment by the mid-1660s, especially in London. The principal supplier during the middle years of the century was William Burroughs, the City's founder, who was said to have made roughly sixty engines by about 1660[2] but there were other makers in London, including John Shaw and Ahaseurus Fromanteel, a clock-maker. The particular advantages claimed for Fromanteel's engines were that they were so tough they would not break 'without extreme violence' and were small enough to be taken into a building. They could even be carried upstairs, an important consideration in the City, where the height of many buildings created problems in fighting fires on the upper floors [. . .] Unless the speedy carriage of water up several flights of stairs was feasible a dangerous delay was likely before a fire could be dealt with [. . .]

Fromanteel's claim for the durability of his fire-engines was designed to allay fears that such appliances were too easily damaged and put out of action. Complaints also show that it was not uncommon for engines to be useless when they were needed because of lack of maintenance. Another feature of the early seventeenth-century engines which restricted their capabilities was that they could throw only an intermittent jet of water. This was overcome by the invention of the air vessel by a Nuremburg maker in 1655, which permitted a continuous stream to be delivered on to the fire, although it may have been some time before this refinement was incorporated into the engines built in London.

Fire-engines represented a major advance in the technology of fire-fighting and they were adopted with considerable enthusiasm, but there were disappointments [. . .] Whatever the quality or efficacy of the equipment, a good supply of water was essential, for the engines contained a tank of only limited capacity that had to be continually topped up by a bucket chain [. . .] By the 1660s London had a good supply of water from [. . .] various sources, but it was not completely reliable. In dry weather the levels in wells and the flow along the aqueducts could both be low, and it was more difficult to draw water from the river at low tide [. . .]

The provision of fire-fighting equipment and an adequate supply of water were of little use unless they were deployed to good effect when a blaze began. Resolute action and good organisation were essential [. . .] In terms of an awareness of the hazards of fire and attempts to reduce them, and the provision of fire-fighting equipment, by the mid-1660s London had achieved a great deal.

Notes

1 R. Jenkins, 'Fire-extinguishing Engines in England, 1625–1725,' *Transactions of the Newcomen Society*, 11, 1930–1, pp. 17-18

2 Thomas Fuller, *The History of the Worthies of England*, P.A. Nuttall, (ed.), 3 Vols, London, Thomas Tegg, 1840, Vol. 2, pp. 334-5

25

TECHNOLOGICAL INNOVATION IN SEVENTEENTH-CENTURY PARIS

by Leon Bernard

Source: Leon Bernard, *The Emerging City: Paris in the age of Louis XIV*, Durham, N.C., Duke University Press, 1970, pp. 56–68, 162–7

Transportation

No aspect of the greatly quickened pace of life in seventeenth-century Paris will strike a more responsive chord among moderns than the appearance of traffic congestion.

Traffic is, of course, a very relative thing. From the earliest beginnings of any urban society, citizens can be suspected of taking a perverse pride in the volume of traffic in the streets. Paris was no exception [. . .] But in the seventeenth century, for the first time, we can take such talk seriously and sympathetically, for this was the century in which the carriage appeared and transportation was consequently revolutionized. [. . .]

At the beginning of the seventeenth century, the august presidents of the Parlement of Paris [. . .] still made their way to the Palais de Justice on the backs of mules, long the favourite conveyance through the mud of Paris [. . .] One authority maintains that in 1594 there were only eight carriages to be seen in and around Paris – great cumbersome vehicles owned only by the very rich.[1]

Under Louis XIII [1610–43], the number of carriages in Paris increased sharply. An official report [. . .] in the 1630s set the number at over 4,000[2] [. . .] According to Voltaire, magnificent new spring-suspension carriages equipped with glass windows began to make their appearance in the streets of Paris shortly after mid-century.[3] He was probably referring to the new French window glass manufactured at the Manufacture des Glaces in the Quartier Saint-Antoine, a factory which by the end of the century employed 400 people and was probably the largest industrial establishment in the city. Voltaire's suspension devices are harder to fathom [. . .] But whatever the nature of the new suspension, its superiority over English vehicles, at least, is well attested by Martin Lister. He wrote that he was less tired after six hours of riding in public carriages in Paris than after one hour in the finest English conveyance.[4] [. . .]

By the end of the seventeenth century Brice estimated there were 20,000 carriages in Paris.[5] A new status symbol had been born [. . .] One of the most striking social developments was the *promenade à carrosse*. Life took on new dimensions, at least for the well-to-do, as carriage outings became the rage [. . .] Far and away the most popular promenade of the fashionable world was the Cours-la-Reine, which ran for almost a mile alongside the Seine just west of the Tuileries gardens. Thanks to a wall on the north side and gate houses at both ends, all but the *beau monde* were theoretically kept out. [. . .]

The seventeenth-century revolution in transportation made an impact on many more than the fortunate few who could afford carriages and gain access to midnight promenades on the Cours-la-Reine. For the first time transportation within the city became an industry [. . .]

The great assortment of rented vehicles which appeared for the first time on the streets of Paris on the Grand Siècle fell into three main categories: the portable *chaises*, the *chaises* on wheels, and the famous *fiacres*. Within these three general types could be found a bewildering assortment of styles and models, many of whose names survived into the era of the horseless carriage. The right to operate these vehicles was earned and precariously protected by hard-won royal privilege. An inventor had no choice but to keep his creation off the streets until such time as he, or more likely, an associate with access to official circles, could obtain letters patent giving him the right of exploitation. [. . .]

Of the many kinds of public conveyances which transformed the streets of Paris in the seventeenth century, the oldest was the litter chair, carried by two men dubbed 'baptized mules' [. . .] Early in the reign of Louis XIV (1644), the Marquis de Montbrun stole from London the idea of a covered chair, and he and his natural son retained their monopoly well into Louis XIV's reign [. . .]

Once the litter chair had been well accepted, the next step, obviously, was to attach wheels to it. But this apparently simple development was an inordinately long time coming. For one thing Parisian streets around mid-century were in a particularly poor state of repair, making any kind of wheeled vehicle impractical. Probably more important, the owners of the monopoly for chairs [. . .] strongly opposed the wheeled vehicles. Although the King to ease their fears of competition gave them the right to exploit the wheeled chairs any time they wished, they made no effort to do so. According to Delamare, they were deterred by their inability to procure a mysterious 'secret [invention] which greatly improves the rolling power' of the vehicles.[6] The 'secret' must have been safeguarded like the crown jewels because not until 1669 did the small two-wheeled contrivances appear [. . .] The *brouette* was simply the old portable chair suspended between a pair of wheels with the fore-and-aft porters replaced by a single human mule, or *brouetteur*, in the front. [. . .]

The *brouette* doubtless had its uses [. . .] but for longer journeys within the city, where more speed and comfort were desirable, still another type of conveyance came into use in the middle of the century, the *fiacre*, or as Englishmen corrupted the word, hack. Until the start of Louis XIV's reign, the *carrosse* had been associated exclusively with the rich and great, but a few years before the outbreak of the Fronde [i.e. in the early 1640s] an enterprising stablemaster by the name of Nicholas Sauvage, residing in the Rue Saint-Martin at the sign of Saint-Fiacre, began to rent carriages and horses to anyone who presented himself. So great was his success that all rental carriages as well as their drivers were henceforth styled *fiacres* [. . .]

Before long the rental carriages were taken out of the stables and made available at designated spots around the city. [. . .]

The most startling of all the many seventeenth-century innovations in Parisian transportation was the system of *carrosses à cinq sous*, the first omnibus system in Paris and probably anywhere else [. . .] According to a well-founded tradition Blaise Pascal was the inventor and moving spirit of the new enterprise[7]. But obviously any plan as daring as this required unusually strong protection at court. This was duly obtained in the persons of the Duc de Roannez, Governor of Poitou, the Marquis de Sourches, Grand Prévôt de l'Hôtel, and the Marquis de Crenan, the King's Cupbearer [. . .] In January 1662 these three gentlemen received permission from the King to operate public carriages on the streets of Paris according to fixed schedules and routes and a uniform fare of 5 sous.

The first line opened in March 1662. Its terminals were the Saint-Antoine Gate and the Luxembourg Palace. Seven *coches*, very much on the order of the vehicles then being used in intercity transportation, were employed, each seating eight passengers. The scheduled time between the vehicles was to be fifteen minutes, and one had the right to hail them at any point along their route [. . .]

Whatever the reaction of the populace, the second omnibus line opened with almost as much fanfare a few weeks later. This one ran on a half-hour schedule from Saint-Roch Church in the Rue Saint-Honoré clear across Paris to the Saint-Antoine Gate. Subsequently, three more routes were established and success seemed assured. Even the King and the Duc d'Enghien are said to have tried the new service [. . .] But without anyone ever being able to ascribe a logical cause, the *carrosses à cinq (six) sous* went out of existence sometime well before the end of the century, not to be revived until 1828, when the price, incidentally, was still the original 5 sous.

Longer lived were the nautical equivalents of the *carroses à cinq sous*, the *bachots*, or small boats which helped meet the seventeenth-century demand for promenades into what Delamare described as 'the enchanted places which gave a new éclat to the city of Paris'[8]. [. . .]

The mass of traffic which came into being in the course of the seventeenth century, alongside the carts, wagons and four-legged conveyances of older times, inevitably created new pressures for street improvements. The latter could no longer be motivated by simple aesthetic and hygienic considerations [. . .] In the seventeenth century, especially in the second half, street improvements became essential to the city's functioning and its citizens' wellbeing [. . .]

From the time of Louis XIV, important changes began to take place in street work.

The many-fold increase in the paving budget was one sign of changing times [. . .] Also noteworthy was the new concern for such matters as the optimum width of streets, an obvious consequence of the many street-construction projects executed in these years. Three classifications were established: the *grandes rues*, which were to be between 42 and 60 feet in width; the *rues de communication et distribution*, from 18 to 30 feet; and the *petites rues* from 6 to 18. By general agreement, 5 *toises* (about 30 feet) was established as the ideal width of the ordinary city street, and this became the size of most of the streets in the new quarters. In the older sections of the city the problem of street modernization was, of course, vastly more complex than in the new neighbourhoods to the west. In many parts of Paris little attention had been paid in the past to the proper alignment of houses along the streets, and some residences by being allowed to jut out into the thoroughfares had become formidable bottlenecks. A creditable beginning was now made in street modernization. The standard collection of royal ordinances dealing with projects for widening and straightening Parisian streets begins for all practical purposes with the reign of Louis XIV [. . .] [From] 1669 to the end of the century, at least fourteen ordinances (most of them dealing with a number of arteries) emanated from Versailles on this subject. Authorization for the construction of streets in new neighbourhoods was the subject of still other ordinances.

The Hôtel de Ville [centre of Paris' government] had the responsibility for carrying out these projects. The well-established rule was that the cost of new pavement fell on the adjoining property owners, but a much thornier question was who would indemnify them for land appropriated for street-widening projects. Delamere wrote piously that 'all private interests had to give way to the public good,'[9] but this, of course, did not negate the right of indemnification. The latter was always carefully provided for in each ordinance ordering a street improvement. [. . .]

Street lighting

[In 1667] the new municipal street-lighting system was inaugurated and Paris finally emerged from its age-old nocturnal gloom. No longer did pedestrians have to rely desperately on moonlight, their flickering hand-lanterns, or stray friendly reflections from the windows of shopkeepers and taverners. Under La Reynie's orders [La Reynie was Chief of Police], up to 6,500 lanterns were strung across the streets, a number which did not increase appreciably until well along in the next reign and probably even declined at times. Lister in his visit of 1698 was all praises for the new lights and thought the 'near 50,000*l*. Sterling' they cost to operate five months each year well worth the price[10]. He particularly admired the policy of keeping the lights burning throughout all the winter nights, contrasting this more open-handed practice with 'the impertinent usage of our People at London to take away the Lights for half of the Month, as though the Moon was certain to shine and light the Streets and that there could be no Cloudy Weather in Winter'[11].

La Reynie's lanterns consisted of simple squares of glass reinforced at the edges by iron frames. They utilized quarter-pound candles made of 'good and trusty tallow of Paris'[12] [. . .] The innovation of 1667 was by no means the first attempt to relieve the darkness of the streets. In the sixteenth century, and probably even earlier, efforts were made to persuade the citizens living on lower stories to keep candles at their window during the early evening hours. In 1588 [. . .] the Parlement of Paris ordered the fabrication of lanterns to be placed at the windows of householders at public expense. None of this worked. The lanterns were soon sold to pay for the cost of manufacture, and nothing more was attempted until Louis XIV's time.

In 1662 an Italian *abbé* of the distinguished Caraffa family obtained a privilege for the rental of lanterns and torches on the principal streets of Paris. The *abbé*'s plan was to establish stations spaced six hundred feet apart at which one could rent a lantern for 1 sou to use from one station to the next. The torches, more expensive but giving off a great deal more light than the lanterns, were pound-and-a-half sticks of tallow inscribed with the arms of Paris and marked off into ten divisions. One paid according to the divisions one consumed. Caraffa's plan was probably quite sound except that it came too late. The Paris of young Louis XIV and Colbert was ready for bolder schemes based on the principle that street lighting was the common responsibility of the bourgeois, just as street cleaning had long been recognized to be. The lighting ordinance of September 1667 provided that the relatively high cost of operating the system should be met by a new tax supplementing the old street-cleaning levy. The resulting *taxe des boues et lanternes* was destined to become very familiar for the balance of the Old Regime as the only direct levy on property owners.

While La Reynie's street lights were a marvelous novelty, the system by which they were operated was strictly traditional in its complexity and excessive decentralization of authority. The financing of the lanterns was placed in the hands of select committees of the bourgeoisie in each of the then seventeen quarters of the city; the purchase of all equipment was entrusted to the Lieutenant of Police; overall supervision of the operation was given to the latter's subordinates, the *commissaires*. The actual nightly lighting of the lanterns was the province of men quite independent of any of the foregoing [. . .]

One of the great difficulties in the operation, the authorities soon discovered, was the failure of residents to lower their lanterns at the moment the *allumeur* [unpaid official elected for the term of one year to light the streets in a given district] appeared in the street below them. La Reynie therefore issued a police regulation in 1671 which called for the ringing of a bell to signal the arrival of the lamplighter on the street. However, the problem was only solved when the lamp-lowering mechanism was enclosed in a locked box set at the street-level, thus obviating the need to rely on the residents. [. . .]

Lighting an area as large as that of Paris by candlelight would seem a difficult undertaking indeed, but – especially on the busier streets where supplementary illumination could be counted on from shops, carriages, and other sources – it seems to have been quite effective [. . .] An apparently reliable census showed 5,580 lanterns in 1715,[13] in close agreement with the figure of 5,522 shown on a map of Paris for 1715. If we accept an estimate of 5,500 and intervals of 60 feet, the result would have been the illumination, all in the early part of Louis XIV's personal reign, of some 65 miles of city streets – a very respectable civic accomplishment certainly.

Fire-fighting

Before the end of Louis XIV's reign, still another municipal service, professional fire-fighting, came into being. Until that time, techniques employed to keep fires in check had changed very little since the Middle Ages. Bucket brigades manned by residents had continued to be the principal weapon against conflagrations. As soon as fire was detected, the *commissaire* of the quarter was supposed to be notified and would take charge at the scene. If the ordinances had been obeyed, buckets, axes and ladders were to be found stored in the houses of the *commissaire*, *quartenier*, or *dizainier* [officers in charge of districts known respectively as *quartiers* and *dizaines*].

The main problem was finding a nearby water supply for the bucket brigade. For some obvious firetraps [. . .] owners provided substantial nearby water reservoirs to be used if the need arose. A private well on the scene was, of course, a great boon for fighting smaller fires, and La Reynie had established a 50-livre fine for householders neglecting to maintain their back-yard wells in good operating condition. Public wells were another obvious source of supply, but if an underground conduit ran close by the burning house, the *commissaire* would likely give the order to *dépaver* [take up the paving stones] and tap the pipe [. . .]

In the last quarter of the [seventeenth] century, hand-powered, wheeled water pumps began to make their appearance in Holland. Jan van der Heyden's device was perhaps the first to come into use there. A syringe-like contraption enabled him to squirt a jet of water to the top floors of city houses. It was probably his machine which furnished the model in 1699 for the *pompes à incendie* promoted by a versatile actor at the Comédie by the name of Dumourier Dupérier. They were successfully demonstrated at several fires, but because of the reluctance of the magistrates to invest in the new devices at a time when the city was on the verge of insolvency, a few years passed before they received official sanction.

A turning point in Dupérier's fortunes came in 1704 when a fire broke out in the Tuileries. Several notables happened to be present, including the Lieutenant of Police, the Superintendent of Buildings, Mansart, and the great Marshal Vauban. Someone sent out a call for 'Dupérier, *comédien*, with his pumps'. His distinguished audience saw him dart water 'wherever he wanted' and he was credited with extinguishing the fire [. . .] Dupérier was put in charge of the city's first fire department the following year, but his troubles were not over. In the financial disasters of the closing years of the reign, fire-fighting was sacrificed on the altar of economy and Dupérier disappeared temporarily from the scene, to be recalled in the last year of the reign when the pumps were revived and fire-fighting established on a permanent basis. In 1716 the actor-turned-fireman was finally recognized as Directeur Général of the city's fire pumps at a salary of 6,000 livres annually. He became responsible for maintaining sixteen pumps (thirty by 1722) stored in strategic locations, mainly convents, about the city. Each contraption was manned by a trained two-man crew, whose professional status was signified by *bonnets particuliers* by which they could be more easily recognized in the hubbub of a street fire. [. . .]

Notes

1 Alfred Franklin, *Dictionnaire historique des arts, métiers et professions exercés dans Paris depuis le treizième siècle*, 1 Vol. in two Pts, Paris, 1905-6, p. 127

2 Henri Sauval, *Histoire et recherches des antiquités de la ville de Paris*, 3 vols, Paris, 1724, 1: p. 24

3 F.M.A. de Voltaire, *The Age of Louis* XIV, Everyman's Library, London and New York, 1926, p. 326

4 Dr Martin Lister, *A Journey to Paris in the Year 1698*, London, 1698, p. 12

5 Germain Brice, *Description nouvelle de ce qu'il y a de plus remarquable dans la ville de Paris*, 2 Vols, Paris, 1698, 1: p. 12

6 Nicolas Delamare, *Traité de la police, ou l'on trouvera l'histoire de son établissement, les fonctions et les prérogatives de ses magistrats, toutes les lois et les règlements qui la concernent*, 2nd edn, 4 Vols, Paris, 1722-38, 4: p. 451

7 Sauval, *Histoire*, (see above note 2), 1: p. 173

8 Delamare, *Traité*, (see above note 6), 4: p. 464

9 Ibid., p. 11

10 Lister, *Journey*, (see above note 4), p. 24

11 Ibid., p. 23

12 Commandant Herlaut, 'L'éclairage des rues de Paris à la fin du XVIIe et au XVIIIe siècles,' *Mémoires de la Société de l'histoire de Paris et de l'Ile de France*, 43, 1916, p. 184

13 Ibid., p. 157

26

REBUILDING LONDON AFTER THE GREAT FIRE OF 1666

by John Evelyn

Source: John Evelyn, 'Londinium Redivivum', in Guy de la Bédoyère, (ed.), *The Writings of John Evelyn*, Woodbridge, Boydell Press, 1995, pp. 337-45

The pretences of the several Proprietors in the fonds [building sites] being first of all secured, and put into such a method, as by the wisdom of his Majesty and Parliament shall be found most conducible to the prevention of future suits and disturbances, interrupting the order of a new designation; the city of London might doubtless be rendered as far superior to any other city in the habitable world for beauty, commodiousness, and magnificence [. . .] as it has hitherto been somewhat inferior to many imperial cities in Europe, for want of improving those advantages, which God and Nature have dignified it withal above them.

In pursuance of this [. . .] I humbly conceive, that an exact plot [. . .] ought in the first place to be taken by some able Artist, and in that accurately to be described all the declivities, eminences, water courses &c. of the whole Area [. . .] After this I conceive there may be delineated some more particular iconographical plan of the whole city [. . .] with the principal streets; where the piazzas, churches, hospitals, courts of justice, halls, markets, quay, exchange, magazines &c. shall be placed. And this ought to be the joint and mature contrivance of the ablest men, Merchants, Architects, and Workmen, in consort; and such as have a true idea what proprieties, and conveniences, belong to so great a city, and which I therefore briefly, but fully, comprehend in these two transcendences, Use and Ornament.

The plan thus prepared, and resolved on, hands must be employed for the speedy removal of the rubbish [. . .] that so the inequalities, and several affections of the surface might be more apparent. But [. . .] this will be found a work of unimaginable difficulty, and require a multitude of hands; nor can it be affected to purpose, without infinite confusion [. . .] till those several plans and types of the future city be first concluded on [. . .] and then they make stake it out, and deliver it to the owners of the ground; provided they exactly conform to the plot, to the shape of the front, and to such other directions for uniformity and solidity, as his Majesty's Surveyors or Commissioners shall appoint. In this work it might haply be thought fit to fill up, or at least give a partial level to some of the deepest valleys, holes, and more sudden declivities, within the city, for the more ease of commerce, carriages, coaches, and people in the streets, and not a little for the more handsome ranging of the buildings [. . .] But here is to be considered the channel running thence through Holborn [the Fleet River], which should be so enlarged as not only to be preserved sweet (by scouring it through flood-gates into the Thames on all occasions) but commodious for the intercourse of considerable vessels [. . .] and which therefore should be accordingly wharfed on both sides to the very quay of the river, and made contiguous to the streets by bridges arched to a due level, as it might easily be contrived (and with passage sufficient for lusty barges and lighters under them) were the valley so elevated as 'tis projected [. . .]

These considerations and employments would greatly forward the prompt and natural disposure of the more useless and cumbersome rubbish, unless it might be thought more expedient [. . .] to design it rather towards the enlargement of a new and ample quay, which I wish might run parallel from the very Tower to the Temple at least, and if it were possible (without augmenting the rapidity of the stream) extend itself even as far as the very low water mark; the basin by this means kept perpetually full, without slub [mire], or annoyance, and to the infinite benefit and ease of access, like that of Constantinople, than which nothing could be imagined more noble. What fractions [fissures] and confusions our ugly stairs, bridges, and causeways, make, and how dirty, and nasty it is at every ebb, we are sufficiently sensible of; so as next to the hellish smoke of the town, there is nothing doubtless, which does more impair the health of its inhabitants.

In the disposure of the streets due consideration should be had, what are competent breadths for commerce and intercourse, cheerfulness and state; and therefore not to pass through the city all in one tenor without varieties, useful breakings, and enlargements into piazzas at competent distances, which ought to be built exactly uniform, strong, and with beautiful fronts. Nor should these be all of them square, but some of them oblong, circular, and oval figures, for their better grace and capacity. I would allow none of the principal streets less than an hundred foot in breadth, nor any of the narrowest than thirty, their openings, and heights proportionable [. . .]

In the piazzas should be kept the several markets, in others the coaches may wait &c. and in some should be public fountains placed; not as formerly immured with blind and melancholy walls, but left free to play, and show their crystal waters, as in most of the best cities of Europe they do, save this of ours, where an officer for a small stipend, might protect them from injury and pollution, till custom has civilised us. [. . .]

I should think the Royal Exchange might front the quay betwixt Queenhithe and the Bridge, about the Steel-yard I conceive were a proper place, respecting the goodliest river in the world, where the traffic, and business is most vigorous [. . .] If it should be erected near the Thames, let there be spacious piazzas about it, either for dwellings, or public warehouses; which yet I should rather advise might be contrived in the vaults under those edifices [. . .] And for such other stores as will not be well preserved under ground, there would be by any means some expedient to be found out, that they might not front the Thames on London side, at least very sparingly; not only for that they may yet become obnoxious to the like accidents (being built contiguous to the rest) but because, they, if there be not ample separations and distances (which would infinitely disfigure and interrupt the face of that quay) they will no where stand commodiously. How greatly therefore were it to be wished, that such a depth of those wretched houses on the opposite side of the water were purchased, and demolished, to make room for those stores? [. . .] Or if needs the warehouses must be on this side, yet that they were made rather to front Thames street than the river, because of the dull and heavy aspect of those kind of erections. [. . .]

For the rest of the necessary evils, the Brewhouses, Bakehouses, Dyers, Salt, Soap, and Sugar-boilers [. . .] I hope his Majesty will now dispose of to some other parts about the river, towards Bow and Wandsworth on the water; Islington, and about Spitalfields &c.

The charge of bringing all their commodities into the city would be very inconsiderable, opposed to the peril of their being continued amongst the inhabitants; and the benefit of the carriage, which would employ a world of people both by land and water, without the least prejudice. [. . .]

One of the last, not least Considerations, will be that of Paving, for which we have a laudable example in those streets of my Lord Treasurer, and Hatton Gardens, which may be imitated. And why may not some of the distorted bricks, to be found amongst the rubbish, be reserved for these purposes; especially the elevations destined for the foot causeys [raised pavements] before the fronts of the houses? Unless they will be at the charge to lay it with Purbeck and flat stones, which indeed were to be preferred. Yet their clinkers in Holland does very well; and, as I remember, the Roman streets are so paved.

I have now no more to add, for the ease, and preservation of the streets, than to wish, that the use of sleds were introduced, and as few heavy carts as might be countenanced. And that for the universal benefit (especially for those who are not born to ride in coaches) that intolerable nuisance of spouts and gutters might be strictly reformed, and the waters so conveyed by close and perpendicular pipes (where they cannot be avoided) or to drop only from above the Modillions [projecting brackets forming part of a cornice], as from Italian roofs. That no pipes for conveyance of waters for domestic uses be derived from the Heads through church-yards, or like unclean places, without being well immured in plaster of Paris. That plain tile may be only employed instead of pan tile. That no Bay windows and uncomely jettings, nor even Balconies (unless made of iron) be for the future permitted. And that for the better expediting of this great design [. . .] store of all materials may be provided betimes (bricks and tiles especially) because all seasons are not fit for it; and that there be a diligent inspection to examine their goodness: but the greatest and almost only desiderate [requirement] will be that of Timber, which peace and industry will furnish. And when all these were prepared, and the Undertakers too as ready, if they be permitted to gratify their own fancies, without religiously intending to pursue the Plan; and that his Majesty (who is best able to judge of it) overrule in this; it may possibly become a new indeed, but a very ugly city, when all is done. Whereas, if they permit themselves to be governed in this, we are not yet to despair of seeing (after a few years) such a city to emerge out of these sad and ruinous heaps, as may dispute it with all the cities of the World; fitter for commerce, apter for government, sweeter for health, more glorious for beauty; and in sum for whatsoever indeed could be desired to render it consummately perfect.

27

TECHNOLOGICAL CHANGE IN A TRADITIONAL SOCIETY: THE CASE OF THE DESAGÜE IN COLONIAL MEXICO

by Louisa Schell Hoberman

Source: Louisa Schell Hoberman, 'Technological Change in a Traditional Society: the case of the *desagüe* in colonial Mexico,' *Technology and Culture*, Vol. 21, University of Chicago Press, 1980, pp. 386–95, 400–6

The drainage canal-tunnel was constructed in response to an environmental problem, the increasingly severe flooding of Mexico City. The city was the capital of the Viceroyalty of New Spain, one of the two major administrative subdivisions of the American empire, and the most important urban center in the New World. Yet it was precariously situated on an island and landfill in the lowest of five shallow lakes in the Valley of Mexico. Three major causeways connected the city to the lakeshore; over them passed the supplies, workers, and business of the capital. Surrounded on all sides by mountains, the lakes of the valley lacked an outlet, and during the rainy season they overflowed onto the city and nearby land. As a result, Mexico City was badly flooded in 1555, 1580, 1604, and 1607. Supplies from the mainland were cut off. People were confined to the upper stories of their houses, if their houses had not been washed away. As flood conditions persisted, some inhabitants died of disease or were forced to migrate (usually temporarily); those who were able to remain suffered from profiteering, property damage, and a slowdown of normal public life.

Flooding had occurred during pre-Colombian times also, as Aztec occupation of the area and the intensive agriculture practiced by that society eroded the valley soil and silted the lakebeds. Both of these processes, however, were accelerated when the Spaniards arrived in the valley in the 1520s. They introduced iron plow agriculture and livestock and stripped the area of trees. In trying to create a familiar setting for themselves by converting the principal canals into streets and building heavy stone houses, the Spaniards unwittingly further disturbed the land–water relationships of the valley. The immediate consequence was flooding; the long-term consequence was the opposite: the drying up of the valley terrain. Only the visible consequences were perceived at the time, however, and all efforts were directed toward preventing floods.

To these environmental causes must be added the political and economic, if we are to grasp the reasons for the building of this immense public work. When the dangers of endemic flooding were apparent one response would have been to move the city out of harm's way by changing the site. In fact, the original foundation of Mexico City on the ruins of the Aztec capital of Tenochtitlan had been a controversial decision. Many of the first settlers had opposed it, alleging, among other drawbacks, the site's vulnerability to flood. Lesser cities had been moved in colonial Latin America. But this response was rejected on several occasions because of the pivotal political and economic role of the city and the fear that moving its site would diminish that role. [. . .]

The pre-eminence of Mexico City was due to the combined power of the state agencies and the economic interests located there. Only the combination produced the demand for a new technology, since neither the market nor the merchant class exerted sufficient influence on its own. The canal-tunnel was built to continue the dominant position of the capital, to perpetuate the control of city groups over the bureaucracy, trade, manufacturing, and cultural institutions by making the site of the city secure. A new technology, therefore, was introduced to maintain the status quo.

The magnitude of the goal of preventing floods, given the environmental obstacles to it, was matched by the magnitude of the effort expended to attain it. The *desagüe* was a monumental structure. Starting from the northernmost of the valley lakes, Lake Zumpango, the *desagüe* stretched a total of 8.02 miles over terrain rising 110 feet. It extended 3.89 miles from Lake Zumpango to the town of Huehuetoca at the northwest corner of the valley. From there it became a tunnel, continuing 3.84 miles to a spot called the Boca de San Gregorio. An exit canal of 0.39 miles conducted the water from that point to the gulch of Nochistongo, then to the Tula River, and finally, out to the Gulf of Mexico. The tunnel was from 5.5 to 8.25 feet wide and about 11 feet high throughout. At its deepest point it lay 149 feet below the surface of the earth. In the 1680s, it was deepened to 198 feet. Another way of appreciating the monumental quality of this work is to consider the resources drawn on to build it. At least 4,700 Indian workers, 15 per cent of the adult male population of the valley, were drafted to dig and haul out earth and water; a minority of these performed the skilled tasks of assembling hoists and pumps

and roofing the tunnel and shafts with wooden beams. The total cost of the *desagüe* to 1789, the year of its completion during the colonial period, was 5,399,869 pesos; the cathedral, begun in 1536 and finished in 1813, cost, at the most, 3,191,313 pesos.

The chief engineer, Enrico Martinez, regarded the *desagüe* as a spectacular achievement, which 'many had believed impossible.'[1] [. . .]

As a canal, the *desagüe* was not an unusual public work in either European or pre-Colombian contexts. The 16th and 17th centuries in Europe were a great era of canal building. Numerous navigational and drainage canals were opened up at this time to meet the demand for cheap bulk transport and cultivatable land. Likewise, the drainage of the Pontine marshes in Italy, the Great Level of the Fens in England, the inland lakes of the Netherlands, and marshes throughout France was undertaken on a scale unprecedented since antiquity. In Spain, irrigation canals from the Ebro and the Tajo were begun and the Tibi reservoir in Alicante was built. The Aztecs had themselves excavated lengthy canals both for irrigation and transport; those near the capital also served to prevent floods.

What was new in both contexts was the nearly 4-mile-long tunnel. The regions drained in 16th- and 17th-century Europe were by and large coastal. There were no topographical obstacles to the expulsion of water drained and no need for tunnels. It is necessary to go back to the Roman Empire to find a comparable structure, the $3\frac{1}{2}$-mile tunnel that Claudius built (AD 41–51) to drain Lake Fucinus into the river Liris. The tunnel of the famous Languedoc Canal, the first canal-tunnel built in modern Europe, was only 0.1 mile long. Nor had the Aztecs built drainage tunnels. They did not practice deep shaft mining, and their flood control system was based on dikes and runoff canals within the valley. These structures were intended to keep as much water out of the city as possible until the water evaporated rather than to drain it away entirely. Some of these flood works were very impressive. The dike of Nezahualcoyotl, built in 1449 and protecting the capital's eastern side, was 10 miles long. There is, however, no evidence of a drainage tunnel. Only after the arrival of the Spaniards were such tunnels proposed, and only in the 17th century was one finally built. The *desagüe*, therefore, represented a new and more effective conception of how to solve the problem of floods.

It also entailed the application of techniques devised for other engineering works to a different structure. These techniques were the excavation of mine tunnels, the laying of foundations in water, and the vaulting of buildings and aqueducts. Although none of the techniques employed to solve these problems was new in itself, when combined and applied to the *desagüe*, they constituted a departure from their previous use. The *desagüe* is a good example of a 'new design with familiar components.'[2] Unfortunately, the success of the new design was short-lived. The history of this public work must be divided into a period of planning and successful execution, 1555–1608, and a period of deterioration, which began in 1609 and lasted throughout the colonial period. During this second period the canal-tunnel functioned but in a sharply reduced capacity, and the problems posed by its maintenance were not systematically solved. The *desagüe* was not restored to dependable working order until 1900 [. . .]

The two major engineering problems of the *desagüe* were the tunnel's tendency to become clogged and the smallness of both tunnel and canal relative to the volume of water that they had to drain. The tunnel became clogged up with earth partly because of the kind of soil it was cut through. About three-fourths of the tunnel and all of the canal were excavated through *marga*, a soft, loamy soil, which was easy to dig out but hard to pack down. The

passage of water through the tunnel and the alternation between wet and dry weather in the valley made this soil crumble easily. This situation was made worse, however, by the head engineer's failure to appreciate it. Underestimating the problem of the soft soil, he did not systematically reinforce the walls as the *desagüe* was being opened. Once it was opened, the reinforcement was much harder to accomplish, for political reasons. Although parts of the tunnel were gradually fortified with well-made masonry vaulting, it was never adequate. The tunnel walls were also vulnerable because Martinez gave the floor too great a gradient. The pressure of fast-moving water, which contained debris from the outside, put additional stress on the walls.

The smallness of the tunnel even before the clogging began was the second impediment. There was no way to calculate accurately the volume of water that had to be drained; no records of annual rainfall were kept. Also, to make the tunnel deep enough, it had to be excavated down to the level of subsoil springs. Animal-powered drainage pumps existed and had been used to good effect in the initial opening. Their clumsiness, cost, and slowness, however, led to the tendency to make the tunnel as shallow as possible. Thus, although the equipment existed to drain the *desagüe*, the tunnel was built too high, and although the techniques existed to wall and anchor it, it became blocked. [. . .]

Not surprisingly, paying for the *desagüe* was disagreeable to all members of colonial society. The Creoles, whether rich or less well-to-do, did not want to pay additional taxes or to contribute their black slaves or Indian peons to work on it. The nonwhites, for their part, were not eager to contribute their labor to a project that, relative to other kinds of employment, paid little and had a reputation for high mortality. The Creoles' opposition was more effective than the nonwhites', but both slowed and sometimes halted work on the *desagüe*.

The three alternatives

The best way to appreciate the obstacles to the successful upkeep of the *desagüe* is to analyze the response of the government to the different plans put forward from 1609 to 1637. These permit us to see how technological progress not only does not necessarily lead to social change but can be undermined by social conservatism.

The first heavy rains since the *desagüe* was built, which occurred in the summer of 1609, revealed its technical weaknesses. The long debate about what should be done and how to pay for it began. What was needed was to continue the work done already in 1607–8 in a more thorough fashion: enlarge the tunnel, especially its depth; fortify its walls and roof with rubble masonry; and set up filters to block the entry of destructive debris. After the huge effort expended in 1607–8, however, it was difficult to find the men and money to execute these measures. Even if the resources had been readily available, the work would have taken a few years to complete. Yet, from the beginning, the correctness of the above strategy was disputed by engineers who opposed Martinez. They proposed three alternatives to the *desagüe*; return to a system of dikes and runoff canals; abandon the tunnel and make the *desagüe* a canal throughout its entire length; exploit the natural tunnel believed to exist in Lake Texcoco. Alonso Arias, a Creole engineer in charge of dike repairs before the *desagüe* was built, and some of the other maestros favored forgetting the *desagüe* entirely. The most persistent advocate of the dike strategy was, however, Adrian Boot. Boot was a Dutch engineer who had worked in France and the Netherlands before being recruited by the Spanish ambassador to come to New Spain in 1614. He was supposed to work with Martinez on improving the tunnel. After inspecting it, however, he dismissed it as impossible and presented his own plan, the protective circle.

The dikes to the east and south of the city were to be fortified, and on the west a 6-mile semi-circular canal, receiving the river water from that quarter, was to be built. The water level within Mexico City was to be regulated by sluices in the dikes and by windmill-driven drainage pumps.

Boot has often been treated as a ridiculous figure in the history of the drainage of the valley. In fact, he was trying to introduce the most advanced European techniques and machines. His protective circle was an adaption of a method used with great success to 'lay dry' the inland lakes of the Netherlands in the 16th and 17th centuries. Unlike the inland dikes of the Netherlands, which surrounded empty swamps, the Mexican system would have to give access to the canoes that brought in the city's food supply. Realizing this, Boot recommended a type of crane to lift the canoes over the dikes when flood conditions made it impossible to open the sluice gates. He also specified the use of scoop dredges. These were probably hand-operated, although the very efficient Amsterdam mud mill, driven by a man- or horse-powered treadmill, had come into use by the late 16th century. Because of the sedimentation that occurs at the foot of a dike, causing the water to rise higher and higher as time goes on, dredges were essential for an effective diking system. This was not understood by the Creole engineers, who concentrated instead on building higher and wider dikes. Nether the sensible scoop dredges nor the more far-fetched canoe-carrying cranes were received with enthusiasm by Boot's colleagues in New Spain. [. . .]

[. . .] Boot's approach was inappropriate in that he did not indicate what would be done with the water that his protective circle would keep away from the city. He suggested only that it be emptied into the far side of Lake Texcoco, which would have put even more pressure on the city's eastern dike. What he should have done was combine his plan for regulating water levels in the city with a proposal for channeling the excess out of the valley. But Boot never really grasped this possibility. Comparing Mexico to 'the cities of our land which are tormented by the waters of the Ocean Sea,'[3] he allowed his European training to get the better of his planning. In the Low Countries the water could never be got rid of completely because the country was below sea level. Dikes were the only answer. With Mexico City located at 7,800 feet above sea level, a gravity drainage canal was a much better solution than dikes. Another feature of New Spain's topography also escaped Boot's attention. The rivers of the valley, descending from the encircling mountains, were fast-flowing, seasonal rivers, much more difficult to dike and channel than the sluggish canal waters of the Netherlands. [. . .]

A longer acquaintance with valley topography than that possessed by Boot was the claim of the engineers who favored the second alternative, the conversion of the tunnel to an open canal. These engineers insisted that the soft soil made the tunnel impossible. An open canal would reduce erosion, since, without a roof, less would fall in and what did could be more easily removed. [Andrés de] San Miguel did not understand that erosion would still be a problem unless the sides of even an open canal had a gentle slope. The *tajo abierto* designed by San Miguel had a slope of 83 degrees; that given the canal in the 18th century, which was still too steep, was 48 degrees. San Miguel also underestimated the time and effort it would take to open the tunnel. Much of it would have to be opened from underneath, 'digging like rabbits'[4] since if the work were started from the top, the whole tunnel might cave in. There was also the difficulty of excavating additional large quantities of earth. Finally, unlike Boot, San Miguel did not appreciate that the retention of a low level of water in Lake Texcoco itself was necessary to keep the subsoil under the city from receding. San Miguel wanted to drain the lake dry,

thinking that this would stop the city from sinking, whereas the opposite was true.

The third alternative, the unstopping of a natural tunnel in Lake Texcoco or in another of the valley lakes, was not a very serious contender for government support, but it was proposed several times. If pumped clean, it was claimed, this natural tunnel would suck the excess water out of the lake and, most important, save the city the expense of elaborate dikes and tunnels. No such natural drain has been found in these lakes [. . .]

Between 1609 and 1629 the choice was seen as that between repairing the *desagüe* and extending it to Lake Texcoco, and adopting the protective circle. The government investigated these options on ten occasions [. . .]

A major reason for this indecisiveness was the disagreement among the engineers themselves [. . .]

The city council, representing most directly the taxpayers and employers of Indian labor and expressing its usual preference for funding public celebrations rather than public works, became more critical of the *desagüe*. [. . .] The viceroy, high court judges, and treasury officials were more steadfast backers of the *desagüe*, but even their support was not sufficient in its scope.

Between 1630 and 1637 the choice of a flood control program was seen as that between the repair of the *desagüe* tunnel and the conversion to an open trench. A devastating flood, which began in the fall of 1629 and lasted through 1634, forced a serious review of all possible programs, and among them the *tajo abierto* or open trench. [. . .]

The plan of opening the tunnel suffered from the same problems that had previously beset its repair. From a technical standpoint, it was difficult to execute. Its proponents had no new ideas about how to pay for it. The posts created to administer it were filled by new men, but the administrative procedures remained the same. The defects of the political system had been overcome to the extent of making a commitment to a partial change in policy but not to the point of being able to implement it with a reasonable degree of speed. Begun in 1637, the conversion of the tunnel was not complete until 1789. [. . .]

When the tunnel was finally converted to a canal in 1789, it was under the auspices of the merchant guild, not of the government. However, the canal was still too small and too prone to landslides to secure the safety of the city, which was again flooded in 1792 and 1795. A fairly effective drainage system had to wait until 1900, when a tunnel at the northeastern corner of the valley was added to the open trench at Huehuetoca to form the network that is still the city's chief protection against floods. Now it is recognized, however, that this

protection is deceptive, since the drainage of water away from the city and, more recently, the drilling of deep wells to pump up water for urban use, have caused an alarming increase in subsidence of the soil. [. . .]

Notes

1 *Memoria histórica, técnica y administrative de las obras del desagüe del Valle de México, 1449-1900*, 2 Vols, Mexico City, 1902, 2, p. 10

2 Edwin Layton, Jr., 'Technology as Knowledge,' *Technology and Culture*, 15, January 1974, p. 38

3 Fernando Cepeda and Fernando Alfonso Carrillo, *Relación universal legítima y verdadera del sitio en que está fundada la muy noble . . . ciudad de México . . .*, 2 Vols, Mexico City, 1637, 1, pt 2, p. 5

4 Juan Gemelli Carreri, *Viaje a la Nueva España*, 2 Vols in 1, Mexico City, 1927, 1, p. 126

Part 3
PRE-INDUSTRIAL CITIES IN CHINA AND AFRICA

The final reading of the previous section opened up vistas of the global dissemination of European cities and technologies, for good or ill. It reinforced the emphasis thus far in this collection on urban and technological traditions that can be seen as running continuously from the ancient Near East urban heartland through various empires, including Islam, into modern Europe. The readings in this section seek to counterbalance that preoccupation, and to explore other ways of building cities, which either developed quite independently, or retained a strong indigenous character.

Readings 28 to 34 are about aspects of cities and technology in imperial China, before the start of the colonial period in 1840. The first extract from the Book of Odes gives some indication of the philosophical framework within which Chinese building and city planning was conducted. China has a canon of classical writings known as the 'Four Books' and the 'Five Classics', brought together in their present form during the time of the Han Dynasty (206 BCE–220 CE). The 'Four Books' contain works by Confucius (the Latinized name of Kong Fuzi, 551–479 BCE), mostly about ethics, manners and the politics of his time. The 'Five Classics' are Confucian in tone, though some date from five centuries before the birth of Confucius. One of the classics, entitled *Li Ji*, the *Book of Rites*, gives advice on house design and the proper layout of a capital city; another, *Yi Jing* (former spelling *I Ching*), the *Book of Changes*, includes the theory of *yin* and *yang*, and other topics discussed by practitioners of *feng shui* or geomancy. Much the most accessible in English translation is *Shi Jing*, the *Book of Odes* (or *Book of Songs*), the contents of which date from the eleventh to the seventh centuries BCE. The passage selected from this classic refers to the way in which the houses in a new capital city are oriented north–south; there are also references to rammed earth construction, timber-framed buildings, and layers of floor matting. Although there is no textbook on architecture or planning in the classics, references like these have given ancient Confucian authority for the work of succeeding ages. The ideological context of Chinese architecture and planning is also the subject of reading 29, by Nelson Wu. He shows how the walls of a square-city plan have many layers of meaning with regard to government, privacy and human ideas of order and control. He also shows how gateways in walls (including city walls) have the same significance as the facades of western buildings.

Readings 30 and 31 deal more directly with technologies linked to urbanization and city-building. The first is from the multi-volume *Science and Civilization in China*, produced under the direction of the Cambridge biologist Joseph Needham. This selection has been edited to avoid some of the complications arising from Needham's tendency to pursue connections between one topic and another across several centuries. The reading, written by Needham and his long-time Chinese collaborator Wang Ling, deals with pipes and water supplies, but includes fountains and water-clocks, many of which could not have functioned without some kind of pipework. The technologies described challenge the common assumption that although China produced practical inventions like the wheelbarrow, the ruling elite and the most imaginative minds took little interest in technical matters. The author of reading 31, Sung Ying-Hsing (Song Yingxing in modern spelling), was born in 1587 and died about 1665. He held relatively low-status posts in provincial government, and eventually in 1636 became Director of Education for Fenyi county in Jiangxi province, a region just south of the Yangzi River with a prosperous agricultural economy as well as coal mines and a copper industry. The post gave him access to a good library and time for writing; his book on technology was intended as a contribution to better informed, more practical government; it was not a manual for engineers, but for civil servants.

The next three readings focus on Hankou, a commercial city close to the geographical centre of China on the Yangzi River. Now part of the major industrial conurbation of Wuhan, Hankou began to grow rapidly during the eighteenth century, and its population probably passed one million sometime between 1800 and 1850. Readings 32 and 33 are taken from William Rowe's comprehensive two-volume social history of the city in the eighteenth and nineteenth centuries. The selection from the first volume provides some of the context of the extensive building stimulated by Hankou's growth, and in particular the role of commercial guilds as agents of city-building. Among the developments they promoted were public and commercial buildings, roads, a pier and a public ferry. A fire-fighting service is also mentioned, and this urban technology, along with the sponsorship of the guilds, is considered in more depth in the following excerpt, from Rowe's second volume. It is at this point that technology transfer from the West, in the form of European and then US fire engines, intrudes into this section. Readers attentive to the previous section will note that the 'pathetic' system of water buckets on which Hankou had relied in the mid-eighteenth century was the principal means of fighting fires in European cities at the beginning of the seventeenth century. Comparisons are made between nineteenth-century Paris and Hankou in the section on public utilities, which deals with street lighting, water supply, sanitation and public transport. The adequacy of these services was what might be expected without a culture of government involvement; an exception was the provision of official ferries from the late 1870s, linking Hankou with Wuchang and Hanyang, the other cities of Wuhan. Reading 34 is taken from an eye-witness account of the bustling commercial activity of nineteenth-century Hankou, written by a French missionary Évariste Régis Huc, who was one of the first Europeans to visit Hankou, and one of the last to write about the city as it had been in the first half of the nineteenth century, prior to its destruction during the Taiping Rebellion. The reading is also a useful source on the construction of China's great system of canals.

The final two readings in this section are taken from J.C. Moughtin's study of the architecture of the mud-built cities of Hausaland, in what is now the north of Nigeria. These sub-Saharan settlements did not spring from a wholly independent building tradition. They were heavily influenced by Islamic culture, because of their involvement in trade across the desert with the cities of north Africa; Moughtin even speculates about a shared ancestry with the trabeated (post-and-lintel) structures of Pharaonic Egypt. The result was a building tradition in which indigenous sub-Saharan African and external north African Islamic cultures interplayed. Earth, the essential material, was not the optimal choice, given the season of torrential rain that is part of Hausaland's climate. A necessary adjunct to this earth-based architecture has therefore been the ingenious development of a range of waterproof finishes, and frequent maintenance by the occupier, resulting in a dynamic, constantly changing built environment.

28

BUILDING IN THE BOOK OF ODES

Source: *The Book of Songs*, trans. A. Waley, London, George Allen and Unwin, 1937, pp. 231–4. The introductory paragraph by Arthur Waley and text use the Old Wade-Giles spelling convention for Chinese names. Waley translated the title as *The Book of Songs*, although *Book of Odes* is more usual

IN 658 BC the people of Wei, continually harassed by the Ti tribes, were forced to abandon their capital north of the Yellow River, in northern Honan, and transfer it to the southern enclave of Hopei that runs in a narrow strip between Shantung and Honan. In their move they were assisted and protected by Duke Huan of Ch'i, who sent a gift of three hundred horses,[1] presumably because most of the Wei people's horses had been captured by the Ti. The following song describes the building of the new capital. We do not know its exact site, nor what is meant by 'T'ang' and the 'Ching hills.'

256   THE Ting-star[2] is in the middle of the sky;
         We begin to build the palace at Ch'u.
         Orientating them by the rays of the sun
         We set to work on the houses at Ch'u,
         By the side of them planting hazels and
             chestnut-trees,
         Catalpas, Pawlownias, lacquer-trees
         That we may make the zitherns great and
             small.

         We climb to that wilderness
         To look down at Ch'u,
         To look upon Ch'u and T'ang,
         Upon the Ching hills and citadel.
         We go down and inspect the mulberry
             orchards,
         We take the omens and they are lucky,
         All of them truly good.

         A magical rain is falling.
         We order our grooms
         By starlight, early, to yoke our steeds;
         We drive to the mulberry-fields and there we
             rest.

Those are men indeed!
They hold hearts that are staunch and true.
They have given us mares three thousand.

257   CEASELESS flows that beck,
         Far stretch the southern hills.
         May you be sturdy as the bamboo,
         May you flourish like the pine,
         May elder brother and younger brother
         Always love one another,
         Never do evil to one another.

         To give continuance to foremothers and
             forefathers
         We build a house, many hundred cubits of
             wall;
         To south and west its doors.
         Here shall we live, here rest,
         Here laugh, here talk.

         We bind the frames, creak, creak;
         We hammer the mud, tap, tap,
         That it may be a place where wind and rain
             cannot enter,
         Nor birds and rats get in,
         But where our lord may dwell.

         As a halberd, even so plumed,
         As an arrow, even so sharp,
         As a bird, even so soaring,
         As wings, even so flying
         Are the halls to which our lord ascends.[3]

         Well levelled is the courtyard,
         Firm are the pillars,
         Cheerful are the rooms by day,
         Softly gloaming by night,
         A place where our lord can be at peace.

         Below, the rush-mats; over them the
             bamboo-mats.
         Comfortably he sleeps,
         He sleeps and wakes

And interprets his dreams.
'Your lucky dreams, what were they?'
'They were of black bears and brown,
Of serpents and snakes.'

The diviner thus interprets it:
'Black bears and brown
Mean men-children.
Snakes and serpents
Mean girl-children.'

So he bears a son,
And puts him to sleep upon a bed,
Clothes him in robes,
Gives him a jade sceptre to play with.
The child's howling is very lusty;[4]
In red greaves shall he flare,
Be lord and king of house and home.

Then he bears a daughter,
And puts her upon the ground,

Clothes her in swaddling-clothes,
Gives her a loom-whorl to play with.
For her no decorations, no emblems;
Her only care, the wine and food,
And how to give no trouble to father and
    mother.

Notes

1 *Kuo Yü* (Ch'i Yü). Multiplied in this song to three thousand
2 Part of Pegasus; also called the Building Star
3 This verse is corrupt and not intelligible with any certainty
4 *Huang*, 'lusty,' suggests the *huang*, 'flare' of the red greaves. These could only be worn by the king's command and constituted a decoration similar to our Garter. Women (see the next verse) received no such marks of distinction

29

MEANINGS OF WALLS AND GATES

by Nelson I. Wu

Source: Nelson I. Wu, *Chinese and Indian Architecture*, New York and London, Prentice-Hall International and George Braziller, 1963, pp. 8-9, 11-12, 29-32, 34-5, 37-8, 42-3. The text uses traditional spelling conventions, and thus Beijing appears as Peking and Chang'an as Ch'ang-an

Colonized by Indian and Chinese architecture, Asia is divided into a Chinese world of walled cities and an Indian world of holy places. From Java to Japan the landscape is shaped by the Chinese ideal of regulated harmony in society and by the Indian concern for eternity.[1] [. . .]

Like two poignant air currents, countering and blending with each other in a stormy atmosphere, the Chinese and Indian traditions affect a large surface of the continent and nearby islands, leaving very few pockets of vacuum between. Today, both traditions are going through fundamental transformations that are probably as significant as any in their entire histories. If architecture is to continue to provide a valid setting for, and to participate in, the new cultural programs, new forms must come forth which will echo the difficult adjustments now being made deep within the very fibers of these cultures - the religious life of India and the family life and social organization of China. The superficial architectural continuity, on the other hand, as seen in such details as the upturned eave lines and stupa [Buddhist monument] motifs on numerous contemporary buildings in China and India, is no solution and brings no rebirth.

Before the next turn of events obliterates the message in these ancient architectures, we should search for the essential meaning behind the true achievements of these two glorious traditions. Different as they are from each other, the traditional Chinese and Indian architectures both refer to programs deeply rooted in everyday life, and are responsible to a collective awareness and need. Both traditions exhibit a strong desire to relate a cosmic ideal with man's own image and role within it. Only in exceptional cases, such as Chinese garden design, was the architecture ever an expression of an individual's artistic whim.

Several additional factors make it profitable to discuss these two traditions in relation to each other. For instance, both India and China, in addition to their architectural remains, have preserved quantities of literature dealing not so much with architecture itself, as with rituals, symbolism, and life in general as it involves architecture. In these writings the popular practices since primeval times and folk beliefs from various components of the nation are no doubt idealized and codified. After tumultuous eras of cultural change, periods of chaos, reunification, or foreign domination, the surviving cultural elements would close ranks and the tradition would take inventory. With all the best intentions, writers in periods of new stability incline to develop a classic-complex - a tendency to glorify old virtues, imaginatively re-creating and streamlining details to enrich the meaning and the form of the old architecture. Han (206 BC–220 AD) and T'ang

(618–906) dynasties of China, as well as Maurya (*c.* 322–185 BC) and particularly Gupta (*c.* 320–600 AD) periods of India, are such classical eras. [. . .]

The design of the Han dynasty tile (Figure 29.1) was not a sudden discovery. Early awareness of the rectangular shape and its orientation was clearly seen in late Shang dynasty tombs (*c.* twelfth-eleventh centuries BC). In the Han image, the world of man is a clearing marked off from the unknown on all four sides by symbols in animal form[2]. Reading these signs in a clockwise manner and oriented to the south, there is first the Blue Dragon of the East, which stands for the blue-green color of vegetation and represents the 'element' of wood and the upreaching tree. Occupying the direction of the rising sun, it is also the symbol of spring. To the south is the Red Phoenix of summer and of fire at the zenith. Next there is the west and the White Tiger of the metallic autumn, symbolic of weapons, war, executions, and harvest; of fruitful conclusion and the calmness of twilight, of mem-

Figure 29.1 'Ssu-shen (Four Deities) Tile,' ink-rubbing, c.200 BC.

ory and regret, and unalterable past mistakes. It is the end of the road, but not the end of the cycle, for the new beginning will have to come from the all-inclusive darkness of the winter. Its position is the cold region of the north; its color, black, and its element, water. There time is immeasurable and elastic. Pictured here is Hsüan-wu, a snake coiling around a turtle, two hibernating reptiles forming a picture behind man's back of life preserved underground.[3]

Facing south, his feet firmly on the fifth element, the earth, is man. Via a negative approach – not knowing how high is up, how deep is down, and how far away is the end of the world in each direction – man fixes his position as equidistant from the end of the universe on all sides, and places himself squarely in the middle. [. . .]

The rectangular Han dynasty tile [. . .] is a rigid, finite, and unnatural design. Wherever circumstances permit, this image is readily translated into the equally unnatural classical city plan of China, sometimes using the same animal symbols to name the gates at corresponding cardinal points. It manifests an intellectual order superimposed upon a natural terrain. The T'ang dynasty capital of Ch'ang-an and its Japanese copies, Nara and Kyoto, are such expressions. This tradition, developed in the north, eventually influenced such southwestern areas as Szechwan and Yunan. Their capitals, Ch'eng-tu and Kunming, interesting variations of this basic design, make a dramatic contrast to a southeastern, naturally shaped river-town like T'ung-lu[4]. [. . .] This rationalized basic design is not only frequently seen in city plans but is also sensed in the layout of houses, palaces, and tombs. Always keeping man in its center, it is an image of man's society, organizing its enclosed space around him. The Chinese designer is continually challenged and inspired by the specific requirements of each social program and by the human relationships in the society which his building serves and portrays.

The engineering problems of construction or the discovery of new materials are relatively unimportant here in the generation of new ideas [. . .] The drama of Chinese architecture's struggle to meet the requirement of its program is one of ingenious exploration of the potentialities of the wall, the height of the platform, the placing of individual buildings, the organization of a compound. It concerns itself above all with the prescribed *position* and *movement* of man in an architectural complex. [. . .] The lasting values in the tradition beginning with the formative period of Han (at the latest) perhaps should be sought in the enduring design that has shaped the humble family house as well as the Chinese city of man. The house is the basic cell in the organism of Chinese architecture, just as the family it houses is the microcosm of the monolithic Chinese society.

From ceramic models found in tombs, and murals of tomb chambers and temples, we have no lack of examples showing the types of houses built from Han to T'ang times. Literary documents supply information about the dwellings of much earlier date. Certain basic techniques of composing a house seem to have survived extremely well and are still found in modern buildings. In one type, a few rooms are connected by walls to form a yard ('t'ing' or 'yüan') utilizing the back wall of the rooms as the exterior wall of the compound. The other type, also well developed and seen frequently in T'ang dynasty murals, has freestanding buildings set within the yard requiring a continuous wall to enclose them. Both clearly demonstrate the significance of the courtyard, which as a negative space plays a prominent role in forming the 'house-yard' complex. The student of Chinese architecture will miss the point if he does not focus his attention on the space and the impalpable relationships between members of this complex, but, rather, fixes his eyes on the solids of the building alone.

In [. . .] houses both in the city and the village, we [. . .] identify quickly the courtyard and certain other aspects of the Chinese house that deserve our special attention (Figure 29.2). [. . .] The privacy here is a partial one; horizontally the yard is separated from the street by the wall or by the surrounding buildings, but it shares both the sky and the elements of the weather with other houses and yards.

As the wall blocks only the view, thus creating a visual privacy, it offers a particularly refreshing experience of communication with the outside through the senses of hearing and smell. Chinese literature abounds with examples describing city life through the sounds of the peddlers on the streets and the scent of the flowering trees coming over such a wall. The implicit paradox of a rigid boundary versus an open sky reminds us of the similar situation in the land use of this ancient agricultural country: while the boundaries of farm land are guarded by everyone, the 'Will of Heaven' is a fate shared by all. [. . .]

Theoretically, the number of courtyards one could have, and the accompanying sense of depth in privacy, was determined by one's status. The situation of a house in a *fang* (a square of houses with its own network of streets), as indicated in Han dynasty sources, was again determined by the status of the owner.[5] [. . .] Houses with one, or only a few courtyards, opened to a side-street or lane. More magnificent houses, therefore having more courtyards, opened to larger streets and main thoroughfares. The entire city was made of groups of *fangs* in which houses of all sizes were organized. In the case of the capital city, the emperor occupied the central and largest house-yard complex, thereby dominating the walled-in town and, symbolically, the nation. Such regularity as that seen in the Sui-T'ang capital of Ch'ang-an, or in the greatest and last of this tradition, Peking, was by no means always a rule or a fact since the earliest times. The development has been a long process and

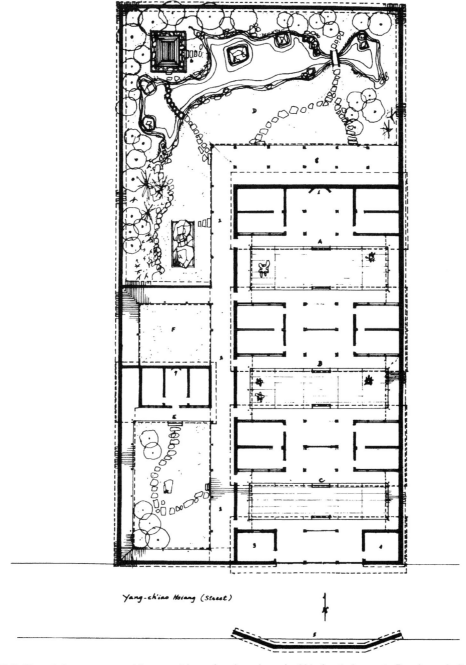

Yang-ch'iao Hsiang (Street)

Figure 29.2 Plan of three-courtyard house with garden, based on the Wu family house in Foochow. A, B and C = family courtyards; D = garden; E = family tutor's courtyard; F = storage and service; 1 = family shrine; 2 = service passageway; 3 = doorman's lodging; 4 = utility and refuse; 5 = pond; 6 = pavilion; 7 = tablet of Confucius; 8 = loggia

the eventual synthesis incorporates elements from many sources. Indeed, when the powerful and the rich vied for splendor, as during Emperor Jentsung's reign (1023–63) in the Sung dynasty, these classical codes of propriety were shattered by the building of magnificent and luxurious structures clustered in the capital city of Pien.[6] As the cities rose and fell on the terrain of China, the Chinese house-yard and its 'graduated privacy' developed an interesting and organic pattern of city life, in which the separation of houses was as clear as the sense of sharing in a community was enduring. Describing a spring night in such a city, the T'ang dynasty poet *Li Po* (701–762) sang:

> *From whose home comes this music of a jade*
> *   flute?*
> *Borne by the spring breeze, it fills the city of*
> *   Lo-yang!*

Sitting in his hall and facing his courtyard to the south, the Chinese sees himself, in an idealized manner of course, not at all fixed in the center of his world, but longingly looking out beyond his walls. In almost every aspect his attitude is different from that of his Indian counterpart. He focuses his eyes on the ground instead of heaven; he is the originator of knowledge and not the seeker of enlightenment who makes the eternal pilgrimage from the periphery toward the center. The Chinese organizes his basic cell in order to organize the world around it. His immediate world, measurable, controllable, is forever encroaching on the Unknown. His kingdom is the Central Kingdom. In adverse times his sphere can be reduced to a hibernating spore, inactive and defensive but still organic. Inside his walls he regulates human relationships to achieve internal harmony, to him the highest goal on earth. As the society became complex and the relationships between men more refined, new architecture was created which further clarified human positions and their movements in space. [. . .]

Preserved in the *Book of Rites* is a 'Code Book of Works' (*K'ao-kung Chi*). The Han dynasty scholars may have edited the work and left their mark on it, but it could not have been a completely new fabrication.[7] In this work a rather idealized plan of a city is given: 'The capital city is a rectangle of nine square *li*. On each side of the wall there are three gates . . . The Altar of Ancestors is to the left (east), and that of Earth, right (west). The court is held in front, and marketing is done in the rear,' forming a Chinese mandala of nine squares with man in the center. [. . .]

[At Chang'an,] as the palace backed against the northern wall, the design provided two market places, one left and one right. If it deviated from the ancient canons in many ways, it set new standards for both its size (5.8 × 5.28 miles) and its rigidity (eleven north-south streets and fourteen east-west streets, the former an odd number to provide a north-south axis). Later in T'ang times (618–906), it developed into a great cosmopolitan city known to many corners of the world, and inspired Japanese copies, Nara in 710 (approximately 2.5 × 3 miles), and Kyoto in 794 (3 × 3.5 miles). [. . .]

[. . .] We are now ready to read the *summa* of the Chinese City of Man in the architecture of Peking, a capital that has served six dynasties in six hundred years and has been modified and maintained continuously.[8]

Superficially, the northern Inner City and the southern Outer City remind us of Ch'ang-an, which also extended its southern approach, and Lo-yang which had a southern region. Against this city of innumerable smaller walled-in courtyards, against its warp and woof of streets, a core of structures stands out more meaningfully in the orderly design: the Forbidden City and its long avenue of approach from the south with all the gates and yards enhancing each significant stage of procession (see Figure 29.3). Unlike a temple which as the house of God can only be read by

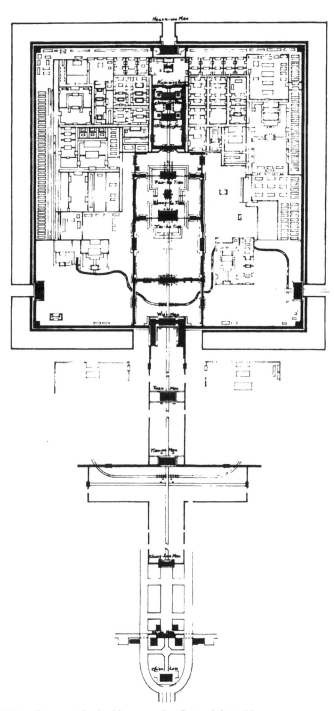

Figure 29.3 The Forbidden City and the buildings on the Central Axis. Map.

man from the outside in, the city of man is to be read both ways: the resident begins from the center, the stranger from the outside gate. Many experiences of this architectural space are preserved in records. Emperor Ch'ien-lung was pictured receiving Amursana.[9] [. . .] at Wu Men, the Gate of the Noonday Sun (see Figure 29.4). The new candidates for Civil Service murmur their thanks to its gates and walls[10] from outside (see Figure 29.5). Therefore, to understand this design as the man in his city understands it, one begins with the central group, immediately grasping the whole composition and with full knowledge and control expanding his view farther and farther out.

[. . .] From here on the world of man is a sequence of rectangular courtyards defined by walls and with graduated privacy. The Palace City is twice enclosed by city walls: first, by those of the Forbidden City with its elegant corner towers; and second by those of the city proper, austere, high, and protective. Then, still defining the sense of privacy, far

beyond the city walls of Peking is the Great Wall of China. Outside of this, the 'barbarians', or the 'outsiders', dwell.

The height of the palace platform [on which the emperor's Audience Hall stands] is nothing to compare with the steep terrace of the Han Palace [at Chang'an]; but leading to this climax is the long southern avenue whose length must be read as height. No matter how the natural terrain of China is formed, one always goes *up* to Peking (see Figure 29.6).

The Chinese doorway (*men*, meaning much more than an opening and indeed an architectural complex in itself) relieves the continuous wall surface by exaggerating the void it creates. Indicating a particular depth along the axis of approach, each doorway gives meaning to the courtyard behind it. The practice of illustrating a Chinese hall with a frontal view, as if it were a Renaissance building introduced by its façade, is most misleading. The Chinese hall, considered singularly, is façadeless. But seen within the house-courtyard composition, it indeed has a

Figure 29.4 Emperor Ch'ien-lung receiving Amursana in 1754, in front of Wu Men. [Engraving of 1774.]

Figure 29.5 Lin Ch'ing at the Wu Men (Gate of the Noonday Sun), thanking the Emperor for the Chin-shih Degree conferred upon him. Wood-block print, nineteenth century.

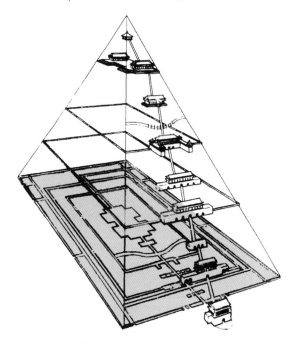

Figure 29.6 Plan of Peking interpreted as volume, with the front and rear courts of the palace superimposed, (shaded area).

façade: the superimposed image of all the doorways leading up to it. [. . .] The emperor's Audience Hall is [. . .] the climax after the sequence of nine important doorways.[11] [. . .]

Notes

The following notes were provided by Arnold Pacey.

1 The influence of Indian architecture can be seen in temples in Cambodia (Angkor Wat), Java, and elsewhere in South-East Asia, while many buildings in Korea and Japan show evidence of Chinese influence. This chapter does not pursue connections outside China, however

2 City walls and gates have several kinds of meaning, the first being expressed by the rectangular plan of the classic city, further articulated in many instances by the animal symbols, with the further idea that the unknown and uncontrollable is excluded from the walled area

3 Practitioners of *feng shui*, the study of natural energies relevant to the design of cities and buildings, often make use of these animal symbols to describe the energies with which they deal. Although Nelson Wu does not mention *feng shui* explicitly, this paragraph could easily be paralleled by similar passages in books on the subject

4 Hankou is another example of a 'naturally-shaped' river

town, with streets parallel to the river bank, and with no attempt to conform to the square plan

5 The different kinds of privacy discussed here provide a second kind of meaning associated with walls and gates

6 The reference here is to Kaifeng, a city which lies on the Pien (modern spelling, Bian) River, and is sometimes referred to by that name

7 *The Book of Rites* (*Li Ji*, traditional spellings *Li Chi* or *Li Ki*) is one of the great Confucian classics compiled under the Han Dynasty (206 BCE–220 CE) from much earlier sources

8 To say that Beijing (Peking) is a capital that has served six dynasties is only true if one includes dynasties such as the Liao which ruled partitioned areas of North China, and then only over a longer time-scale

9 Amursana was a Mongol chieftain seeking aid from the Emperor in a local civil war

10 Another meaning of city walls is that they are symbolic of order and good government. Hence candidates successful in government examinations which qualified them for jobs in the civil service could very appropriately signify their appreciation before the walls and gates of the Forbidden City

11 The 'nine important doorways' are mostly gates in walls, and this last paragraph indicates a fourth (or fifth) meaning of the wall penetrated by a gate. It is the public face of a city or a building, and on passing through it into the enclosed space, one enters an area that is both more private and more elevated above the public domain

30

WATER-PIPES, FOUNTAINS AND CLOCKS

by Joseph Needham and Wang Ling

Source: Joseph Needham and Wang Ling, *Science and Civilization in China*, Vol. 4, Pt. 2, *Mechanical Engineering*, Cambridge, Cambridge University Press, 1965, pp. 129, 133–4, 507–8. The spelling convention used for Chinese names is peculiar to Needham's work; the Tang Dynasty, for example, is referred to as the Thang; − and + are used instead of BCE and CE, or BC and AD

A transfer of technology from western China to Hangzhou (Hangchow)

We are not able to find any instance of the use of metal piping in [early] China,[1] but Nature offered there a material which was admirably adapted for the same purpose, and unexpectedly strong, though perishable, namely the stems of bamboo. It may well be that the earliest large-scale use of this took place in the Szechuanese salt-fields, for brine, unlike fresh water, will not permit the growth of algae and consequent rotting of the tubes.[2] [. . .] The joints are sealed with a mixture of tung-oil and lime. From rubbings of Han bricks which show the salt industry it seems certain enough that the bamboo pipe-lines (*lien thung*) were already in full use at that time. For agricultural purposes also bamboo piping was used (cf. a drawing which has been noticed by several Western historians of technology),[3] but it needed frequent replacement. References to piped water-supplies for palaces, houses, farms and villages are not uncommon. But the largest systems of this kind seem to have been due to the great poet-official Su Tung-Pho, who as a Szechuanese knew of the brine pipe-lines in his own province. Under his inspiration, water-mains made of large bamboo trunks were installed at Hangchow in + 1089 and at Canton in + 1096, caulked with the usual composition and lacquered on the outside.[4] In the latter system there were five parallel mains. Holes were provided at intervals for freeing blockages, and ventilator taps for the removal of trapped air. Significantly, Su Tung-Pho had the help of a Taoist, Têng Shou-An, in planning and executing these works.[5] [. . .]

Exotic technology in Kaifeng (Khaifeng) and Chang'an

[. . .] Mêng Yuan-Lao, describing in + 1148 the glories of Khaifêng, the capital lost to the Chin Tartars, tells us that at a certain temple there were

> two statues of the Buddhas Mañjuśrī and Samantabhadra riding on white lions. From the five fingers of each of their outstretched hands, which quivered all the time, streams of water poured in all directions. For this purpose wheels were used[6] to hoist the water up to the top of the high hill behind, where there was a wooden cistern. At the appointed times this was released (through pipes) so that it sprayed like a waterfall.

This must have been quite worth seeing.

Four hundred years earlier [. . .] the great worthies of the Thang had been equally interested in fountains and similar means of cooling halls and pavilions in summer. The *Thang Yü Lin* says: [7]

After the empress Wu Hou died, the mansions of the princes, princesses and notables in the capital grew daily more magnificent and imposing.[8] During the Thien-Pao reign-period (+ 742 to + 755) the Grand Censor Wang Hung was found guilty of crimes and sentenced to death, so his mansion in Thai-phing-fang was confiscated by the district officials. Several days were not enough for this. In the grounds there was a pavilion called the Lodge of Artificial Rain (Tzu Yü Thing), from the roof of which cascades of water ran down in all directions. If one was there at midsummer one felt as cool as if it were mid-autumn.

This passage does not in itself imply upward-shooting fountains but something perhaps rather more like those lodges or bathing pavilions in Indian lands in which the bathers could sit surrounded by sheets of water descending on all sides.[9] One such remains to this day in the royal gardens at Anurādhapura in Ceylon[9]. But another text which refers to about the same date indicates true fountains rather clearly. The *Thang Yü Lin* again says:

When the emperor Ming Huang (Hsüan Tsung) built the Cool Hall (Liang Tien) (about +747), the Remonstrator Chhen Chih-Chieh, submitting a memorial to the throne, admonished most severely against it (on grounds of extravagance).[10] At the request of the emperor, (Kao) Li-Shih summoned him to court. It was when the heat was really extreme. The emperor was in the Cool Hall, and behind his seat the water struck the fan-wheels while cool air played around one's neck and clothes. Chhen Chih-Chieh arrived and was given a seat on a stone chair. A low thunder growled. The sun was hidden from sight. Water rose in the four corners and forming screens fell again with a splash (*ssu yü chi shui chhêng lien fei sa*). The seats were cooled with ice,[11] and Chhen was served with marrow-chilling drinks, so that he began to shiver and his belly was filled with rumblings. Again and again he begged permission to leave, though the emperor never stopped perspiring, and at last Chhen could hardly get as far as the gate before stopping to relieve nature in the most embarrassing way. Next day he recovered his equanimity. But people said that 'when one discussed affairs one

should deliberate thoroughly on them first, and not put oneself in the emperor's place'.

For fountains this must suffice. But what were the fan-wheels kept rotating behind the emperor's seat? [. . .]

Fountains and clocks in Beijing

[Six hundred years later] the last emperor of the Yuan dynasty, Toghan Timur,[12] [. . .] surrounded himself with mechanical toys of all kinds, clocks with elaborate jackwork, and fountains of several different sorts. We know about them from the description of Hsiao Hsün, who left a vivid record of the architecture and contents of these Yuan palaces which it had been his duty, as an official, to destroy upon the orders of the first Ming emperor in + 1368.

[. . .] Hsiao Hsün was a Divisional Director of the Ministry of Works, and thus had the opportunity to see in detail the beauty of the buildings which had housed more than a thousand concubines, as well as the arrangements of the workshops in which so many ingenious mechanisms had been made by the emperor himself and his artisans for their delectation. Hsiao Hsün describes dragon-fountains with balls kept dancing on the jets, tiger robots, dragons spouting perfumed mist, and several dragon-headed boats full of mechanical figures. [. . .] It was indeed a shame to destroy them, to 'break down the carved work with axes and hammers', but the demands of a demagogic asceticism were in the ascendant, [and it is here that we] [. . .] take leave of the millennial indigenous tradition of water-powered clockwork. In order to do this, we must enter the private apartments of the imperial palace, where about the middle of the + 14th century we find the last emperor of the Yuan [. . .] busied – like Louis XVI – in making clocks himself. The following excerpt will show the style of his work:[13]

The emperor (Shun Ti) himself made (in his workshops, in + 1354) a boat [. . .] which sailed about on the lake between the Front and the Back Palaces, (having mechanical arrangements so that the) dragon (figurehead's) head and tail could move about, while it rolled its eyes, opened its mouth and waved its claws.

He himself also constructed a Palace Clock (*kung lou*) 6 to 7 ft. high and half as wide. A wooden casing hid many scoops (*hu*) which made the water circulate up and down within (*yün shui shang hsia*). On the casing there was a 'Hall of the Three Sages of the Western Paradise',[14] at the side of which there was a Jade Girl holding an indicator-rod for the (double-)hours and quarters. When the time arrived this figure rose up on a float, and to left and right appeared two genii in golden armour, one with a bell and the other with a gong. At night these jacks struck the night-watches automatically without the slightest mistake.[15] When the bells and gongs sounded, lions and phoenixes on each side all danced and flew around. East and west of the casing there were 'Palaces of the Sun and Moon', in front of which stood six flying immortals (*fei hsien*). Whenever the noon and midnight (double-)hours arrived, these figures went in procession two by two across a 'Bridge of Salvation' so as to reach the 'Hall of the Three Sages', but afterwards they withdrew and returned to their original positions.

The ingenuity of all this was beyond belief, and people said that surely nothing like it had ever been seen before.

If this judgment was more flattering than accurate, one need not doubt that the imperial jack-work, though quite in the tradition of I-Hsing and Su Sung[16] was impressive enough. More interesting and notable is the fact that this clock, though almost certainly without dial and pointer, had lost practically all trace of the original astronomical components. Yet in becoming a purely time-keeping machine it had not acquired any new name, and was called merely a *lou*, a 'leaker', like the simplest clepsydras of two thousand years before. It was of course far more, as we can deduce from the water-wheel escapement hinted at by the 'scoops'. By a curious coincidence, this clock was almost contemporary with the wonderful astronomical time-piece of Giovanni de Dondi in Italy (+ 1364). [. . .]

The final blow to the indigenous tradition might well be dated about + 1368, when the new forces of the Ming dynasty captured Peking and ended the Mongol domination. Hsiao Hsün, about twenty years later, in his *Ku Kung I Lu*,[17] left a striking account of the architecture and contents of the Yuan palaces which were destroyed by the first Ming emperor's order. [. . .] The destruction of all these things, understandable though it was as the act of a dynasty representing the people's resentment at economic exploitation, and able to direct it against quasi-alien overlords, was nevertheless very unfortunate. Much that could have been put to better use probably perished – the Ming, like another revolution later, 'had no use for' clock-makers.[18] No doubt the Chinese horological tradition had become smothered in its own jack-work, and was hopelessly identified with the 'conspicuous waste' of the Mongol court. But its death (if indeed it did quite die at that time) was a circumstance of peculiar historical importance, for it meant that when the Jesuits arrived two hundred and fifty years later there was extremely little to show them that mechanical clocks had ever been known in China.[19]

Notes

These notes, provided by Arnold Pacey, are closely based on those by Joseph Needham and Wang Ling, but with amplification in places, and some obscure bibliographical references removed.

1 But from the fifteenth century onwards, pipes of copper and bronze were used in the Ming palaces (in Beijing, etc.)

2 Szechuan (now rendered Sichuan) is a province in West China, its nearest boundaries being over 1,000 km west of Hangzhou (Hangchow). Bamboo pipes and the Szechuan brine wells were described by Song Yingxing in 1637, and one of the illustrations in extract 31 by that author shows a bamboo being lowered into a well, probably for use as a pipe, see Figure 31.6, p. 197

3 The drawing is given by Wang Zhen in his agricultural encyclopedia of 1313 entitled *Nong Shu* (*Nung Shu*),

and the westerners who comment on the illustration of pipes include B. Buffet and R. Evrard, *L'eau potable à travers les âges*, Liège, Solédi, 1950

4 The system at Hangzhou (or Hangchow) was reconstructed in 1270 by the then governor who gave a detailed description in a book he wrote. The 'usual composition' for caulking joints was putty made from tung-oil and lime as described earlier, tung-oil being a vegetable oil pressed from seeds of a tree

5 Archaeologists have found remains of bamboo and earthenware piping in several cities, so this technology may have been more widely used than written sources indicate. Earthenware joints and bends were sometimes used where the main lengths were bamboo, and there were lengths made wholly of earthenware pipes with interlocking joints

6 The word used here implies pulleys and buckets, and the circumstances seem rather to suggest a chain pump in which an endless chain of buckets (or pallets in a channel) was running on a wheel

7 The *Than Yü Lin* is a book whose title translates as: *Miscellanea of the Tang Dynasty*. It was compiled by Wang Tang in 1107, that is, two centuries after the fall of the dynasty

8 The capital referred to here is Chang'an. The Empress Wu (the only woman to have ruled with full powers of an emperor) died in 704 CE (+ 704)

9 The Lodge of Artificial Rain is referred to in several other Tang Dynasty records. At the date of its construction, there was a good deal of contact between Chang'an and India, so the idea could have come from there

10 Puritanical officials who rail against the extravagance of the court, or over-ambitious engineering works, are a recurrent theme in Chinese history, and at the end of this extract, we will note that the overthrow of the Yuan Dynasty by the Ming gave an opportunity for destruction of another emperor's extravagant indulgences

11 Snow and ice, collected in winter and packed into a deep underground storage chamber would remain fro-

zen for several months; thus iced drinks were a real option in hot weather

12 Toghan Timur, born in 1320, became emperor in 1333 at a very young age. He died in Mongolia in 1370, two years after being deposed. His name as emperor was Shun Ti

13 The long quotation which follows comes from a book called *Yuan Shih*, which translates as *The History of the Yuan Dynasty*, written about 1370, by Sung Lien and others

14 The three sages of the western paradise would be Confucius, Lao Tzu, and Buddha. Lao Tzu (modern spelling, Laotze) was the founder of Daoism (Taoism)

15 This implies great ingenuity, because the night-watches were unequal, varying with the season of the year

16 Su Song (Su Sung) was the designer of the great astronomical clock erected in Kaifeng during the 1080s. I Hsing was an 8th-century astronomer and clock-designer

17 This book title translates as *Description of the Palaces*

18 One should not suppose that this great nationalist uprising was in all respects anti-technological. On the contrary there is much evidence connecting its success with the first adequate large-scale tactical use of metal-barrel guns, then a new invention

19 Moreover, for reasons not yet at all clear, there was in Ming times a general decline in most of the indigenous traditions of physical science and technology (with some exceptions such as the ceramics industry). There was thus almost no one who could explain Chinese mathematics, astronomy or other sciences to the Jesuit missionaries (*c.* 1600 and later). This situation they naturally exploited in several ways. To the Chinese they emphasised as much as possible the supposed superiority of the natural science of Renaissance Europe, because by the aid of it they hoped to convince them of the superiority of European religion. To the Europeans they praised the ethical and social philosophy of China as much as possible in order to raise the prestige of the Jesuit mission, which was converting not savages but highly civilised people

31

TILES, BRICKS AND COAL

by Sung Ying-Hsing

Source: Sung Ying-Hsing, *T'ien-kung K'ai-wu: Chinese technology in the seventeenth century*, trans. Sun E-tu Zen and Sun Shiou-chuan, University Park and London, Pennsylvania State University Press, 1966, pp. 135–42, 204–5. This text renders Chinese words using the old Wade–Giles spelling convention. Where it is important to know the modern spelling, this is indicated in the footnotes. The author's name and the title of the book in modern spelling are: Song Yingxing, *Tiangong kaiwu*. The title may be translated as: 'The exploitation of the works of Nature'

Master Sung has observed that, with earthenware brought about through the use of water and fire on clay, even the daily labor of a thousand workers is not enough to fill the demands of a large country, so numerous are the people's needs for pottery. As shelter against wind and rain, houses are built, and so there is need for tiles. In order to defend the country the rulers must construct strategic defense works, hence city walls and ramparts are built of bricks, and invaders are kept out. Sturdy earthen crocks preserve wine to a good vintage, while clean pottery vessels are instruments for containing the sacrificial offerings of wines and bean sauces. The sacrificial dishes of Shang and Chou times were made of wood; was it not because the people then wanted to show great respect [towards the spirits]? In later times, however, ingenious designs began to appear in various localities, human craftsmanship exerted its specialities, and superior ceramic wares were produced, beautiful as a woman endowed with fair complexion and delicate bones. [These wares] sparkle in quiet retreats or at festive boards, a concrete sign of civilized life. It is hardly necessary to adhere [to the ways of Chou and Shang] forever.[1]

Tiles

The material for making tiles is a sandless clayey earth, which is obtained by digging about two *ch'ih* under the earth's surface.[2] Within a hundred *li* [of any place][3] there is sure to be a suitable clay for the construction of houses. The tiles used for the houses of the ordinary people are all made from the quarter sections [of a cylinder]. The process of making such tiles involves the preparation of a cylindrical core mold, on the surface of which are four ribbed demarcation lines. The clay is mixed with water and made into a high rectangular pile. Then the moist clay pile is sliced by the iron wire of a bow, with the wire located at 0.3 inch away from the back of the bow.[4] The sliced clay layer is lifted like a piece of paper, and wrapped around the cylindrical core mold. When slightly dried, the clay is removed from the mold, and naturally falls apart into four pieces (see Figure 31.1). There is no standard size for tiles, which range from a measurement of eight to nine inches on each side for the large ones to seven-tenths of the above size for the smaller ones. Those used to catch rainwater along house roofs, called drain tiles, must be the largest in size, so that they can sustain a continued rainfall without overflowing.

Figure 31.1 Making tiles

When the green tiles are made and dried, they are piled in the kilns and fired. The firing may last either one day and night or two days and nights, depending upon the number of tiles placed in the kiln. The methods used for the water-tempering and glazing of tiles are the same as those for bricks. There are 'water drip tiles,' to be placed at the eaves; 'cloud tiles,' below the roof ridge; 'curved cover tiles,' over the roof ridge; and tiles variously shaped like birds or beasts, to be affixed at the two ends of the ridge. These latter tiles have to be handmade, piece by piece; but in so far as they are prepared through the use of water and fire and firing in kilns, they are made by the same method [used for making other kinds of tiles].

Tiles used by the Imperial Household, however, are quite different from the above. The *lazuli*[5] tiles are made piece by piece, either flat or rounded in shape, with the aid of round bamboo or polished wood molds. Only clay from T'ai-ping prefecture [in Anhui province] is used for [making imperial tiles]. (This clay must be transported 3,000 *li* to Peking,[6] in the course of which indescribable damage is caused by adulteration with sand, and by the lawless behavior of the transport laborers and boat crews. Even the imperial mausolea are built with such tiles, and no one has dared give advice to the contrary.) When the tiles have been shaped, they are first placed in the 'glazed tile' kilns, in which [they are fired] to the proportion of 5,000 catties of fuel wood per 100 pieces of tiles. The fired tiles are painted with a mixture of pyrolusite [MnO_2] and palm hairs for green color, and with [a paste of] ochre, rosin and rushes [juice] for yellow. The painted tiles are then placed in another kiln to be fired again at a lower temperature. The gemlike brilliance of *lazuli* colors will then be achieved. Occasionally this kind of tile is used in the palaces of princes (living) in the provinces as well as in temples and monasteries, although the materials and methods of glazing may be somewhat different from one place to another. Ordinary people, on the other hand, are forbidden to use [*lazuli* tiles] for their houses.

Bricks

Clay for making bricks is also obtained from the ground, in color either blue, white, red, or yellow (that found in Fukien and Kwangtung is mostly red; blue clay is produced chiefly in Kiangsu and Chekiang[7] and is called 'good clay'), the best being the kind that is adhesive, fine-textured and sandless. The clay is mixed with water and trampled over by several oxen driven by a man, so that it becomes a thick paste. After that, wooden frames are filled with this clay paste, the surface is smoothed with wire-strung bows, and so the green or unbaked bricks are formed (Figure 31.2). For the construction of city walls and house walls, there are two kinds of bricks known as 'recumbent brick' and 'side brick.' The former are oblong pieces, used in building city walls and the houses of well-to-do people; they are placed solidly one upon another. Those people who wish to economize, however, usually put down alternate layers of 'recumbent bricks' and 'side bricks,' and using stone fragments and bits of earth they fill the inner spaces; this is done to reduce the cost. Aside from building walls, the bricks used for making foundations are called 'square-frame bricks'; for supporting tiles at corners of eaves, '*Huang-pan* bricks'; for constructing curved arches of small bridges, rounded doorways, or graves, 'knife bricks' or 'curved bricks.' One side of the 'knife brick' is cut off wedgewise; when several of these are packed closely together and constructed into a dome, the structure will not collapse even when trodden and weighed down by horses and carriages.

To make 'square-frame brick,' the clay paste

Figure 31.2 Making bricks

is first put into the square frame, and then a smooth board is placed over it; two persons step on the board in order to pack the mass and ensure the solidity of the green brick. [After being air dried,] the green brick is then fired to harden it. The bricks used for laying foundations are first trimmed around the edges by stone workers. The knife bricks are larger than the wall bricks by one-tenth of an inch; the *huang-pan* brick is one-tenth the size of the wall brick, while the square-frame brick is ten times as large as the wall brick. After the green bricks are made and [air-dried], they are placed inside a kiln and are fired for one complete day and night for 3,000 catties [of bricks],[8] and twice that length of time for 6,000 catties. The fuel for firing bricks is either wood or coal; when the former is used the brick produced will be bluish gray in color, while the latter will make white bricks.[9] At the top of a wood-fuel kiln three holes are opened on one side for the emission of smoke. When the process of firing is completed and the fire is withdrawn, these holes are sealed with mud, and then water is employed for the superficial glazing or quenching [of the bricks].

[As to the amount of heat applied in the firing process], if it is one ounce [i.e. degree] less than the proper temperature, the glaze will have no luster; if three ounces less, the product is known as 'low temperature brick,' with the clay colors plainly showing.[10] This brick will disintegrate immediately after exposure to frost and snow and change back to clay. On the other hand, if the heat applied is one ounce more than the proper temperature, cracks will appear on the brick; if three ounces more, the size of the bricks will shrink and they will be full of crevices. The whole piece is warped, breaks like iron fragments when struck, and is altogether unsuitable for use. Skillful [builders], however, utilize these pieces by burying them underground as foundations for walls, thus putting them to the same kind of use as good bricks. The way to check the temperature in the kilns is to watch the clay through the kiln doors; under the attack of fire the clay will manifest an attitude of uncertainty, and appear similar to gold and silver at their melting point. The pottery works foremen will recognize [the proper temperature].

The superficial glazing or quenching of [bricks and tiles] is done by pouring water onto the level space atop the kiln which is surrounded by a raised wall on all four sides. Forty *tan* of water are needed per 3,000 catties of bricks or tiles in the kiln (see Figure 31.3). The water, once having permeated the earthen covering [of the kiln], will react with the heat [to form superficial glaze on the brick surfaces]. When the right proportion of water and heat is achieved, then the quality and durability of the [bricks] are assured.[11]

Figure 31.3 Superficial glazing of bricks and tiles by water-quenching

The coal-fuel kilns are twice as high as the wood-fuel ones, having tops that are domed but not sealed. Inside the kilns are laid round cakes [briquettes] of coal one and one-half *ch'ih* in diameter, and each layer of coal is alternated with a layer of bricks (see Figure 31.4). The bottom layer consists of reeds that serve as kindling.

A large factory for the making of bricks for Imperial Palaces is situated at Lin-ch'ing [in Shantung], and is operated under the direction of an official of the Board of Works. Previously many different kinds of bricks were produced here, such as 'secondary brick,' 'tally brick,' 'flat-bodied brick,' 'lookboard brick,' 'axe-blade

窯磚燒炭煤

Figure 31.4 Coal-fired brick kiln

brick,' 'square brick,' and so forth, but half of these varieties were eliminated in later times. The bricks are transported to Peking by loading forty pieces per rice-tribute boat, and half as many pieces per ordinary private boat. The fine-quality square bricks, used in building the central halls [in the Palace], are made in Soochow and transported [to the capital].[12] As to the making of *lazuli* bricks, we have already mentioned it in the section on tiles.

Coal

Coal is obtainable everywhere, and is used for the smelting and calcination of metals and stones. South of the Yangtse River, coal is found in mountains that are bare of trees. We need not discuss the situation in North China. There are three kinds of coal: anthracite, bituminous, and powdered. The large pieces of anthracite coal are about the size of a bushel. The coal is produced in places like Yen, Ch'i, Ch'in, and Chin [that is, the provinces of Hopei, Shantung, Shensi, and Shansi in North China].[13] It is kindled with a little charcoal, and can burn for a whole day without the use of bellows. The fragments beside the large pieces can be used as fuel after being mixed with clean yellow mud and made into cake-shaped briquettes.

Of the bituminous coal, which is mostly produced in Wu and Ch'u [i.e. in the middle and lower Yangtse region], there are two types: the high volatile type is known as 'rice coal' and is used in cooking, while the low volatile is called 'iron coal' and is used in smelting and forging metals. The coal is first dampened with water before being placed in the furnace, and the bellows must be used to bring it up to red heat. While the fire is going, coal should be added repeatedly. The small fragments of this coal, fine as flour, are called 'automatic wind.' These fragments, when made into briquettes with the addition of mud and water, will burn constantly throughout the day and night similar to anthracite coal. Half of the briquettes are used for cooking and half for smelting, calcination, and the manufacture of cinnabar. As for burning lime, alum, or making sulphur, all three kinds of coal can be employed.

Experienced coal miners are able to find underground coal by the color of the earth. Coal is reached in pits about fifty *ch'ih* deep.[14] The first appearance of the coal seam is accompanied by strong poisonous gas; some people therefore erect a thick hollow bamboo pipe on the coal, thus drawing up the poisonous gas[15] and enabling the men to shovel and pick the coal underneath. Sometimes the coal seams extend in several directions from a shaft; in which case the miners simply follow the seams, with timber built overhead [in the mine] to prevent collapse[16] (see Figure 31.5). After a coal mine has been depleted, the shafts are then filled with earth, and more coal will appear there again after twenty or thirty years – it is inexhaustible [*sic*].[17]

The round stones at the bottom of and around the coal seams, called 'copper coal' [i.e. carbonaceous shale containing cupriferous pyritic minerals] by the local people, are mined as raw material for making black vitriol and sulphur. This kind of round stone, used only for making sulphur, has a partly oxidized, penetrating odor and is called 'smelly coal' [i.e. carbonaceous shale containing pyrites and native sulphur]. It is found occasionally in

Figure 31.5 Coal mining

Fang-shan and Ku-an in Yen-ching province [modern Hopei], and in Ching-chou in Hu-kuang province [modern Hunan and Hupei].

The essence of coal disappears with the element of fire after burning, and leaves no residue or ash. This means that it is a special manifestation of Nature placed between the species of metal and that of earth and stone.[18]

Notes

The following notes were provided by Arnold Pacey.

1 The reference to Chou and Shang times denotes a period before 1000 BCE when the pottery being made was earthenware, some of it painted. Glazed pottery first appeared in China during the Han dynasty, that is, after 200 BCE, and 'superior ceramic ware' whose white, delicate appearance Song so vividly describes became famous only in the Tang dynasty (618–906 CE). Fired bricks were occasionally made from 200 BCE onwards, but glazed tiles and bricks were probably first regularly used during the Tang dynasty. When the capital city at what is now Beijing was being rebuilt for Kublai Khan, four kilns were established in 1276 for making plain, white-glazed bricks and tiles, and brick-making developed further during the Ming dynasty, up to the time when Song was writing

2 The *ch'ih* (modern spelling, *chi*), often referred to as the Chinese foot, was standardised at 311 mm during the Ming dynasty (standards were markedly different in other periods)

3 A distance of 100 *li* is about 30 miles or 50 kilometres

4 The Chinese foot or *chi* was divided into smaller units somewhat larger than the western inch. The bow with iron wire used for slicing through soft clay is illustrated in Figure 31.2. Similar bows were used for the same purpose in English brickworks during the eighteenth and nineteenth centuries

5 *Lazuli* is short for the European term *lapis lazuli*, which refers to a blue, silicate mineral, used in this case for making glazes of several colours

6 This refers to a journey of a thousand miles (1,600 km) by canal to Beijing (Peking)

7 Fujian (Fukien) and Guangdong (Kwangtong) provinces where red clay is found are coastal provinces in South China. The other two regions referred to, where blue clay is found, are the next two coastal provinces to the north, around the mouth of the Yangzi River

8 A *catty* was a weight of 600 g, so 3,000 catties of brick would weigh 1,800 kg (1.8 tonnes)

9 The editors of the English translation, Sun and Sun, comment as follows: 'The bluish-gray colour is apparently caused by the deposition of fine smoke particles from the wood fuel in the pores of the bricks. These

smoke particles are subsequently fixed by the fusible matter on the surface of the brick. The blueing of bricks and tiles by smoke is practically eliminated when the wood fuel is replaced by coal, particularly anthracite, since a coal fire gives a cleaner atmosphere and a higher temperature. The term 'white' is often used very loosely in clay working and is used to include all shades from a true white to a distinct cream or even a slight reddish or pale yellow colour'

10 There were no precise units for measuring temperature at this time. A small unit of weight, the *liang* or ounce, is being used figuratively. The temperature within a kiln was judged visually, mainly by colour

11 The editors of the English translation, Sun and Sun, have a footnote here to the following effect: 'The water-quenching (i.e. with steam) of red-hot bricks and tiles requires great skill and care. An improper cooling of kilns is the source of many defects, especially crazing, cracks, dunts, and feathering or crystallization'

12 Soochow, now written Suzhou, is the famous 'canal city' south of the Yangzi. Apparently, this specialist type of brick was transported from there (or from brickyards in that district) by canal, a distance of nearly 1,500 km. The large factory at Lin-ch'ing (now written Linqing) mentioned earlier in the same paragraph was much nearer Beijing (perhaps 300 km by water), but still a long way to transport bricks. Since there are reports of brick kilns close to Beijing during the Yuan and early Ming periods, these more distant works must have all been for quality products

13 Anthracite was found in relatively few places, mostly in Shanxi (Shansi) province, where it was used in iron smelting; see Donald Wagner, *The Traditional Chinese Iron Industry*, Richmond (Surrey), Curzon, 1997, pp. 48, 53

14 This mine is only about 50 feet or 15 metres deep

15 A rotary fan, first used to provide a draught in machines for winnowing rice after threshing, was adapted for blowing fresh air into mines. See Mark Elvin, 'Skills and resources in late traditional China', in Dwight Perkins, *China's Modern Economy in Historical Perspective*, Stanford, Stanford University Press, 1975, pp. 85–113, esp. pp. 87–9

16 The illustration of pit props, ventilation pipe and the method of hauling coal to the surface (Figure 31.5) is too rudimentary to be realistic. Some editors of Song's work find 'simplicity and clarity' in his illustrations, but it is clear from other drawings added to later editions that mines, like salt wells, could have headstocks constructed of bamboo which would allow containers of coal to be hauled up using a buffalo-powered capstan without contacting the sides of the shaft: see Figure 31.6.

Figure 31.6 Raising brine from the bottom of a salt well [Ch'ing addition]

17 An organic theory of matter seems to be implied here, but note that when 'exhausted' coal mines were opened up after a long interval, it would often be found that more coal could be extracted simply because mining technology had developed since the mine closed, and coal that was previously out of reach could now be worked

18 Master Song is referring here to the 'five phase' theory of matter widely accepted at the time in China. The 'element of fire' is one *phase* and not an element in the western (Aristotelian) sense. Earth and metal were two other phases, with coal somewhere between them. For a discussion of how Song used the five-phase theory, see Christopher Cullen, 'The science/technology interface in seventeenth-century China: Song Yingxing on *qi* and the *wu xing*', *Bulletin of the School of Oriental and African Studies, University of London*, 53, 1990, pp. 295-318

32

GUILDS AND PROPERTY DEVELOPMENT IN HANKOU

by William T. Rowe

Source: William T. Rowe, *Hankow: commerce and society in a Chinese city, 1796–1889*, Stanford, Stanford University Press, 1984, pp. 303–9. Like most US scholars, William Rowe uses the old Wade–Giles spelling convention, most conspicuously in his spelling of Hankow (modern Hankou). Where it is important to know the modern spelling, this is indicated in an end note

Guilds made capital investments of two general kinds. One, [. . .] was investment in business enterprises, particularly financial ones. Occasional reports of local authorities forcing guilds to stand behind failing credit institutions (for example, the Shaohsing Guild for pawnshops and the Kwangtung Guild for native banks)[1] suggest that the guilds were major investors in these shops or banks. Later, in 1891 the Tea Guild backed Hankow's first Chinese-owned modern bank – an investment that demonstrates a guild's willingness to invest in profit-making ventures outside its own direct purview. Unfortunately, few materials on this type of investment have survived.

The second and more common type of investment was in real property – outside the guild's own precincts and specifically to generate income. Such real estate could be already developed, developed by the guild, or left undeveloped. Occasionally it could be cultivated land, but the surviving documentation attests that even this was almost always urban. Virtually since their inception, guilds had invested in urban land; however, such investment intensified beginning in the late eighteenth century to become very widespread in the second half of the nineteenth. Thus the guilds, as corporate groups, became increas-ingly important as urban landlords. Moreover, the remarkable lack of guild investment in rural agricultural land both reinforces the impression of Hankow's detachment from its rural surroundings and casts doubt on general postulations of the extractive role of late imperial cities vis-à-vis their hinterlands. Fortunately, considerable documentation on real estate investment is available; it figures prominently, for example, in the records of the Hui-chou and Shansi-Shensi guilds.

The Tzu-yang Academy (Hui-chou Guild). During the early Ch'ing,[2] the Hui-chou merchant community at Hankow was headquartered and partially lodged at two small temples located in an idyllic suburban setting on the landward side of the city. The prosperity and increasing size of this group in the late K'ang-hsi period urged a greater visibility, however, and in 1694 three leaders of the compatriot association called a meeting of the group's 24 most influential members to deliberate on plans for a grand Chu Hsi academy (and guild-hall), to be located in the middle of town and to take as its centerpiece an elaborate meeting hall to be known as the 'Respect the Way Lodge.' One of these leaders, Wu Chi-lung, was asked to draft a 'Letter to My Compatriots,' soliciting contributions for the project. The response, apparently voluntary, provided sufficient funds to cover the cost of the building. Four project managers and 24 associate managers were selected by the fraternity as a whole; the three men who had convened the original meeting were among the former. The group acquired four *mu* of land (slightly more than half an acre) in a downtown area of Hsün-li ward, within a neighborhood already heavily populated by Hui-chou men. Several prominent group members, voluntarily but apparently with some hesitation, sold their houses to the group and moved into temporary quarters elsewhere in order to vacate the proposed guildhall site. The final complex, comprising some one hundred separate chambers and constructed at a cost approaching ten thousand taels,[3] was completed in 1704, eleven years after the plan had originally been conceived. Among its attractions were accommodations for some members and a great ceremonial gate upon which was mounted a stage for theatrical performances. The project was a massive undertaking, which reflected the dominant position of the Hui-chou community in salt and in other leading trades.[4]

In the eighteenth century, the corporate property and assets of the academy grew. The guild built an adjoining West Hall in 1717, a large lecture hall in 1721, a dormitory for transient compatriot merchants,[5] in 1743, and several subsidiary chapels in 1775; several times throughout the century it also expanded lodging and library facilities. Although the guild also purchased and developed for collective use property in other sections of town (for example, a small temple on the shores of Back Lake), most of its holdings were in or around the sprawling complex, popularly known as 'Hui-chou ward' (Hsin-an fang), which dominated one portion of the city's central district. At some time in the early part of the century, the whole area was enclosed by a wall.

The Hui-chou complex stretched transversely across the city's major traffic arteries,

which paralleled the Han River. The guildhall's central hall was north of Chung chieh (Center Street) and faced south toward the Han. Across the street, running southward as far as Han-kow's principal thoroughfare, Cheng chieh (Main Street), was a street enclosed at either end by ceremonial gates and joined along the way by several radiating alleys. This street had been built largely by the guild and was lined by its properties, but during daylight hours it was open to the public and constituted one of the most bustling shopping areas of Hankow. Known originally as Hsin-an Alley (*hsiang*), the street soon outgrew this designation and became known as Hsin-an Street (*chieh*), one of the few transverse thoroughfares given this name. In 1775 Hsin-an Street was completely reconstructed by the guild as a 'major highway' (*k'ang-ch'ü*). Across Main Street, Hsin-an Street was again begun by a gate and continued its course south to the riverbank, where in 1734 the guild had constructed one of the major pier complexes in the city, known as Hsin-an ma-t'ou, or Hsin ma-t'ou. Although intended primarily for guild members, the pier was described as a 'public wharf' (*i-pu*) and may have been open for general use upon payment of rental and service charges.

In addition to the pier, in these years the Hui-chou men engaged in several other projects deliberately intended for the public good, including a major public road, a school, and a public ferry. [. . .] The guild had also accumulated many rental properties, the income from which was used to underwrite its continuous building projects. These properties were generally commercial, and although guild members were given preference in renting them, guild membership was apparently not a necessary condition of lease: the primary goal was corporate profit. A document of 1734 describes an example of such rental holdings, a single block containing sixteen shophouses, two kitchens, and a larger hall or warehouse. The guild valued this block at one thousand taels –

whereas its annual rent was 360 taels. Thus, even in the early eighteenth century investment in such property seems to have repaid capital outlay in something less than three years, indicating the immense opportunities open to groups that could generate significant capital for investment.

In 1788 the academy and many other holdings were badly damaged by flood. After several attempts at piecemeal repair, a plenary meeting was convened in 1796, and the guild decided to undertake a mammoth reconstruction of the entire complex. Accomplished between 1798 and 1805, this reconstruction affected all parts of the academy, as well as most of the guild's rental holdings. Special attention was lavished on the roads within and around the compound, some of which, it was claimed, were transformed from 'winding, twisted, stagnant, muddy, and foul-smelling' puddles into broad, paved thoroughfares. As for the shops that lined these streets, 'the bamboo huts of old, which were crowded together like the teeth of a dog, can now be seen to form a straight, even line.' Several related projects, like establishing fire-fighting companies and acquiring and clearing a suburban burial site, were accomplished at the same time. The group celebrated the completion of the project in 1806 by publishing its own gazetteer.

How was this undertaking financed and managed? The 1796 plenary meeting of the guild delegated responsibility to one of the current general managers of the organization, Wang Heng-shih, who was to be assisted by 26 specially designated project managers. Almost immediately Wang ran into problems with the other members. Several expressed doubt that the guild's present capital was sufficient to see such grand plans through to completion. To be forced to give up the project halfway, they argued, would be to dishonor their forebears in the organization. In particular, it seems, they feared that it might eventually be necessary to borrow from outside sources. Wang rebuked

them for their timidity and issued a general call for contributions. Two years later the guild decided that a sufficient amount had been collected, and the project was begun. The managers reportedly exercised great care to check the economic feasibility of each step before it was commenced.

The skepticism, however, proved to have been well founded. Shortly after the project was begun, funds had to be diverted to hire a paramilitary defense force when White Lotus rebels threatened the city;[6] moreover, expenses in general tended to exceed initial estimates. By 1804 the project was close to four thousand taels in arrears, and when a related street renovation estimated at a cost of eleven thousand taels was proposed, the guild found itself in need of more than fifteen thousand taels in order to go on building. By this time rental incomes were substantial, but they could not produce so large a sum in short order. Consequently, the guild for the first time voted to levy a 'likin' charge on all commercial transactions by member merchants to cover future lacks, and in addition took up a strikingly innovative subscription to supply the fifteen-thousand-tael deficit.

The new subscription was a bond issue floated among the guild's membership. One hundred and fifty bonds (*ch'ou*) were printed up at one hundred taels per bond, to total the fifteen-thousand-tael deficit. These were issued to members, apparently after some arm-twisting, according to the assessed financial capability of each individual. Each bond was to be paid off by the guild over a ten-year period in annual installments of sixteen taels, thus yielding a total interest payment (*tzu-chin*) to the bondholder of 60 taels, or 6 per cent per annum. The feasibility of this bond flotation was calculated precisely. At the time of issue the surplus of total rental income over guild operational expenses was averaging between 2,000 and 2,300 taels per year; however, it was anticipated that the income generated by the projects completed via the bond issue would raise the annual surplus to 2,400 taels, the amount required to repay the borrowed capital plus interest (16 taels per bond × 150 bonds = 2,400 taels). Any additional surplus revenue accrued over the ten-year course of repayment, plus all surplus revenue after the repayment had been completed, would be reinvested in property development, with profits to go to the guild's entire membership.

By the beginning of the nineteenth century, the Hui-chou Guild had accumulated a staggering amount of real estate. In order to safeguard the guild's title to these properties, its managers decided to petition the Hanyang prefect:

> Over the years, the Hui-chou scholar–merchant guild has bought up various properties and market buildings. Our managers in yearly succession assume control over the collection of rental fees, which are used to meet the expenses of our spring and autumn sacrifices. Between the seventh year of K'ang-hsi [1668] and the sixtieth year of Ch'ien-lung [1795] we have purchased collective property for which we now hold a total of 67 title deeds [*ch'i-yüeh*]. Our only fear is that in the years to come, since these deeds are so numerous and so frequently passed from one custodian to another, they may be lost or scattered, and we would then have no record upon which to base our claim of ownership.
>
> Thus we have collected these deeds and recorded them in a single register, and we request that you officially seal them and proclaim them, so that they will become a matter of formal public record. Moreover, we beseech you to permit them to be engraved in stone, so that there will be no fear of their becoming lost. Thereupon we guest people[7] will be forever the recipients of your great kindness.

The prefect agreed to this, and ordered that a copy of the register be retained at his yamen as an authoritative source to resolve any future litigation.[8]

Fortunately, the guild's 1806 gazetteer reprints the texts of all 67 deeds, which reveal a great deal not only about the Hui-chou Guild's holdings, but also about the concept of urban property during the Ch'ing. The property that

they represent shows clearly the capital-investment potential of large Hankow guilds by the early 1800s. In addition to the various properties the guild occupied and used itself (which involve but 5 of the 67 deeds), the Hui-chou organization rented out: (1) two large blocks along Hsin-an Street, incorporating 36 shophouses; (2) 31 'foundationless [*wu-chi*] properties,' including shops and residences constructed of tile, earth, bamboo, and rushes, in the immediate neighborhood of the academy and Hsin-an Street; (3) ten shops on the Hsin-an pier; (4) a large, enclosed market on Hou chieh (Back Street), elsewhere in Hsün-li ward; (5) a large, enclosed market on Main Street, not far from Hsin-an Street; (6) a row of shophouses along Hsiung-chia Alley, in Ta-chih ward; (7) eight additional, extensively developed properties scattered throughout Hankow proper; (8) a large and important block of shophouses and residences across the river in the county seat, Hanyang city; and (9) many smaller holdings, including urban garden plots and some rural paddy fields and wheatfields. Most of these properties had been purchased out of capital funds, and many had been developed by the guild itself as investments. They were managed by specially elected officers of the guild known as *szu-shih* (overseers), who were replaced annually following an audit of their accounts by the guild's general managerial board.

One acquisition-of-title document printed in the gazetteer sheds light on the conduct of land transactions by such groups and provides evidence of cooperation between guilds. [. . .] This document is a 'Certificate of Exchange' drawn up jointly in 1804 by the Hui-chou Guild and the Che-Ning (Ningpo) Guild:

> It is here recorded: We natives of these two regions, gathered together in Hankow, have both founded *hui-kuan* to worship our former worthies and honor our native-place ties. In its new reconstruction, the Che-Ning Guild is hampered by the narrowness of its site and hence

cannot expand to the desired dimensions. Behind the Temple of the Triple Origin lies a vacant property owned by the Hsin-an Academy, upon which the Ningpo Guild might expand. Since no reliable market exists by which to estimate a value so that the land might be purchased, the Ningpo men have decided to request the intercession of a go-between [*chü-chien*], through whom they have approached the Hsin-an Academy about the property. The latter graciously convened for a private deliberation [*meng-i*], at which it agreed to the proposal presented.

The Che-Ning Guild has long owned one residence and one shop in front of the Nanking guildhall. They have further recently acquired from the Cheng family three shops located in Hsün-li ward. By the signing of this contract [*ch'i-chü*], these properties are ceded to the Hsin-an Academy to be used as rental properties, in exchange for which the Hsin-an Academy will cede a portion of the vacant land southwest of the front wall of the Temple of the Triple Origin, four *chang* deep and three *chang* wide [approximately forty by thirty feet] . . . and will transfer it to the Che-Ning Guild. The latter may never expand beyond the temple wall, and the rest of this vacant land remains the property of the Hsin-an Academy.

The document was signed by three Managers of the Che-Ning Guild and four of the Hui-chou Guild, as well as by twelve other members of the two clubs as witnesses. Although the Hui-chou Guild seems to have driven a rather hard bargain, the incident demonstrates that other expansive guilds, like that of Ningpo, engaged in both major reconstruction and real estate investment during this period.

The Hui-chou Guild left a detailed statement of its corporate finances for the year 1806, a statement that reveals the profitability of collective enterprise in a prosperous and growing commercial city. All income listed in this statement is from rental properties – that is, the guild seems to have run no business directly and not to have invested in commercial enterprises per se. Total income for the year was listed as 4,404 taels. (Elsewhere, the average annual income from rents is placed at over 4,300 taels, so this year's revenue was not

exceptional.) In 1806 the total income from rural property owned and rented out by the group was only six taels, indicating not only the small percentage of the guild's total capital that was invested in this type of land, but also probably why so little was thus invested. In contrast, the amount of rental income derived from the two rows of shops along Hsin-an Street (an area probably equal to or less than the total rural holdings in square footage) was 2,249 taels. Rural investment simply could not compete with urban property in profitability, particularly when an investor could afford to develop the property attractively.

Balanced against this gross income were the guild's total annual operating expenses, which for 1806 were 1,830 taels, or only about 42 per cent of the gross revenues listed. Where did the remaining 58 per cent go? The accounts indicate that an unspecified but substantial portion went for the 'repair and new acquisition of various shops,' that is, for reinvestment in profit-making ventures. The remainder was applied to paying off the bonds issued in 1804.

After 1806, our information on the Hui-chou Guild is fragmentary. A book-length travelogue on Hankow printed in 1822 indicates that at this time Hui-chou natives were still both socially and economically the dominant local-origin community, but after the Taiping interregnum and the subsequent changes in the salt trade, the Hui-chou merchants seem to have declined, and they are no longer mentioned among the leading commercial powers of the port.

Notes

The following notes were provided by Arnold Pacey.

1 When the author refers to 'native banks', he is discussing the *traditional* system of finance and banking which developed very strongly during the eighteenth and nineteenth centuries for the purposes of extending credit facilities, transferring funds from place to place and changing currencies. When the author refers to 'modern banks', he is referring to banks set up on the basis of western models

2 The Tzu-yang Academy (modern spelling Ziyang) was an alternative name for the Hui-chou (or Huizhou) Guild, because the school run by the guild was regarded as so important. The 'early Ch'ing period' is the period after the founding of the Ch'ing (or Qing) dynasty in 1644. In the next sentence, the 'late K'ang-hsi period' refers to an individual emperor's reign in the later seventeenth century

3 'Tael' is a Portuguese word corresponding to the Chinese *liang* and the English 'ounce'. In this context it refers to 37 grams of silver, nominally equivalent in value to a string of copper cash

4 The author here indicates that the source for most of what he says in this extract is the *Gazetteer* published by the Hui-chou guild in 1806. Nearly all of what follows comes from this source, which will not be cited again

5 The 'compatriot merchants' are merchants from the Huizhou (Hui-chou) area in South China, as this guild was a 'native-place association' for people from that area

6 The White Lotus Society was a secret society which fomented a rebellion during the late 1790s. It affected much of Central China, and was suppressed with considerable loss of life. The economic consequences are often seen as ending (or beginning to end) the long eighteenth-century phase of commercial expansion and prosperity

7 Members of the guild refer to themselves as 'guest people' because they think of themselves as natives of Huizhou (Hui-chou), not Hankou, despite their large stake in the latter city

8 The prefect was a civil servant appointed by the central government as the senior official responsible for a district or 'county'. The 'yamen' was the prefect's office building

33

FIRE BRIGADES AND FERRIES IN HANKOU

by William T. Rowe

Source: William T. Rowe, *Hankow: conflict and community in a Chinese city, 1796-1895*, Stanford, Stanford University Press, 1989, pp. 163-73. Again Rowe uses the Wade–Giles spelling convention, notably in defining his period of study as 'the late Ch'ing.' This refers to the Qing Dynasty whose emperors ruled China from 1644 to 1911

Fire Brigades Given the capricious nature of fire prevention in nineteenth-century Hankow, it is not surprising that considerable energy was expended in the area of fire-fighting. Legally, responsibility for such matters lay with the prefect and the magistrate. But in fact evidence from Hankow shows these officials' role to have been negligible. In the eighteenth century, a certain degree of leadership did come from the *provincial* bureaucracy.[1] For example, of the 12,000 taels contributed by the salt merchant Ch'iu Chien to the viceroy's office in 1752, 10,000 was placed by that official in a special interest-bearing account with the Hankow salt merchant treasury, the proceeds to be released annually for 'various fire-fighting items.' The exact nature of these were unspecified. A clue, however, is provided in a 1747 proclamation by the exemplary 'statecraft' official Ch'en Hung-mou, then serving as Hupeh governor. Ch'en reveals that the sole means of fighting fires at Hankow at that time was some 300 water buckets, distributed along the major streets! Moreover, even this pathetic system was not well maintained; upon inspection, Ch'en found most of the buckets to be empty. The governor thus ordered a wholesale intensification of the system, with buckets placed at *each shop* along the town's major thorough-fares, and at prescribed intervals along side-streets. Old buckets were to be repaired and kept watertight, and the water level regularly monitored by local military personnel.

At the end of the eighteenth century a revolution took place in the technology of fire-fighting at Hankow with the advent of the 'water dragon' (*shui-lung*), or hand-drawn fire engine. Mark Elvin describes this ingenious piece of gear as a 'cylinder and piston pump with inlet and outlet valves for projecting a jet of water,' and notes that it was introduced to China by the Jesuits in the seventeenth century.[2] All *shui-lung* in operation at Hankow were of Chinese manufacture (referred to uncharitably by a Western observer as 'the Chinese bathtub model') until the 1870s, when they began to be displaced by lighter and more efficient American products such as that pictured in Figure 33.1, manufactured by the Gould Company of Seneca Falls, New York.[3] Apparently some five *shui-lung* brigades had been established at Hankow in the 1790s under local military management. These quickly proved insufficient, and at the turn of the century local officials called for private groups to sponsor additional units; almost immediately the official *shui-lung* operations disintegrated and fire-fighting in practice became almost totally a

Figure 33.1 An American-made fire engine (*shui-lung*) used in China, 1879. From an advertisement in *Shen-pao*, Kuang-hsü 5/1/15.

local societal task. The first private group to respond was the pre-eminent Hui-chou Guild, which was surely thinking first of protecting its magnificent new guildhall complex, but which nonetheless claimed the interests of the broader neighborhood as the guiding principle for its action. The Guild held a plenary meeting in 1801, voting to import two fire engines from Kiangsu[4] and operate them out of guildhall precincts. The Hanyang magistrate, in a stele announcing his legal patronage, noted with approval the 'central location' of the Hui-chou Guildhall as the site for such a broad community service, and added that inasmuch as the brigades were funded solely out of guild members' 'contributions,' there was no need for any 'official management.' (One may probably interpret this both as a local official abrogation of direct financial responsibility for fire-fighting matters, and as a warning to yamen functionaires to leave the merchant-sponsored fire brigades alone.) Other guilds soon followed the Hui-chou example.

In the aftermath of the Taiping devastation,

six fire brigades (*shui-chü, lung-chü*) were quickly re-established at Hankow. The number had grown to twenty within a decade, and had doubled again by 1875 [. . .]

The great majority were sponsored by benevolent halls, and operated either out of the hall itself or out of a more centrally located temple. However, several groups such as the Grain Dealers' Guild and the Medicinal Herbs Guild also sponsored brigades at their guild headquarters. The six *shui-chü* of the late 1850s were said to have been initially funded by a broad-based subscription drive among the entire urban population, under the titular leadership of Prefect Liu Ch'ang; their locations and managers were selected in accordance with a single, municipality-wide plan. Those founded later were created by independent initiative of the sponsoring agency, and were financed out of organizational funds.

Our best picture of the management of fire brigades comes from the detailed accounts and regulations left by the Hui-chou and the Shansi-Shensi guilds,[5] in their respective organizational histories of 1806 and 1896 (the time differential conveniently confirming for us the basic continuity of managerial style over the course of the century). The two guilds each operated two fire engines, and their guild regulations contained strict instructions regarding maintenance of equipment and operator readiness. The fire brigades employed a system of headmen (*fut'ou*) comparable to that of labor gangs operating elsewhere in the local economy [. . .] These headmen were fully responsible for the training and operational command of their crews. The Hui-chou Guild's two brigades were each made up of one headman and 21 firemen; [. . .] guild documents made it plain that firemen were full-time careerists. While undoubtedly drawn from the same lower-class labor force as Hankow's porters, carters, warehouse stockboys, and so on, they were comparatively well-paid. The Hui-chou Guild allocated 84 taels per year for the wages of its firemen, an average of nearly two taels per man, far more than it paid its guildhall watchmen, gatekeepers, and custodians, none of whom received more than a quarter of a tael per year apiece. Shansi-Shensi Guild paid its fire headmen 280 cash per year apiece, its skilled firemen 140 cash, and its unskilled men 80 cash, but all received an additional bounty for each time they were called out, as well as occasional bonuses of 'drink money' (*chiu-tzu*).

The rules of the Shansi-Shensi fire brigades also suggest another element of fire-fighting management in the post-rebellion years: a system of assigned precincts. The Guild's two *shui-chü* were each assigned specific territorial jurisdictions, but at the same time it was stipulated that they could go anywhere in the city to fight a fire, if necessity so dictated. Indeed, mutual assistance among fire-fighting companies at Hankow was routine, and numerous newspaper accounts tell of several brigades from various parts of the city cooperating to quell a major blaze, with neighborhood residents manning water buckets to help out. This enthusiasm to assist could be a mixed blessing, however, as the account of a bemused foreign reporter for the *North-China Daily News* makes clear:

No sooner had the first cloud of smoke burst from the upper story of the house in which it had originated, than the alarm was sounded, and almost instantly the first fire brigade appeared on the scene. Not a moment had been lost in coming to the rescue. One would hardly have credited that the twenty or thirty men composing each brigade could have assembled at the headquarters of the corps, donned their uniform, carried out their engines, buckets, pipes, ladders, etc., and have appeared, with banners flying and gongs beating, in so short a time. Brigade after brigade, in distinct uniform, came rushing on in rapid succession, and had there been sufficient room to work all the engines the fire might have been put out in a very short time. But the streets were so narrow and so crowded that only a few of the engines could be got forward, and in the eagerness and rivalry of each group to obtain a vantage ground, collisions were inevitable. Several engines were dashed into each other, water coolies were knocked over, buckets and all, several men received bruises on the head and legs, and the water intended for the fire was in the general melee poured out into the street.

The courage and dash of the firemen was beyond all praise . . . The confusion and the uproar, however, was something indescribable. Everyone was shouting a different order, and loud above the crackling of timbers and the roar of the flames rose the deafening clamours of the multitude, and the incessant din of gongs as if leading an army into battle. By and by some military Mandarins appeared on the scene, and quickly restored order. Detachments of troops cleared the streets of spectators, the different brigades wee better distributed, and within two hours the fire had been extinguished; but not before it had completely destroyed some twenty shops and houses.[6]

The cumulative impression left by Chinese press accounts, though hardly so picturesque, is essentially the same. The fire brigades are greatly appreciated for their energy and bravery, and their successes are far from negligible.

But largely for the reasons of congestion and confusion depicted by the *Daily News* reporter – especially the brigades' 'struggling to be first' (*cheng-hsien*) – all too frequently a major disaster cannot be averted.

The problem of coordination was a natural consequence of the piecemeal establishment of the fire brigades. Operational command was usually exercised by the gentry-directors of the brigade's parent benevolent hall – Yeh Tiao-yuan in 1850 left an amusing picture of these worthies being hastily conveyed to the scene of a blaze in their sedan chairs – but in major fires some local bureaucrat such as the Hankow subprefect would personally assume charge of the operation. The five official brigades of the late eighteenth century were at least in theory centrally administered, and the six revived units of the late 1850s were likewise set up with some intention of central planning. But throughout most of the century, apart from the coordination expressed in their compartmentalized jurisdictions and their commitment to mutual assistance, the various *shui-chü* operated as fully autonomous units, as symbolized by their colorful diversity of uniforms. Only in the early twentieth century was a concerted effort undertaken to bring them under central control. At first an informal agency was established in the Little Kuan-ti Temple to provide operational coordination for the brigades, which remained separately administered. Then, in 1910, the new Hankow Fire Department (Han-k'ou hsiao-fang hui) finally brought all fire-fighting activities under the control of a single, municipality-wide (but still extrabureaucratic) association. The department was headed on a voluntary basis by a wealthy medicinal herbs merchant, Hsu Jung-t'ing.

The emergence of Hankow's fire brigades in the nineteenth century was a significant development in the city's social history. For one thing, of course, the harnessing of private muscle under the aegis of neighborhood or commercial elites had potential consequences for class relations in the city. Imahori Seiji's famous study[7] of neighborhood fire-fighting associations in Peking showed that by the late Ch'ing units in that city had already assumed permanent police or paramilitary functions. The potential for this clearly existed in Hankow as well. [. . .] However, though the presence of this cadre of relatively well disciplined and well paid stalwarts must have afforded their patrons some margin of personal security, the temptation to routinely turn the fire fighters into an elite-controlled local militia was resisted in Hankow. It was resisted, that is, until the *shui-chü* metamorphosed into 'Peace Preservation Associations' (*pao-an-hui*) on the eve of the 1911 Revolution.[8]

More importantly in the present context, fire-fighting was simply one more of the self-nurturance functions that the local urban society came to undertake on its own behalf. Though it remained one of the few areas explicitly delegated by statute to the responsibility of local magistrates, as fire-fighting became ever more sophisticated at Hankow it grew ever more independent of bureaucratic influence. And although the scale of organization clung rather stubbornly to the neighborhood or the guild, increasingly systematic coordination between neighborhood fire-fighting groups reflected a clear rise in the level of urban consciousness and community spirit, and of the municipal public sphere.

Public utilities

At the beginning of the nineteenth century, according to Louis Chevalier, the great city of Paris found itself in a 'pathological' state. Both objectively and subjectively the city's population had outgrown its physical plant. Housing was inadequate, but even more visibly so were such basic utilities as water and sanitation facilities (recall, for example, the famous sewers haunted by Jean Valjean and other unfortunates). The resulting condition was unhealthy both physically and psychologically.[9] Despite the differing cultural perceptions that in France and China governed the distribution of responsibility for such matters, the similarities already noted between early modern Paris and Hankow[10] prompt us to ask whether public utilities in the latter city had, like those in the former, failed disastrously to keep pace with demographic change, with critical results for urban social relations.

The area of public utilities, like those of social welfare, civil construction, and many others, shows the local agents of the imperial bureaucracy at Hankow playing a sharply restricted role, one that was determined by a consistent set of principles but that, in practice, often appears capricious. The question of street lighting is a good example. This remained almost exclusively the domain of societal initiative, the few areas of the city that were effectively lit being so through the efforts of neighborhood philanthropists, street associations, and commercial managers such as those of the night market at Ma-wang Temple. In one recorded case, for example, popular fears of the ghost of a recently drowned neighbor induced a prosperous merchant to finance the lighting of a dark alleyway..Only when the administration felt social order at risk, as it did during the perceived crime wave of early 1880, did it enter the picture even so far as to make it mandatory for neighborhoods to *keep themselves* lit. When a reasonably comprehensive system of municipal lighting did appear in Hankow during the early twentieth century, it was not the government but rather the umbrella association of the city's benevolent halls that sponsored the project.

A similar pattern appears in the history of water supply. During the Ch'ing, Hankow residents had several options in procuring water. Wells were relatively easy to dig in Hankow's sandy foundation, but well water tended to have a high iron content and carried bacteria, so it was generally used only for industrial purposes, for example in dyeshops. One alternative, adopted by some neighborhoods adjacent to the Yü-tai Canal, was to dig a channel diverting river water for local supply. Most citizens, however, relied upon water carried up by bucket from the Han. At one time most households seem to have drawn their own water, making morning sorties for this purpose down to the riverbank, but by the late eighteenth century a class of professional water-carriers (*t'iao-shui*) had succeeded in professionalizing this task, selling water door-to-door by the bucketful. The administration's only role in this process was that of promulgating zoning regulations restricting water-carriers to stipulated 'water-lanes' (*shui-hsiang*). Even this was done not through codified statute, but merely through proclamations issued in individual cases. Usually state intervention came at the request of some important property holder, but officials seem nevertheless to have sought to balance the interests of property with those of consumers and the water-carriers themselves. Even after a private entrepreneur had inaugurated Hankow's first mechanized source of water supply, the Chi-chi Water and Electric Company of 1901, the state continued to restrict itself to providing indirect coordination.

Sanitation facilities, not surprisingly, were grossly inadequate. This was remarked upon not only by foreign residents, who noted 'heaps of filth accumulated in the native streets,' but also by Chinese visitors like the Chekiangese Yeh Tiao-yuan, who wrote in 1850 of the filthy and unsanitary streets in the butchers' quarter, covered with blood draining from fresh-killed animals and fish.[11] Garbage collection and street cleaning, which in some eighteenth-century European cities had already begun to come under municipal control, were left in Hankow to the neighborhoods. The inevitable result was an inconsistency based on the individual neighborhood's financial capabilities; an early Methodist missionary commented on 'the foulness which riotously reigns in the dark [read: poorer] places of this wealthy mart.'[12]

Sewage – removing human excrement from households and the city's several large public latrines – was an even greater problem. One answer was found in professional nightsoil-carriers (*t'iao-fen* or *p'a-ni*), who hauled buckets of excrement to farmers in the suburban green belt. Paid by cultivators for each bucket they delivered, these men also received periodic gratuities from street associations in the urban areas they served. The administration's role was merely regulatory. It limited the hours at which public latrines might be emptied, and in 1880, following the model of Shanghai and Foochow, it ordered that nightsoil pots be covered during transport. As with water-carriers, the administration zoned the routes that nightsoil-carriers might follow, again usually acting on petition from property holders who sought to divert this odoriferous traffic from 'better' neighborhoods. Here too, however, the state's role as coordinator of private and subcommunal interests and as preserver of Confucian harmony dictated that, when called upon to arbitrate specific cases, it could not simply find for property. In one 1874 case, for example, the Taotai, after lengthy investigation, called the nightsoil-carriers' headman and neighborhood propertied interests to his yamen and painstakingly negotiated the most direct and least offensive route for the porters to follow.[13]

The other means for removing excrement was via sewage ditches, leading into the larger flood-water drainage channels, and ultimately to the rivers. These ditches were dug alongside or sometimes down the middle of streets, and their maintenance was left entirely to the individual neighborhood. Such decentralization again led to wide variations in levels of upkeep. In wealthier neighborhoods the sewers were covered with stone and periodically dredged; in the poorer districts they were open and

perpetually blocked up, 'cesspools which have no outlet save in the bubbles of gas which slowly and loudly distill from the brooding filth' and 'fill the air with their concentrated and choice effluvia.'[14] Never in our period did the state evince a systematic concern for the severe problems of health and sanitation at Hankow. Such efforts as did appear – for example, a pamphlet and broadside campaign in 1869 urging greater community action and personal thoughtfulness in these matters – came from concerned local elites or from crusading foreigners such as the missionary doctor A.G. Reid.[15]

The one public utility in which official and societal efforts combined to prove largely equal to their task was public transport, in the form of a ferry service linking the three Wuhan cities. Most of the landmark temple-piers that dotted the Hankow shoreline[16] served as ferry termini to Hanyang (the Ts'ung-san, Wu-hsien, Lao-kuan, and Chieh-chia-tsui piers) or Wuchang (the Lung-wang and Ssu-kuan piers). A growing number of petty entrepreneurs – by the late 1870s several *thousand* of them – had set up commercial operations here, carrying anywhere from four or five to more than a score of passengers, plus freight, per crossing. Community leaders saw chronic problems, however, with leaving this vital element of the area's transport infrastructure to private enterprise. One problem was the unacceptably high number of collisions and capsizings, due to incompetent seamanship, unregulated hustling for fares, and – especially during threats of flood when the demand for escape from Hankow was intense – the reckless disregard by fare-hungry ferrymen for government-issued foul weather warnings. Above all, poor people were excluded by the high fares that ferry services commanded.

Consequently, the felt need for 'charitable ferries' (*i-tu*) was met from a number of sources. Establishing a free ferry, either in flood emergency or on a more permanent basis, became a favored form of private philanthropy. Corporate donors followed the lead of wealthy individuals, from the Hui-chou Guild in 1806 to the Wei-sheng and Jen-chi benevolent halls in the 1870s and 1880s. The most ambitious scheme, however, came from an administrative source, the Hankow Taotai. In August 1877 Taotai Ho Wei-chien established a system of ten 'official ferries' (*kuan-tu*) plying between the Lung-wang Temple and the facing Wuchang and Hanyang shores. The boats were large and unusually sturdy, carrying twenty passengers apiece at the considerably under-cost fare of two cash. Later, the system was modified into a two-class system, with more luxurious ferries charging elite passengers slightly more, thus underwriting the cost of lower-class ferries that operated free of charge. As to the ferries' financing, we are only told that Taotai Ho 'raised funds' (*ch'ou-k'uan*) for their operation, probably by imposing a customs surtax on major parties to the foreign trade (the Shanghai Taotai just a few years earlier had used this strategy to finance public ferries at his port). Management was to rest with a specially created Official Ferry Bureau (Kuan-tu chü), headed by two gentry-directors appointed and paid by the Taotai. Each ferry would be commanded by a master-seaman (*ch'uan-hu*), who would recruit his four-man crew and post bond (*pao-chieh*) for their good conduct.

The new official ferries drew wide acclaim during their first year of operation, but problems set in soon thereafter. In heavy weather the official ferries proved subject to their own share of collisions with other harbor craft, and the Ferry Bureau found itself the target of lawsuits by injured parties. Some of its boatmen, moreover, developed the habit of extorting what fares they could from the supposedly 'free' passengers. Both problems stemmed in part from the growing number of sailors employed part-time (in high-water months only) by the Bureau, over whose competence and conduct it exercised ever less control. In October 1879, therefore, the Bureau's two directors took advantage of a spell of calm weather to hold trials in the harbor for aspiring ferrymen, passing forty out of some seventy applications. *Shen-pao* greeted this innovation with a front-page editorial, praising the Hankow system as a model for official ferries elsewhere in China.[17] Significantly, the editors assumed that the Western presence at Hankow had influenced the system's development (although there is no direct evidence that it in fact did so), and strongly endorsed the Western view of governmental responsibility in public safety and social services.[18] A new conception of the state–society relationship was dawning, which would have very far-reaching implications in the decades to come.

In the overall context of government involvement in public utilities management at Hankow, of course, the energy expended in ferry service after 1877 stands out as exceptional. Overall, the low level of attention paid to utilities does not seem much different from the situation portrayed for early-nineteenth-century Paris by Chevalier. Indeed, there are striking similarities, such as between Yeh Tiao-yuan's description of the grossly unsanitary butchers' quarter and Chevalier's of the Parisian meat market at Les Halles.[19] Yet it would be very far from the truth to say that Hankow at any time in the nineteenth century had been reduced to a 'pathological' state – indeed, it seems overall a city intoxicated with its own youthful vitality. Nor was there any broad popular feeling of entrapment within the confines of a physically outmoded physical environment, or of the critical inadequacy of the available social services. Although these services were for the most part no better than those in early modern European cities, it seems likely that the level of cultural expectation toward them was somewhat lower in Hankow. More positively, we may attribute the lack of a sense of crisis in such matters to the fact that, whereas a commitment to systematic infrastructural improvement was largely absent (it was beginning to dawn in the last years of our period in the pages of the urban-reformist *Shen-pao*), we can clearly see throughout the period a remarkable degree of flexible responsiveness – on the part of the state, the elite, and the community as a whole – to *particular*, acutely felt needs for improvement in the urban environment and facilities.

Notes

The following notes were provided by Arnold Pacey.

1 The author is making a distinction here between *local* officials (notably the magistrate and prefect), and the government of the *province* (the governor of Hubeh – or Hupeh – Province is mentioned later, and there was a viceroy in charge of two provinces)

2 Mark Elvin, 'Skills and resources in late traditional China', in: Dwight Perkins (ed.), *China's Modern Economy in Historical Perspective*, Stanford, Stanford University Press, 1975, pp. 85–113, esp. p. 99

3 We do not have a picture of the Chinese fire engines but they would have been made of wood rather than iron, though otherwise working on the same principle as their American counterparts. Each fire engine would be worked by two men, one using the handle at the back (obscured by the suction pipe in its folded position), and the other working the front handle. There would be another pipe to direct the jet of water onto the fire, and in many situations, it would be necessary to organize a large team with buckets to keep the machine supplied with water

4 Kiangsu (now Jiangsu) Province at the mouth of the Yangzi is where fire engines had been made since the seventeenth century

5 The Hui-chou (Huizhou) Guild is discussed at greater length in the previous extract also by William Rowe

6 *North-China Daily News*, 7 June 1876 (this paper was published in English and circulated among Europeans in the Treaty Ports, including the British Concession in Hankou)

7 Imahori Seiji, *Peipin shimin no jiji kosei*, Tokyo, 1947, pp. 149–54 (this title translates as 'The self-governing organizations of the people of Beijing')

8 The Wuhan cities played a central part in the revolution which overthrew the Empire in October, 1911 and replaced it by a Republic in 1912. While the capture of revolutionaries in Hankou, followed by mutiny of the army in Wuchang, are said to have sparked off the revolution, it appears that the discipline of the fire brigades or *shui-chü* was seen as a source of stability. But there

were extensive fires in Hankou following the army take-over in Wuchang

9 Louis Chevalier, *Labouring Classes and Dangerous Classes in Paris*, Princeton, Princeton University Press, 1973

10 Comparisons between Hankou and early-nineteenth-century Paris (and sometimes London) are drawn at several points in William Rowe's study, e.g. in his first volume, *Hankou: Commerce and Society*, 1984, p. 41

11 The author quoted is described as Chekiangese because he came from Zhejiang (Chekiang) Province, the region around Hangzhou

12 Methodist Missionary Society Archives, School of Oriental and African Studies, London, *Five Annual Reports*, fifth report, 1869, p. 70. Missionaries were always prone to emphasise insanitary conditions, partly because they sought out the poorer sections of society, whereas merchants and diplomats mainly saw the prosperous shopping streets. But missionaries also tended to regard dirty living conditions as symptomatic of a people's spiritual condition – cleanliness really was near to godliness for some of them. Rowe quotes a European non-missionary who thought Hankou 'fairly clean, considering how crowded its streets are' (see Rowe's *Hankou: Commerce and Society*, 1984, p. 25)

13 The Taotai or prefect was the most senior government official in Hankou during the latter half of the nineteenth century. He was responsible for customs, and for diplomatic relations with the British Concession. In the negotiations referred to here, the Taotai aimed to route nightsoil-carriers away from the rice market, to avoid contamination, while respecting property rights in 'private lanes'. Most tran.sport of nightsoil to agricultural areas was ultimately by boat once it had been carried to the shore line or to a jetty

14 London Missionary Society Archives, *First Report of the Hankow Hospital*, p. 6

15 Methodist Missionary Society Archives, fifth report (listed in note 12 above), p. 78

16 Landing stages were often associated with temples to local deities (which might also be loosely Taoist or Buddhist in affiliation). Such temples were popular meeting places, and sometimes functioned as public buildings. A ferry terminal might be approached through an archway belonging to a temple, and in the case of the Wu-hsien Temple (modern spelling, Wuxian), there was also a teashop nearby

17 *Shen-pao* was a Chinese-language newspaper published in Shanghai. It carried advertisements for the American fire engine mentioned earlier. While its staff of Chinese journalists reflected an authentic current of opinion in China among those who advocated modernisation along western lines, the newpaper's standpoint was also that of a 'treaty port' publication with British owners. Thus in praising improvements made in Hankou, it would inevitably be biased toward see-

ing the western presence in that city as a beneficial influence

18 Although newpapers such as *Shen-pao* and *North-China Daily News* are important sources for Rowe's account of Hankou, as are missionary records, he counter-balances the Eurocentric bias of such information by extensive use of local Chinese sources such as Guild records, Benevolent Hall records, and gazetteers. However, references to these documents are omitted here because of their inaccessibility to most readers

19 Chevalier, *Labouring Classes* (see note 9 above), pp. 210-11

34

A DESCRIPTION OF HANKOU IN 1850

by Évariste Régis Huc

Source: M. Huc (Évariste Régis Huc) *The Chinese Empire: a sequel to recollections of a journey through Tartary and Thibet*, trans. J. Sinnett, London, Longman, new edn, 1859, pp. 362-7. The author spells Chinese words in a non-standard manner, with Hankou rendered as Han-keou. Where it is important to know the modern spelling, this is given in the notes.

It is so common for people in Europe to form their opinions of the Chinese from the drawings on screens and fans, and to regard them merely as more or less civilised baboons, that we were glad to have an opportunity of showing how they treat questions of policy and social economy [. . .]

The immense population of China, the richness of its soil, the variety of its products, the vast extent of its territory, and the facility of communication by land and water, the activity of its inhabitants, its laws and public usages, all unite to render this nation the most commercial in the world.

On whichever side a stranger enters China, whatever point may first meet his eye, he is sure to be struck, above all else, by the prodigious bustle and movement going on everywhere under the stimulus of the thirst of gain, and the desire of traffic by which this people is incessantly tormented. From north to south, from east to west, the whole country is like a perpetual fair, and a fair that lasts the whole year without any interruption.

And yet, when one has not penetrated to the centre of the Empire, when one has not seen the great towns, Han-yang, Ou-tchang-fou, and Han-keou, facing one another,[1] it is impossible to form an adequate idea of the amount of the internal trade.

Han-keou especially, 'The Mouth of Commercial Marts,' must be visited, for it is one great shop; and every production has its street or quarter particularly devoted to it. In all parts of the city you meet with a concourse of passengers, often pressed so compactly together, that you have the greatest difficulty to make your way through them. Long lines of porters stretch through every street; and, as they proceed with a peculiar gymnastic step, they utter a measured monotonous cry, whose sharp sound is heard above all the clamours of the multitude. In the midst of this crowded vortex of men, there prevails, nevertheless, a very fair amount of order and tranquillity; there are few quarrels, much less fights, although the police is far from being as numerous as in most of our cities in Europe. The Chinese are always restrained by a salutary fear of compromising themselves that acts like an instinct; and though they are easily excited, and induced to vociferate, they are soon quiet again, and things return to their usual course.

In seeing the streets thus constantly thronged with people, you might be apt to think that all the inhabitants of the town must be out, and the houses empty. But just cast a glance into the shops and you will see they are crowded with buyers and sellers. The factories also contain a considerable number of workmen and artisans;[2] and if to these you add the old men, women, and children, you will not be surprised to hear the population of Han-keou, Han-yang, and Ou-tchang-fou, taken together, estimated at eight millions.[3] We do not know whether the inhabitants of the boats are included in this calculation, but the great port of Han-keou is literally a forest of masts, and it is quite astonishing to see vessels of such a size, in such numbers, in the very middle of China.

We have said that Han-keou is in some measure the general mart for the eighteen provinces;[4] since it is there the goods arrive, and thence depart, which are intended to supply all the internal trade. Perhaps the world could not show a town more favourably situated, and possessing a greater number of natural advantages. Placed in the very centre of the Empire, it is in some measure surrounded by the Blue River,[5] and brought into direct communication with the provinces of the east and the west. This same river, on leaving Han-keou, describes two curves, to the right and left, and bears the great trading junks towards the south as far as the bosom of the lakes Pou-yang and Thoungting, which are like two inland seas. An immense number of rivers, which fall into these two lakes, receive in small boats the merchandise brought from Han-keou, and distribute it through all the provinces of the south. Towards the north the natural communications are less easy, but gigantic and ingenious labours have come to the aid of nature, in the numerous artificial canals with which the north of China is intersected, and which, by marvellous and skilful contrivances, establish a communication between all the lakes and navigable rivers of the Empire, so that you might traverse its entire extent without ever getting out of your boat.

The Annals of China show that at every period each successive dynasty has paid great attention to the canal system; but no other work is comparable to that which was executed by the Emperor Yang-ti, of the dynasty of Tsin, who ascended the throne in the year 605 of the Christian era.[6]

In the first year of his reign he dug many new canals, and enlarged the old ones, so that vessels could pass from the Hoang-Ho[7] to the Yang-tse-kiang, and from these great rivers to the principal smaller ones. A learned man named Siao-hoai, presented to him a plan for rendering all the rivers navigable throughout their entire course, and making them communicate, one with another, by canals of a new invention. His project was adopted and executed, and there were consequently made, remade, and repaired, more than four thousand eight hundred miles of canals.[8] This great enterprise of course required a vast amount of labour, which was divided between the soldiers and the people. Every family was required to furnish one man, between the ages of fifteen and fifty, to whom the Government gave nothing but his food. The soldiers, who had to do the most painful part of the work, received a small increase of pay. Some of these canals were lined with freestone throughout their entire length, and during our various journeys we saw remains enough to attest the beauty of the works. That which ran from the northern to the southern court[9] was forty feet wide, and had on its banks plantations of elm and willow. That from the eastern to the western court was less magnificent, but also bordered with a double line of trees.

Chinese historians have branded the memory of the Emperor Yang-ti, because during his reign he never ceased to oppress the people by these *corvées*, to satisfy his own caprice, and his taste for luxury, but they acknowledged that he deserved well of the Empire for the benefits conferred by his canals on the internal trade.

The wealth of China, its system of canals, and the other causes already assigned, have doubtless contributed much to develop in the country the prodigious commercial activity that has been remarked in it at all epochs; but it must be acknowledged also that the character and genius of its inhabitants has always disposed them to traffic. The Chinese has a passionate love of lucre; he is fond of all kinds of speculation and stock-jobbing, and his mind, full of finesse and cunning, takes delight in combining and calculating the chances of a commercial operation.

The Chinese, *par excellence*, is a man installed behind the counter of a shop, waiting for his customers with patience and resignation; and in the intervals of their arrival pondering in his head, and casting up on his little arithmetical machine,[10] the means of increasing his fortune. Whatever may be the nature and importance of his business, he neglects not the smallest profit; the least gain is always welcome, and he accepts it eagerly: greatest of all is his enjoyment, when, in the evening, having well closed and barricaded his shop, he can retire into some corner, and there count up religiously the number of his sapecks, and reckon the earnings of the day. [. . .][11]

The only legal coinage existing in China is a little round piece made of a mixture of copper and pewter, and called by the Chinese *tsien*, and by the Europeans sapecks. They are pierced through the middle with a square hole, in order that they may easily be passed on a thread. A string of a thousand of these pieces is equivalent usually to a Chinese ounce of silver: gold and silver are never coined in China; when employed for larger purchases than can be paid for in sapecks, they are weighed like any other commodity; sapecks are used for all small transactions, and agreements are also made in strings of sapecks.[12]

The Chinese in the towns generally carry with them little scales for buying and selling, and weigh all the money they give or receive. Bank notes, payable to the bearer, are in use throughout the whole empire; they are issued by the great houses of business, and accepted in all the principal towns.[13]

The value of a sapeck is about half a French centime; and this small coinage is àn incalcul-able advantage to small dealings. Thanks to the sapeck, one may traffic in China on very small means. One may buy a slice of pear, a dozen of fried beans, a few melon seeds, or one walnut, or one may also drink a cup of tea, or smoke some pipes for a sapeck; and a citizen who is not rich enough to afford himself a whole orange will often purchase a half. This extreme division of Chinese coinage has given birth to an infinity of small occupations that afford a subsistence to thousands of persons. With a capital of two hundred sapecks (ten pence English), a Chinese will not hesitate to commence some mercantile speculation. The sapeck is especially an immense resource for those who are asking alms, for a man must be poor indeed not to be able to give a beggar a sapeck.

Notes

The following notes were provided by Arnold Pacey.

1 This sentence refers to the three cities now known collectively as WUHAN facing each other across the Yangzi (Yangtse) and Han Rivers. Wuchang is here written Ou-tchang-fou (a fair representation of the pronunciation), with the syllable *-fou* or *-fu* added as was often done to denote cities with the status of a provincial capital

2 'Workshops' would be a better word than 'factories', since the dyeworks and other textile enterprises were not mechanized, nor were the sack and basket-making shops, nor yet the iron foundry (or foundries). The power of a donkey walking round and round in a gin may sometimes have been used to drive a lathe or other machine, however

3 This population figure is about five times too large. By contrast, official estimates tended to *underestimate* population, and a figure of 800,000 (pronounced 'ba shi wan') may have been quoted to Huc. Since he was already impressed by how crowded Hankou was, he may have misheard this as eight million (pronounced 'ba yi bai wan'). Rhoads Murphey discusses the statistic and comments that European critics regarded Huc as a liar, or teller of tall stories. They were not prepared for the notion of China as a major commercial economy. See Murphey's fn. 78 in Mark Elvin and G. William Skinner, (eds), *The Chinese City Between Two Worlds*, Stanford, Stanford University Press, 1974, p. 396

4 The 'eighteen provinces' represent the whole extent of China from the Great Wall in the north to the borders of

Vietnam and Burma in the south, but exclude the outlying empire in Manchuria, Mongolia, Xinjiang and Tibet

5 Travelling downstream from Hankou, the Yangzi, here referred to as the Blue River, takes a south-easterly direction as far as Lake Poyang, and the lake extends a further 100 km. to the south

6 Huc's historical information is correct. The Emperor Yangdi (Yang-ti) of the Sui dynasty ruled from 604 to 617 CE, and an edict of his dated 605 did indeed initiate a major phase of canal-building

7 The Huang He (Hoang-Ho) is what westerners call the Yellow River

8 This figure may represent the total extent of the waterway *system*, including navigable rivers, but the length of *canal* constructed during Yangdi's time would be somewhat less than a quarter of the distance given

9 The original edition has a fn. here commenting that the imperial court during the Sui dynasty moved between four different cities

10 The 'little arithmetical machine' is the abacus, which is still extensively used for adding up bills and other calculations

11 Huc's generalizations here may apply to some *men* he met in the commercially-oriented context of Hankou, but cannot be true of all Chinese women and men. So while Huc dismissed one stereotype of the Chinese people in the first sentence, he is here constructing another

12 Sapeks were also referred to by Europeans as 'copper cash', or just 'cash', and a thousand made a 'string of cash'. Think of one 'cash' as a penny and a string of cash, worth an ounce of silver (or tael), as a £10 note

13 The Chinese banking system had developed rapidly during the period of commercial expansion in the eighteenth and early nineteenth centuries, and Huc is correct in saying that notes were issued by individual banks and not by the government. See Susan Mann-Jones on Ningpo and Shanghai banking houses in Mark Elvin and G. William Skinner (eds), *The Chinese City Between Two Worlds* (see above note 3), pp. 73-96

35

HAUSA BUILDING TECHNIQUES

by J.C. Moughtin

Source: J.C. Moughtin, *Hausa Architecture*, London, Ethnographica, 1985, pp. 99–115.

Introduction

The elements of Hausa material culture have been developed from the natural products in the local environment, until quite recently, supplemented in a limited way only with imported goods. But under the growing influence of industrialized nations, the number and variety of materials and techniques for constructional purposes has increased dramatically. The changes in building technology caused by external contacts are most apparent in the work of government agencies and commercial concerns, but have been introduced into the traditional structure of the old cities by the growing middle class: those who can afford to now build in concrete blocks, and the tin roof is preferred to thatch or the mud dome. Yet despite these rapidly accelerating changes, which may bring about the demise of a once great architectural tradition, Hausa builders are still only marginally affected by the new technology. Most constructional work outside the scope of the formal building industry is still of materials found in the local environment.

Laterite, used with various additives for walls, roofs and finishes, is still the most important material used by the traditional Hausa builders. There is an abundance of stone in the many stark, black granite outcrops – those marvellous inselbergs that contrast so sharply to the flat savanna landscape of the Hausa plains – yet its use, as a traditional building material is confined to a few of the smaller ethnic groups of the Jos plateau. The primitive stone walls of these peoples are a reminder of earlier great builders responsible for the abandoned and enigmatic dry stone walls of the plateau. Evidence of earlier great periods of building is to be found in travellers' descriptions of traces of ancient burnt brick buildings in areas adjacent to Hausaland.[1] Such archaeological remains may indeed be the last vestiges of a building technique introduced by Islamic scholars such as al Saheli many centuries before.[2] Rejection of diverse forms of structural techniques, and concentration on the development of one system of construction and the almost universal use of one building material has produced a unified architectural composition for complete cities. The uncompromising nature of the architectural pattern derived from constantly repeated structural forms expresses the social and spiritual needs of the Hausa through the symbolic use of those forms and the spaces they create. It is because of this unity of form and the meaning attached to it that Hausa building can be classified and studied as architecture.

The main structural types

There are three main traditional structural types: the circular room made entirely from vegetable material (*dakuna*); the circular, fig-ure-of-eight or rectangular room with mud walls, thatched roof and sometimes a verandah (*adada*), and the building constructed entirely of mud (*soro*) reinforced with beams split from the *deleb* palm. Mud buildings (*soro*) take a number of forms: they may be circular or rec-tangular and have a roof supported on corbels, pillars and beams or arches; some are elabo-rate, two-storey structures (*bene*). In addition to these traditional building types others have been introduced over the last thirty years which are made of more permanent materials. For example, the tin roof may be used with any combination of the former structures, repla-cing the mud or thatched roof and sometimes in addition to an existing mud roof to increase its weather resistance. The latest and most desirable structure, to which many Hausa peo-ple aspire, consists of concrete-block walls, tin roof and verandah. These are the principal building units which now make up the city; in this study, however, only those of mud will be discussed in detail.

The structural process

The constructional process for a mud building is a long one: too little time spent in preparing the materials or in the construction of the building results in repeated and costly mainte-nance. Preparation of the building takes two to three weeks: the minimum period for building the walls of a small room from foundation to wall-plate is ten days. Since there is no way of speeding up the process, building starts at the beginning of the dry season and ends just before the rains; a period of three months is necessary to build the normal house.

The mud walls are made from regular courses of unbaked bricks laid in mud mortar, (see Figure 35.1). The mud brick is made from earth (*kasa*), preferably red laterite which is thoroughly soaked in water, left for twenty-four hours, again soaked, trampled and kneaded. This process is repeated a number of times before the earth is rolled into bricks, usually circular cones varying from 5cm. to 15cm. in diameter, depending on the district

Figure 35.1 Mud wall construction 1

in which they are made. Making the building mud is graphically described in *Labarun Al'Adun Hausawa Da Zantatukansu*:

> Then they started work and dug up the earth on the surface. It was soaked with the water which they brought in water-pots, from morning till about three p.m. – their time for leaving off work. They left the earth to soak for twenty-four hours. On their arrival at daybreak they turn it over with all their might, mixing it thoroughly with water till it is properly mixed. When it has been properly mixed they leave it for another twenty-four hours . . . This is the work they will go on doing every day without interruption for fifteen days.[3]

Before they can be used structurally, the bricks are allowed to dry out thoroughly, which takes about ten days. Old bricks taken from ruins are sometimes re-used in the top courses of new buildings, but better practice dictates breaking them down into earth and repeating the process of soaking and trampling.

The method of making mortar is similar to that used for bricks: after soaking and trampling the earth it is covered with horse manure and continually soaked with water for several days. The mixture is then trampled to get a thorough mixing of earth and dung; and the process is repeated using a fresh layer of dung. This material is mixed three or four times and is ready for use in two to three weeks.

The first step in erecting a building is to clear the ground (*schema*) of stones, vegetable matter and topsoil. Then the building is set out using pegs, ropes and hoes. The plan of the house is drawn on the ground by the chief builder with his foot:

> They came in the morning, and Tanko gave them a rope, hoe and 'pegs'; they marked out the house exactly rectangular, with its entrance facing south; they marked out four huts, a square house, a mud-roofed house, and an entrance-hut.[4]

Shallow trenches (0.45m.) are dug along the line of the plan just beneath the loose topsoil: mud foundations without footings are constructed in the trenches, the load of the roof

and wall being distributed over a large area of subsoil because of the large batter given to the wall. Foundations of important buildings are protected from erosion by the construction of a wide external plinth at the base.

Wall construction

When constructing a wall only two or three courses of mud bricks are laid in one day, and on reaching the height of the door lintel the work is suspended for twenty-four hours so that the walls are thoroughly dry before completing the top courses. The normal method of building is for the builder to sit astride the top of the unfinished wall; mud bricks and pats of mortar are thrown up to him, (see Figure 35.2).

Figure 35.2 Mud wall construction 2

On finishing the day's courses within his reach, the builder moves backwards, away from this new work, and sits on that part of the wall built the previous day. So scaffolding and ladders are used in the construction of large buildings only, such as the emir's palace and important mosques, but ladders are sometimes used for repairs, maintenance and decoration.

Hausa builders have learned, through the accumulated experience of many generations, that the stability of a clay wall is increased by decreasing its thickness towards the top. Consequently, the battered wall [a wall tapering in width as it rises] is a universal and very beautiful feature of Hausa architecture. The use of the vertical line and plumbob is unknown to traditional builders: they achieve structural stability by the mass of the wall acting vertically downwards. Buttresses are not important elements and are used only to prop failing walls.

Because mud roofs are heavy, walls supporting them must be extremely thick. Then they are often strengthened by timber reinforcement taken from the *deleb* palm, the *dumi* palm or *kurna*, none of which is attacked or destroyed by the white ant, a termite found in large numbers in many parts of West Africa including Hausaland. Timber taken from the *deleb* palm for building purposes is called *azara*. It is placed in walls about half to one metre above the ground and again at a level just above the height of the door head. This reinforcement consists of *azarori* laid transversely across the width of the wall, on top of which are placed additional *azarori* running longtitudinally. Walls vary in thickness from 15cm. for a small partition to 1.20m. for two-storey buildings (see Table 35.1). Piers and columns are also reinforced using groups of bonded *azarori* bound together with mud mortar and surrounded by a thick coating of mud: their size varies from 0.9m square to 1.8m. square.

Table 35.1 Wall thickness at ground level

Thickness	Type of Wall
1.2m.	for two-storey buildings
0.9m.	for rooms 4.5 × 4.5m. and over
0.75m.	for rooms 3.9 × 3.9m. to 4.5 × 4.5m.
0.6m.	for rooms less than 3.9 × 3.9m.
0.45m.	for partition walls
0.3m.	for low partition walls and small grass-roofed huts
0.15m.	for partition walls less than 1.8m. high

Source: Based on table in A.F. Daldy, *Temporary Buildings in Northern Nigeria*, Nigeria, Public Works Department, 1945, pp. 4–5

Roof construction

There are three main types of traditional roofing. The cheapest and most common is conical and thatched; then comes the flat, mud roof and finally, until very recently, the most expensive and prestigious, the domed mud roof supported on mud arches. Corrugated iron is now becoming the most popular roofing material and the most prestigious.

Whenever possible, the conical thatched roof is made from the fronds of the raphia palm. But this is expensive and it is more usual for raphia palm fronds to be used for the main rafters, with sticks (*kara*) or guinea-corn stalks (*karan dawa*) between them. For small huts, the whole roof is generally constructed on the ground then lifted into place on the mud walls; roofs of huts with a diameter greater than 4m. are constructed *in situ*.

The wall plate for the roof is made first and to it the thickest ends of the palm stalks are fixed; their thin ends meet at the top where they are securely bound with rope. Bands of corn stalks are tied around the sloping rafters at centres of between 0.2m. and 0.3m. Thatch is brought up to the site in bundles of about 8m. long and up to 1.5m. wide, depending on the length of grass available. The grass is held

together near the thick ends with one row of
sewing, and is unrolled from the bottom of the
roof towards the apex. The thick ends of the
grass are at the bottom, each layer of grass
overlapping the one below by a few centimetres less than its total length. All the layers
of grass are sewn to the framework and the
whole roof is held down by a net which is
made by tying rope over the thatch at about
0.6m. to 0.9m. centres, (see Figures 35.3 to
35.7).

The simplest mud roof is formed by spanning
a space with *azara* joists. The economic span
of *azara* is about 1.8 metres; in other words,

Figure 35.4 Construction of building with thatched
roof 2

Figure 35.3 Construction of building with thatched
roof 1

Figure 35.5 Construction of building with thatched
roof 3

Figure 35.6 Construction of building with thatched roof 4

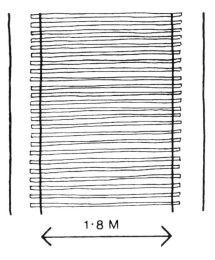

Figure 35.8 Spanning a space of 1.8m

Figure 35.7 Construction of building with thatched roof 5

the weight of the usual thickness of mud required to weatherproof the roof can be held up by closely spaced *azara* joists without an intermediate support over a space of just under 2 metres. It is usual to place the *azara* beams less than 2.5 centimetres apart, but for the sake of economy they may be at intervals of 15 to 30 centimetres, with sticks placed over them at right angles to the main span. The whole structure is covered with grass matting on which the mud roof rests. An understanding of the structural limitations of this simple roof form is the key to analysing and interpreting the form of Hausa architecture from the simplest mud cell to the most

complex roof patterning adopted by Mallam Mikaila in the Friday Mosque, Zaria[5] (see Figure 35.8). Spaces larger than the economic span for *azara* may be roofed by a complex system of corbelling and coffering; the division of the space by pillars, or the use of arches. Sometimes a combination of such devices is employed.

When corbels are used they are reinforced with several layers of *azara* projecting 0.45m. from the face of the wall at about 2.1m. centres. The space between the corbels is spanned by a beam made of several layers of *azara* from which *azara* joists span across to the other wall in the usual way. This extension of the constructional system allows a room to be increased in width from 1.8m. to 2.7m., (see Figure 35.9).

Roofing a rectangular room larger than 1.8m. square requires the use of *azara* placed diagonally across the corners of the space. From these triangular platforms at the corners of the room additional beams span parallel to the walls. By repeating this process a shallow reinforced dome is formed covering a room about 3.4m. square: the free space to be bridged by

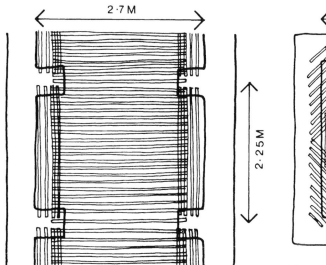

Figure 35.9 Spanning a space of 2.7m

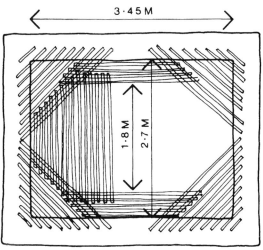

Figure 35.11 Spanning a space of 2.7m × 3.45m

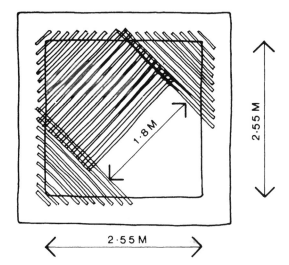

Figure 35.10 Spanning a space of 2.55m square

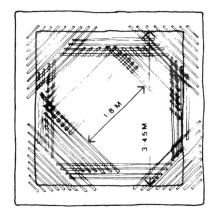

Figure 35.12 Spanning a space of 3.45m square

the joists is gradually reduced to 1.8m. by the clever system of corbelling shown in Figure 35.12.

For very large rooms, the roof may be sup-

ported on a series of columns connected by beams. The column is usually surmounted by a simple capital, consisting of two or four *azara* corbels which make it possible to increase the spacing between columns from 1.8m. to 2.7m. The spaces between the columns are spanned by beams reinforced in the usual way with *azara*, the roof joists spanning between the beams. Figure 35.13 is the draw-

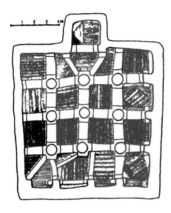

Figure 35.13 Roof plan of the Mosque at Kazaure

ing of the roof plan and Figure 35.14 is a detail of the column of the Friday Mosque at Kazaure, a simple trabeated structure in which some of the corbelling techniques have been used.

Arch construction

Forming spaces larger than 3.5m. square requires the use of the mud arch: with this structural system it is possible to construct rooms 8m. square. The 'mud arch' as used by the Hausa builders is not a true arch in the structural sense of the word, but simply a series of reinforced mud corbels placed one on top of the other until they meet at the centre. They are coated with mud to take on the shape and outward appearance of the arch.

Figure 35.15 is a section through such an arch showing a typical arrangement of *azara* reinforcement. In good construction, the layers of *azara* should not project more than about 0.7m., nor should the change in angle between succeeding *kafi* be too great. For these reasons the arch should start quite low down as in the Friday Mosque, Zaria.

Figure 35.14 Column and capital, Mosque at Kazaure

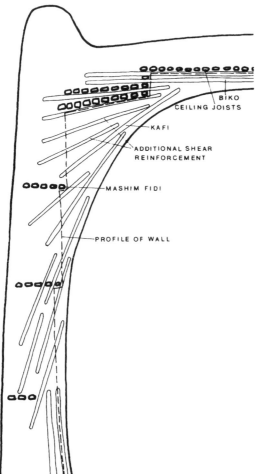

Figure 35.15 Section through an arch showing reinforcement

Arches are normally constructed in the following manner: each layer of reinforcement is tied back to the preceding one, beginning from both walls and working upwards and outwards towards the centre of the room. When the gap between the two halves of the arch is small enough, horizontal *azara* called *biko* are used to complete the arch. Each corbel is allowed to dry overnight before the next one is constructed: in this way the arch can be built without centring or scaffolding. Additional *azara* for sheer reinforcement are placed at right angles to the wall and project into the body of the arch. Lengths of *azara* put in the wall at the back of the arch at right angles to it distribute thrusts through a large area of wall and prevent cracking. In a rectangular room where the shortest side is less than about 4.5m., the room is usually divided into bays of 2.1m. and simple arches span across the room parallel to the shortest side. *Azara* covers the area between the arches which usually conforms to the standard 1.8m. (see Figure 35.16).

A room having all walls longer than 4.5m. may be roofed using arches in three ways: internal pillars with arches spanning from them to the walls (see Figure 35.17); arches spanning from wall to wall, all passing through one central point (*kafin laima*) like a tent construction (see Figure 35.18); arches spanning from wall to wall, but all being parallel to one or other wall called *daurin guga* (see Figure 35.19); a

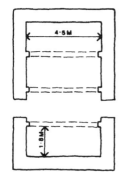

Figure 35.16 Arch construction: room 4.5m wide

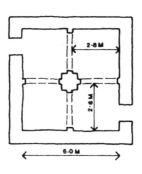

Figure 35.17 Arch construction: room with central pillar

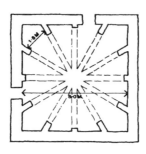

Figure 35.18 Arch construction: *kafin laima* construction

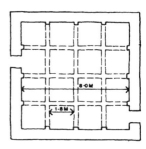

Figure 35.19 Arch construction: *daurin guga* construction

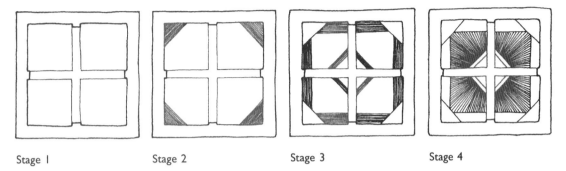

Stage 1   Stage 2   Stage 3   Stage 4

Figure 35.20 Process of arch construction

variation of this latter type of roof consists of three half arches meeting at mid span [. . .]

Figure 35.20 illustrates the process of building a simple, domed roof supported on four half arches [. . .] The arches are built out from the walls until they meet at the summit as previously described. Then diagonal lengths of *azara* are laid across the corners of the room, forming triangular platforms. From these platforms *azara* beams are carried over the backs of the arches. Similar but lighter beams are formed near the apex of the vault from which the *azara* joists span to the wall beam. Using this form of construction most of the weight of the roof is placed directly on the wall, reducing the loading on the centre of the arches which is the most vulnerable part of the structure (see Figure 35.21).

Figure 35.21 Arch: structural failure

The Friday Mosque, Zaria

The six domes of the Friday Mosque, Zaria show all the subtleties of structural technique available to the Hausa architect. Here Mallam Mikaila has exhibited his greatest skill as a builder. Compare the size of the main hall (20m. × 23m.) with that of the *Shari'a* court (7m. × 7m.) which pushes *kafin laima* structure to its limits. Using arches springing from all four walls in the *daurin guga* style produces a space of about 5.5m. × 12.5m., far too small for the purposes of Mallam Mikaila. Instead, he created six main spaces, each approximately 7.0m. square, using two main structural walls, twenty-one piers and one isolated wall at the centre of the composition. At first sight the pattern of the coffering on the ceiling may look arbitrary, but it represents the culmination of many generations of structural experiment. The domes and the arches that support them are bound together into a rigid and monolithic structure by a series of rising beams along the haunches of the arches so that the great mass of the clay domes is transferred to pillar and wall with the least possible eccentric loading.

An important visual feature of the mosque, which perhaps gives it much of its character, is the repetitive use of the double arch supported on twin piers. This device is a simple effort to increase both the free space between the piers and the stability of the structure. But the twin arch was an important structural innovation: it was for mud buildings as important a development as was the flying buttress for medieval European architecture. Look at the roof plan of the mosque and imagine the two main spaces constructed using three arches in both directions (see Figures 35.22 and 35.23). This roofing solution is the obvious one conforming to the modular discipline of *azara* construc-

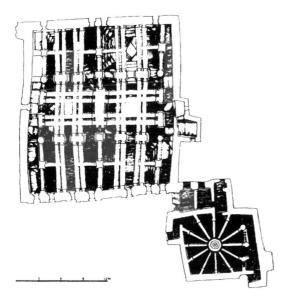

Figure 35.22 Friday Mosque, Zaria, roof plan

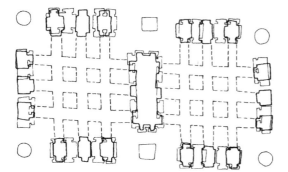

Figure 35.23 Friday Mosque, Zaria, alternative planning solution

tion, but it leads to the use of more piers, one of which is badly placed in relation to the *mihrab* [the devotional focus of the mosque]. Following this system each dome is supported by only six instead of eight arches. By doubling up the arches it is possible to create voids in the centre of each face of the space, which most architectural critics would consider aesthetically desirable. Mallam Mikaila therefore designed a stronger solution and one that is visually more pleasing, while maintaining a distance between arches within the limits of the economic span of *azara*.

Modular design

The structural technology employed by Hausa builders is one important factor in determining the architectural form of buildings. Heights and plan sizes, while not rigidly conforming to a strict set of rules, nevertheless provide the designer with an almost modular discipline based upon a set of preferred dimensions determined largely by the constructional process. Heights, to a large degree, are decided by simple reference to the human body. According to the account in *Labarun Al'Adun Hausawa*, the arches spring from about shoulder or head height: 'When the building of the mud-roofed room reaches the height of a man's shoulder, or his full height, short *azaras* are put in to form the base of the arch'.[6] The height of a simple flat-roofed domestic building is about that of a man with upraised hand; the height of a room with a roof supported on arches is entirely dependent on the span, which in turn determines the height of the arch.

Strictly speaking the 'mud arch' is a curved beam made from a series of cantilevered corbels reinforced with *azara*: it is not a true arch, which is an arrangement of wedge-shaped blocks mutually supporting each other. The shape of the Hausa arch is decorative: it takes on an approximately semicircular form, with the base of the semicircle at the level where the first layer of reinforcement cuts the face of the wall. The usual practice is to continue the arch shape downwards for some distance, giving it the appearance of a semi-eclipse; sometimes it is narrowed into the 'horse shoe' shape so common in the medieval Islamic architecture of North Africa. In yet other examples, including the Friday Mosque, Zaria, the arches spring from the ground. This is stronger not only visually, but structurally, particularly if *azara* is carried down this extra length of arch.[7] Despite its structural ambiguity in visual terms, the preferred shape of the arch determines the height of the building. [. . .]

There is no rigid system by which building heights are determined; this is done empirically, where necessary making reference to the height of a person, or simply working within the structural limits of the building material and the preferred constructional technology. The building technique is such that ladders and scaffolding are used only on important edifices and mainly for maintenance. So the maximum height of a single-storey wall in those buildings which comprise most of the settlement is determined by the height to which a clay brick can be thrown to the builder who sits astride the top of the wall.[8] The height of the wall plate in a circular thatched room is also standardized at approximately the height of a man.

Although in Hausa settlements buildings vary in height, the variations are all within close limits of the standard norm for the particular structural type and its plan form. The account of the building process in *Labarun Al'Adun Hausawa* shows clearly that the Hausa are interested in the appearance of their buildings and conscious of the fact that height has an effect on appearance:

> One must take care about the height, lest one hut be higher than another. It is better to arrange it so that they are all of the same height.[9]

The diameter of the round hut with a thatched conical roof is determined largely by the lengths of available rafters; most round huts are between 3 and 4m. in diameter. Sizes outside this range do occur, but it is more usual to join two round huts together to increase the living area. In such cases, each hut has an independent conical frame; the frames are joined together at the points with a ridge pole and the whole frame, so formed, is thatched as one roof.

Plan dimensions of mud buildings are governed by the system of dimensions that results from the use of economically sized balks of *azara* as both joists and reinforcement (see Figures 35.8 to 35.12 above, pp. 223–224). The maximum room size using this construction is about 3.5m. square at roof level, which after allowing for the batter on the walls gives a floor space of approximately 3m. square. Increasing room sizes further requires the use of columns or piers whose spacing is again determined by the effective span of *azara* beams which, including two mud corbels, gives a clear space between columns at the capital of about 2.7m. Based on these dimensions the ideal arrangement of columns is at approximately 3m. centres, after allowing for a column size of 1m. square with a batter on each face. The form of the building resulting from such a structural system is very similar to the hypostyle hall of Pharaonic Egypt, the product of a similar trabeated constructional system but in stone. In both cases the roof is supported on a forest of columns which in Hausaland vary in size from the huge polished clay pillars of the Bauchi Mosque (about 2.7m. × 2.1m. at the base) to the slender cylindrical columns of the Shehu's Mosque in Sokoto which are only about 1m. in diameter at the base.

Arch construction is also based on the set of dimensions which determines bay sizes in flat roof construction. Rooms where one dimension is less than 4.5m. can be formed using arches spanning in the shorter direction only. The arches are built at centres of 2.1m. so that the roof is divided into bays of 1.8m., making the length of the room 1.8 + 0.3 + 1.8 = 3.9m., or 1.8 + 0.3 + 1.8 + 0.3 + 1.8 = 6m., and so on to a maximum size of 12.3m.

Rooms greater than 4.5m. wide require arches spanning in two directions. The limit of the structural system using either *kafin laima* or *daurin guga* appears to be the construction of a space 8.0m. square. At this size great care must be taken with the construction and first class supervision is required. As shown earlier, the Friday Mosque in Zaria is particularly interesting in its layout of the main spaces where the architect has extended the system to its limits by the technique of twinning the structural arches. Using normal methods, with two single arches in both directions, a floor plan of approximately 5.4m. is created. However, by using sets of twin arches the architect has been able to increase the free floor area to 7.2m. square while maintaining the discipline of spaces between main structural members conforming to the requirements of *azara*. [. . .]

Protecting the building from the weather

A steeply sloping thatched roof with large overhang protects the walls of the Hausa peasant building from rain. But protection from rain of the mud roofed building is a much more complicated process of design.

It is important that rainwater is removed from the building as quickly and efficiently as possible. For this purpose all roofs are designed with a minimum fall of 1 in 15, which is achieved by laying the *azara* joists at this angle and not building up the thickness of clay at one side of the roof. The roof, as was mentioned above, consists of mats placed on the joists followed by 15cm. of mud then 5cm. of waterproofing compound. The rainwater is conducted from the roof by spouts projecting from 0.6m. to 0.9m. from the face of the wall. At one time the spouts were made of pottery, but now they are entirely of beaten tin. At the foot of the wall is a protective plinth which prevents the rainwater damaging the main wall and its foundations.

In addition to these design precautions, the external surfaces of the building are protected with various waterproof finishes. The finish used on a particular building depends upon its type and purpose and the prestige of its owner; on how much is to be spent upon this element of the building and the orientation and exposure of the surface to be treated. *Laso* is the most important waterproof cement in use; its main ingredient is *katsi*, a by-product of the dyeing trade. The slurry at the bottom of a worn out dye pit is removed, partially dried, then moulded into lumps. These lumps are put between layers of wood in an open-air kiln and thoroughly baked. After baking, the lumps are beaten into powder: it is this powder which is known as *katsi*. The description of the manufacture of *laso* is given in *Labarun Al'Adun Hausawa* and is as follows:

> Cement cannot be prepared quickly. A hole is dug first, about four feet [1.21m.] deep. Then *katsi* is taken and poured into the hole together with old indigo liquid. It is stirred in the hole. When it has been properly stirred, it is covered and left for two or three days. Then hair is got from a tannery, and the roots of the wild vine, as well as horse manure without any straw in it. These vine roots are thoroughly beaten with a stone till they are soft. They are poured into large earthen pots which are filled to their brims with water. The horse manure is thoroughly broken up until it becomes powder. Then sticks, such as pestles, are obtained and the *katsi* is pounded in the hole. It is taken out, then put back in again, and so on until it is ready. Then it is left for two days.
>
> One continues to treat the mortar in this way until it is ready, i.e. it begins to smell, and becomes sticky and soft; then the cement is quite ready for use.[10]

Laso has a life expectancy of five or six years when used as a finishing coat for external walls and roofs. Before it is applied to the roof, the roof is twice plastered with good building clay: during the last plastering the parapet walls and pinnacles are built. On top of the mud is sprinkled powdered *katsi*, or a mixture of *katsi* and old *laso* from a demolished building. The *katsi* is watered, and for two days is stamped upon until it becomes very hard. The *laso* is taken to the site where it is mixed with water until it becomes the correct consistency, after which it is applied to the roof in two coats. On flatter surfaces it is covered with a protective layer of red earth mixed with dung which is washed off during the rains, by which time the *laso* has hardened. *Laso* is the best finish for roofs and parapets and if used on walls produces a lovely silvery white building weathering to darker greys.

The most expensive and by far the best finish for external walls is *cafe* which is reputed to last for many years without maintenance. This wall finish varies from place to place in Hausaland, but usually consists of black earth collected from a borrow pit, which is mixed with a solution of the pounded seeds of the *bagaruwa* tree. The black mud is then plastered on to the wall and allowed to partly dry. Pebbles, sifted to remove fine particles, are pressed then beaten into the wall surface. This hardens for two days, then on successive days it is twice wetted with *bagaruwa* fluid and twice with *makuba* solution. *Cafe* is a dark earthen colour with a rough texture which is extremely durable.

The most beautiful wall finish is perhaps *makuba* itself. It is a deep purple brown applied with bold circular sweeps of the arm which gives it a distinctive patterning (see Figure 35.24). *Makuba* is made from the fruit pod of the locust bean tree. The husk of the fruit is ground to a powder which is then mixed with water before it is added to red earth. Unfortu-

Figure 35.24 Makuba

nately, the waterproofing effects of *makuba* last for only about two years and the plaster is prone to attack by white ants.

The most expensive and best wall finishes are normally reserved for entrance huts or for important buildings such as the palace or the Friday mosque. Usually the bricks are simply covered with mortar as the wall is built and then given a further finishing coat of ordinary building mortar. This is often delicately decorated with trailing finger patterns, the colour changing from wall to wall, from silver brown to dark earth orange depending on the age of the structure (see Figures 35.25 and 35.26). It is the finish which predominates in the old cities.

Figure 35.25 Finger patterning in mortar

Figure 35.26 Finger patterning in mortar

Internal surface finish

Internal plastering consists of two coats applied over rough mortar [. . .] First a coat of mud mortar is applied, followed by a finishing coat of red earth and sand which can be brought to a smooth finish on which to whitewash. For ceilings, a special plaster is used which has greater adhesive properties than normal plasters: the undercoat is made from black earth and grass; the final coat is the usual finishing plaster. There are special plasters too, such as the one used occasionally in emirs' palaces, which consists of small fragments of quartz set in gum arabic, or the one used in Wusasa church, Zaria, which contains mica flakes and makes the whole church glisten with a golden or silver sheen depending upon the quality of light which filters through the narrow slit windows (see Figures 35.27 to 35.30).

Floors are made from a mixture of laterite and gravel. The following is a Hausaman's description of the process of floor making:

> Then some other earth called *birji* is dug and put on top of the floor. Then they bring gravel and enough water for their needs. When they have put the gravel on to the laterite, scattered all over the place and not left in a heap, they then start to do the work with the proper sticks; they go on beating until all the gravel has become embedded in the *birji* and it is strong. Pods of the acacia tree are got and pounded up, or the husks of the locust-bean tree are soaked. This liquid is sprinkled all over the floor, which is again beaten, but lightly, till it penetrates the whole floor. It becomes black when it dries. If this is done, it will prevent the floor from cracking.[11]

Conclusion

Although the climate of Hausaland is not particularly suitable for buildings made entirely from mud, an essentially alien style of architecture has been cleverly adapted to the environment through the development and use of

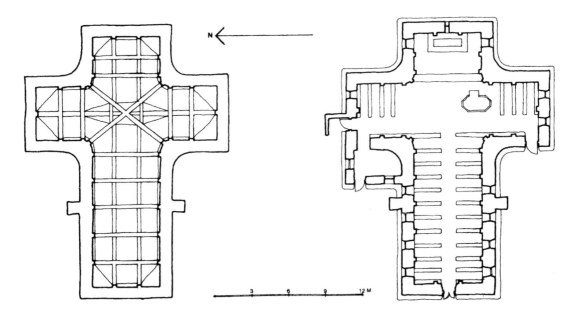

Figure 35.27 Church at Wusasa, Zaria, plan and roof plan

special external finishes. However, despite the ingenuity of the Hausa builder the soft lines of the final building forms are as much the product of nature as of man. When newly built, the Hausa building is crisp, neat, efficient in appearance and based upon a rigorous structural discipline. It is the erosive effect of the rain which produces those softened moulded forms and incidental charm associated with mud buildings. The Hausa builder has learned to accept this outside interference with his designs: each dry season he repairs parts of walls, and even whole buildings when necessary.

Figure 35.28 Church at Wusasa, Zaria, Interior

Figure 35.29 Church at Wusasa, Zaria, Interior

Figure 35.30 Church at Wusasa, Zaria, Interior

The profiles of the buildings vary with great rapidity: a whole street changes from season to season and from year to year, producing what is essentially an organic architecture in which static form is unknown. Such changes take place within a highly disciplined architectural context, where elements added, removed or changed are part of a unified system of design based upon a set of preferred dimensions which maintain continuity and coherence within a continually evolving built environment.

Notes

1 Heinrich Barth, *Travels and Discoveries in North and Central Africa*, London, Longman, 1857-8, Vol. 3, p. 389; Vol. 2, p. 226

2 In 1962 there was a brickworks between Maiduguri and Dikway. Bricks were burnt in the same way as pots, stacked beneath a bonfire of wood

3 F.W. Taylor and A.G.G. Webb, trans., *Labarun Al'Adun Hausawa Da Zantatukansu*, London, Oxford University Press, 1932, pp. 171-2

4 Ibid., p. 171

5 J.C. Moughtin, 'The Friday Mosque, Zaria City', *Savanna*, Vol. 1, Dec. 1972, no.2, pp. 143-63

6 Taylor and Webb, *Labarun Al'Adun Hausawa Da Zantatukansu* (see above note 3), p. 181

7 A.F. Daldy states that: 'This continuation of the arch is done partly for effect, but mainly it is a legacy from the past. Until the last fifty years there was always a pillar projecting from the wall under each arch; then it was realised that this pillar had no structural value, and it was gradually left out; the lengthening of the arch is the last trace of this pillar.' Daldy, *Temporary Buildings in Northern Nigeria*, Nigeria, Public Works Department, 1945, pp. 11, 12. In the Friday Mosque, Zaria, built some hundred years before Daldy wrote his text, Mallam Mikaila was using the type of arch described by him as a relatively recent innovation

8 J.C. Moughtin, 'The Traditional Settlements of the Hausa People', *Town Planning Review*, Vol. 35, April 1964, no.1, p. 29

9 Taylor and Webb, *Labarun Al'Adun Hausawa Da Zantatukansu* p. 177

10 Ibid., pp. 185-7

11 Taylor and Webb, *Labarun Al'Adun Hausawa Da Zantatukansu* pp. 187-9

36

BUILDING-TYPES OF THE HAUSA PEOPLE: CHARACTERISTICS AND FORMATIVE INFLUENCES

by J.C. Moughtin

Source: J.C. Moughtin, *Hausa Architecture*, London, Ethnographica, 1985, pp. 149-55

Introduction

The Hausa system of mud construction is based upon an empirical knowledge of the strength of materials and a method of sizing building elements based on the scale of the human body and building techniques. The limit set by these design parameters has resulted in a series of building sizes that correspond closely to a broad band of normal dimensions which could almost be described as modular. This does not mean that each dimension of a building is determined accurately according to some system of proportion, but rather that sizes are an inherent part of the system, and failure to observe normal rules of good building results in collapse or requires a change of technology.

The environment and technology set the limits within which Hausa architecture developed. Social and economic requirements influenced the spatial arrangement of the various architectural elements to serve the needs of the community. The organic nature of Hausa architecture is a reflection of the structure of society. In the past, the impermanent nature of the building materials made it possible for the total amount of accommodation and its disposition within the settlements to vary rapidly, keeping pace with the organic nature of the extended family system. So far this

quality has not been disrupted by the introduction of more permanent structures: such structures which are still a small proportion of the total building stock.

The Hausa have developed an architecture using materials found in their locality and have adopted, adapted or invented systems of constructions that take these materials to their structural limits. In adapting impermanent and relatively unsuitable building materials to a climate which has periods of heavy rainfall the Hausa have created an extremely attractive built environment composed of buildings whose sculptural forms give expression to the plastic nature of mud. Within the limits set by the structural possibilities of the materials the buildings provide shelter from the extreme conditions of the climate.

Although there are similarities between Hausa architecture and that of other parts of West and North Africa it has a distinct character of its own which is the product of a cohesive and definable culture. The form of Hausa architecture, like the structure of Hausa society, has undergone a long process of change and development with new ideas from outside the area being introduced from time to time. The present building forms cannot be fully appreciated without an understanding of the effect of culture contact between the

Hausa and neighbouring states. The process of acculturation has been a long one; the early influences on the Nok who occupied the area prior to the establishment of the Hausa states are unknown, but it is possible to be more certain of the effects of culture contact closer to modern times.

It is interesting to speculate on the influence of Pharaonic Egypt on West Africa. But speculation it must be, because there is very little evidence. We may note the similarities in character between the building forms of the Egyptians and the older mosques of Hausaland, but this could be explained simply as similar roofing problems giving rise quite independently to similar structural solutions.

Building types in West Africa

In the area of West Africa between the Sahara and the forest zone the chief building type is made entirely from vegetable material - that is some sort of thatching fixed on to a timber frame. The predominance of this building type indicates that it may be the indigenous form of structure for this climatic zone. In the West African Sudan there are three areas where this possible indigenous structure does not predominate: around the great northern bend of the Niger, Hausaland and the Chad area - the West African termini of the Saharan caravan routes and consequently the areas that had most contact with North African and Nilotic cultures. In the Chad area and Hausaland the main hut type is the structure with mud walls and thatched roof (*dakuna*); close to the northern bend of the Niger the all-mud structure (*soro*) predominates.

The mud wall may have been developed in the West African Sudan but is more likely to have been imported from more advanced cultures. As a technique it may even have been introduced by the legendary first rulers of the Hausa states who, according to custom, arrived in their present location after travelling from

the east through Bornu, but coming originally from further north. There is less doubt about the origin of the rectangular room with mud walls and mud roof. Its introduction was late: in the fifteenth century Leo Africanus described Timbuktu as consisting of thatched houses, and there is a description of the building of a mud-roofed mosque in Bornu in the twelfth century which seems to indicate that it was something of an extraordinary occurrence in that area too at that time.[1]

It seems that the rectangular room of mud walls and mud roof was an unusual structural form around Timbuktu until the fifteenth century, and that in the Chad area it has never replaced the older huts as the dominant house form. The rectangular mud structure was possibly introduced into Hausaland from Mali at about the same time as Islam, in the fourteenth century. However, until the seventeenth century this architecture remained an alien and exotic form, used only for mosques and the houses of foreign traders. From about this time direct contact between Hausaland and North Africa was first established yet travellers in Hausaland as late as the nineteenth century still described the Hausa settlements as a mixture of *soro* and *dakuna* building types. Lugard, writing in 1904, described Kano as the southern limit of the *soro*, or northern type of architecture, and he wrote that *soro* buildings were used in Zaria only by the chiefs and the Emir.[2]

Barth's portrait of Kano in the 1850s reveals a mixture of *soro* and *dakuna* building types. He gave detailed descriptions of the various quarters of the town which show quite clearly that the majority of the new ruling class, the Fulani, lived in the simpler *dakuna*. Barth found one exception to this rule, the Emir, who occupied a large mud palace. According to him the defeated Habe still occupied their *soro* buildings, but as the century wore on the Fulani were influenced by the Arabs and by the defeated Habe and they too began to live in houses built in the *soro* style.[3] It was not until the British occupation of Hausaland, which broke the autocratic power of the Fulani, that greater numbers of Hausa traders of the *talakawa* class became wealthy and were permitted to display their wealth by the building of *soro* houses. The British occupation and the new emphasis on trade encouraged the rapid growth of a class of wealthy merchants, who were both willing and able to invest in building forms which were formerly beyond their means and probably regarded as the prerogative of the ruling class.

Before discussing possible alien influences on the form, layout and detail of buildings, it is necessary to trace the origin and development of the two main building types, the *soro* and the *dakuna*. It is not known for certain how or when the *dakuna* hut was first introduced into the Sudan and Hausaland. It may have been developed in this region, but it is more likely that the all-grass hut is the indigenous form of semi-permanent structure associated with the area and that the *dakuna* hut was introduced with other technological innovations such as smelting at the time of the Nok culture. New building techniques were brought to the eastern and western Sudan from North Africa between the fourth and twelfth centuries, among them the use of burnt bricks, introduced into Mali during the fourteenth century by Al Saheli. This material was used at Gambaru in the Bornu-Kanem area in the sixteenth century by Mairam Aisa Kili N'girmaram.

The brick buildings in Mali and Bornu-Kanem were unusual and exotic structures unsuitable for wholesale use in the Western Sudan, but they may, by their form have acted as a stimulus to the development of a rectilinear architecture in mud, a material more suited to curvilinear shapes. However, the evidence suggests that few if any of these ideas reached Hausaland or had any significant impact upon local building technology until the seventeenth century. Then Hausaland was in direct contact with the Arab world of North Africa and came under its influence; the Habe, or ruling class, adopted the North African style of architecture. During the nineteenth century the Fulani conquerors of Hausaland adopted the architectural style of the defeated Habe as they too came within the sphere of influence of North Africa.[4] It was due to the constant cultural contact beween Hausaland and North Africa in the three centuries before the British occupation, that the art of mud building became established among the leading families of the emirate cities of Hausaland. But it is only since the beginning of this century that the *soro* style has become available for all those who can afford it.

The *dakuna* remains the building type in general use by the *talakawa*, especially in rural areas, although many of the peasant class, particularly in the urban centres, still build the *zaure* from mud and may indeed have other rooms in this style. Nowadays they tend to invest in more permanent modern structures. The compounds of the ruling families contain *dakuna*, *soro* and 'modern' buildings, so that in Hausaland building styles are freely mixed. A man decides which type of hut to build after considering such factors as the use to which it will be put, its position in relation to the main courtyard and the cost.

The courtyard plan

The courtyard, in one form or another, is a universal plan type in West Africa. At its most simple form it is either a group of huts forming the walls of an irregular enclosed space, or a group of huts scattered within a circular containing wall. It probably evolved in the same natural way as the round hut itself, and may well be the product of the culture of early sedentary agriculturalists. As such it probably developed quite naturally or independently in Hausaland.

The rectangular courtyard figures in many

pre-dynastic Egyptian hieroglyphics, and together with the rectangular room is either a product of a technological advance requiring the use of tools to shape and frame timber members, or the natural outcome of building with burnt bricks.[5] The rectangular courtyard house was used throughout Egyptian times and was in use in the Middle East when Mohammed built his home at Mecca. Today the courtyard house, using rectangular rooms and rectangular external spaces, is to be found in many areas of the Middle East (see Figure 36.1).

The rectangular courtyard house was probably introduced into Hausaland at the same

time as *soro* building. In a Hausaman's description of the setting out of a building, the rectangular compound is shown to be an ideal type and many houses do indeed have courtyards that approximate to a rectangular shape:

> When they are about to lay the foundations of the wall, it would be best if Tanko gave them ropes and pegs to set up. They should tie the rope to the pegs and align the sides lest they be crooked, or lest one be longer than the other. It is best that the compound should be exactly rectangular. The plan of the house is best if laid out in this way.[6]

To judge from the buildings I have measured, the right angle appears to be unknown in Hausaland, although most buildings have floor plans that are trapezoidal or, as in the case of the Friday Mosque in Zaria, approximating to a parallelogram. Despite the probability that Al Saheli brought the knowledge of the right angle to the Sudan from Egypt there seems little evidence of its use, although the opposite sides of buildings and spaces are often approximately equal, possibly due to their being set out using ropes of the same length. This lack of knowledge, or at least use of the right angle, is most surprising considering the close contact between Hausaland and North Africa. When Clapperton visited Sokoto in the 1820s he noted that Sultan Bello had a copy of Euclid, and he saw a mosque being constructed which had been designed by a man from Zaria whose father, also a builder, had studied 'Moorish' architecture in Egypt.[7] It appears that if the right angle was known to the Habe builders then they saw no good reason to use it.

Although the right angle was not used by the Hausa, the main buildings are formal compositions. For example, the visual effect of the Friday Mosque in Zaria is one of sobriety, and this is confirmed by a study of its plan and roof plan. The regularity of the plan is self evident, most elements correspond about a series of axes and sub-axes. The symmetrical composition of the Mosque, however, is not that artificial symmetry associated with the 'Beaux Arts'

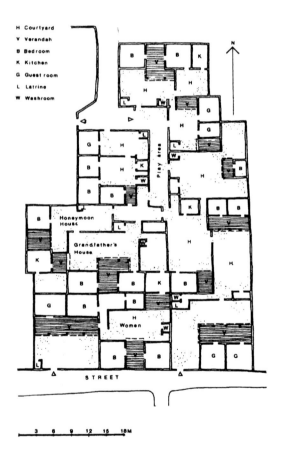

H Courtyard
V Verandah
B Bedroom
K Kitchen
G Guest room
L Latrine
W Washroom

Figure 36.1 Sudanese house plan. The house of the Awad El-Kerimi family. Khartoum

period in Europe, but a lively discipline where corresponding forms vary slightly, either by accident or design or because of the nature of the building materials. Such accidents of detail add charm to the design: because the visual structure of the composition is strong they are able to fall into place without disrupting the overall effect. In designing the Mosque in this orderly way, the architect, Mallam Mikaila, was following in the tradition of Hausa building, for as we have seen it is Hausa custom to build as regularly as building and constructional techniques permit. This tradition is probably a result of the ideas that came to West Africa from North Africa or Egypt where regularity of design was the result of a more precise building technology.

The development of the mosque

The first mosque in Islam, according to Cresswell, was the converted home of Mohammed at Madikia.[8] As in other cultures the religious buildings of Islam initially took the form of the standard home, and the mosques in North Africa still show the effect of this early influence of the house plan. Mosques in Hausaland have taken the standard Islamic form of a building within a walled area, but unlike those in North Africa the courtyards are not surrounded by covered arcades. The Hausa mosque may be free-standing within the courtyard but is entered through small gatehouses used as ablution chambers, a form closer to the house plan of the Hausa than to the fully-developed Egyptian mosques. It is interesting to note the marked differences between the present day Sudanese house in Khartoum and the Hausa house (see Figure 36.1). Although both have been designed to achieve privacy, particularly for the women of the household, the Hausa link between private world of the house and the public world of the street is an entrance hut (*zaure*); in Khartoum the link is the entrance courtyard (*hosh*).

The Friday Mosque in Zaria, for example, uses ablution chambers as a buffer between the courtyard of the mosque and the busy street; the Emir's entrance through the *Shari'a* court and a series of other rooms and courtyards is a planning solution very close to that of the normal Hausa home. [. . .]

Since the beginning of the nineteenth century the Hausa mosque has changed in two important ways. In Habe times, prior to the jihad, there was a tower attached to each mosque. Towers from this period can still be seen in Katsina and Bauchi, although there is evidence that the present tower in Katsina was rehabilitated at the beginning of the present century. In Habe times the mud tower was a feature both of the mosque and the houses of emirs and chiefs.

According to Clapperton, such towers were built for defence.[9] Since Leo Africanus did not comment on mud towers in Hausaland, it is safe to assume that they were first introduced directly from North Africa in the seventeenth century or later. It seems possible that the tower was introduced to improve the defences of Kano and Katsina in their struggle to control the southern terminus of the central Saharan trade route. From old sketches and photographs, it can be seen that the now demolished *sumi'a* of Kano mosque had many small, slit-like windows. It appears ideally suitable as a retreat from which to make a last stand against the enemy. Whatever the reason for building towers in the seventeenth and eighteenth centuries, they are similar in form to the minarets of such mosques as that of Ahmad Ibn Tulun (built AD 876–9) in Cairo, or the Great Mosque in Qairawan, and to the defensive towers of the Dades Valley in Morocco. The Friday Mosques in Zaria and Kazaure and others built since the jihad have no tower which may indicate that it had less significance during the comparatively peaceful times in the latter part of the nineteenth century.

The development of arch construction

Another change in the form of the mud mosque was the replacement of the trabeated system of construction by the use of arches. The first account of such a mosque is given in Clapperton's journal, a record of his second visit to Hausaland in 1828. As mentioned above, the mosque being built in Sokoto was designed by a man from Zaria whose father, also a builder, had studied 'Moorish' architecture in Egypt.[10] Clapperton noted that the design of Bello's main suite of reception rooms was the same as the houses of all the most important families in Hausaland. He described one of Bello's rooms as being about 9m. square and requiring eight arches to support the domed roof. This does not represent an early formative stage in the development of the arch, but is the maximum size for a square room using this system of construction. So it may be argued that by the end of the second decade in the eighteenth century the fully developed Hausa architectural style already existed. It is possible to form some idea of the type of houses in which the Habe rulers of Hausaland lived by studying the *zaure* of the Emir's palace in Daura. This simple structure, which is whitewashed and has very little decoration, is a very beautiful example of the type of building which was to be found in Hausaland in the late eighteenth century.

Since the arch is a latent possibility of the Hausa structural system and is in effect a series of corbels it may well be a product of a slow development. On the other hand, its introduction may be due to the skill and knowledge of the Zaria builder who had studied in Egypt, returning home perhaps full of new ideas with which he experimented. Even if the invention of the arch is not due entirely to the efforts of one man, it seems likely, since his son was invited to Sokoto to build an important mosque, that he was a leading exponent of its use. He may well have been the first builder to realize the full potential of the arch and to use it in more daring ways.

The Friday Mosque in Zaria, built by Babban Gwani Mallam Mikaila for Emir Abdulkarim in the 1830s, is very similar to the one described by Clapperton, and so close in time to it that it is more than likely that the two mosques were built by the same man, or at least by close relatives of the same family of builders. The post of chief builder is still held by the direct male descendant of Babban Gwani, which seems a good reason to believe that the chief builder of Zaria before Babban Gwani was a more senior male member of his family.[11]

Structurally and from an aesthetic point of view, the Mosque in Zaria went further than most contemporary buildings. Its main arches are coupled and sweep right down to the floor; those of the Daura Palace and the Sokoto Mosque, which Clapperton described, rest on columns, a solution which represents a transitional stage between the spanning of a space using straightforward corbels and a lintel, and Babban Gwani's complete arch form. The Zaria Mosque is perhaps the high point of Hausa architecture, but it is said that Babban Gwani built one more mosque for the Emir of Birnin Gwari which, although smaller, was as fine as his earlier building.[12] On completion of his mosque, the Emir of Birnin Gwari seized Babban Gwani and had him executed so that no mosque would ever be built to equal the one in Birnin Gwari. Judging Babban Gwani only on the one work in Zaria, discounting both this last folk tale and the possibility that he may have built an equally fine mosque in Sokoto, it is evident that he made a major contribution to the art of Hausa building.

The roots of Hausa architecture are lost in antiquity, but it is possible that it shares a common ancestry with the great buildings of Pharaonic Egypt. Some of the early pre-dynastic hieroglyphics depict houses with small pinnacles similar to the *zankwaye* that decorate Hausa buildings. The whole character of the

Egyptian house drawings resembles that of the architecture of both present day Nubia and Hausaland (see Figure 36.2). Pre-dynastic reed architecture of Egypt similar to that of the Marsh Arabs of Iraq may have been the forerunner for both the mud architecture of Sudan and that of Pharaonic Egypt.[13]

The *zankwaye* or pinnacles constructed on top of Hausa walls are found in the architecture

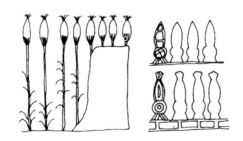

Figure 36.2 House forms and hut shrines in pre-dynastic Egypt. (E.B. Smith, *Egyptian Architecture* New York 1938)

of many other parts of the Sudan. They may have originated in North Africa and found their way into Hausaland from the seventeenth century onwards. However, between Wa in the north of Ghana and Timbuktu in Mali there are many mosques whose walls are a striking series of pointed buttresses (see Figure 36.3). The buttresses are totally unnecessary to resist lateral thrust, since the structure consists of simple columns and beams. This particular building form is similar to the huge termite nests in the same region, a natural form for mud structures, their battered wall surfaces following the line of slip for laterite. Since they have no horizontal surfaces the effect of erosion by rainfall is minimized. It is not suggested that the mosques of Northern Ghana are a conscious copy of the termite nest, but their forms are similar and may have been developed partly for climatic reasons. The buttresses are possibly the mud versions of an older structural form associated with buildings made from local vegetation. Whatever the origin of these intriguing mosques, the projecting buttresses in some cases become simple pinnacles similar to the Hausa *zankwaye*.

The mud arch and the arch used in Egyptian Islamic architecture, from which it was derived, are the same shape as the reed arches of the Marsh Arabs. When loaded, the bundles of reeds which form the arches take the shape of a horseshoe and again suggest a link between mud architecture and an earlier form of reed structure. Reed arches are not unknown in the Western Sudan, for they are used by the Buduma people who inhabit the islands of Lake Chad and have a similar technology to the Marsh Arabs.[14] A less likely reason for the shape of the mud arch may be the simple practicalities of building an arch on to a battered wall; the arch may simply follow the line of the wall until the first *azara* is placed in position, when the angle of the arch automatically changes. However, because of its horseshoe shape the arch merges beautifully with

Figure 36.3 Mosque, Northern Ghana

the general line of the wall and is far more pleasing in its monolithic mud form than its more sophisticated cousin made in stonework where the voussoirs are emphasized. The horseshoe arch may in fact have undergone three phases of development from a reed, to a mud and finally to a stone structure.

The male descendants of Babban Gwani are today the leading family of builders in Zaria, their knowledge having been passed down to them from their ancestors. Clapperton's account of his meeting with the builder of the Sokoto Mosque shows that this family tradition of building existed early in the last century. The Hausa builders are members of the old Habe families. So it would seem quite wrong to assume, as Crowder has, that 'this restraining influence [the simplicity of the Fulani] on the crude art of the Hausas combined with wider Arab influence to produce an unusual style of architecture'.[15] On the contrary, it seems that the structural techniques developed by the Habe have given the archi-

tecture its simplicity and dignity. The Habe, over many hundreds of years, have assimilated the North African influence, producing an architecture quite different from that of the Songhai, the Bornu, or the North African states. The period of contact with other cultures was long and the process of adaptation of indigenous architecture slow, being the work of families of builders who passed their skills from father to son, and who adapted new ideas as and when necessary in the developing society. Such ideas were not altogether alien, and were probably a development of the same original culture, and thus quite compatible with the forms and ideas existing in the West African Sudan.

Notes

1 Leo Africanus, *The History and Description of Africa*, London, Hakluyt Society, 1896, Vol. 3, p. 824, and H.R. Palmer, *The Bornu Sahara and Sudan*, London, Murray, 1936, p. 9

2 F.D. Lugard, 'Northern Nigeria', *Geographical Journal*, Vol. 23, 1904.

3 Heinrich Barth, *Travels and Discoveries in North and Central Africa*, 2nd edn, London, Longmans, 1857-8, Vol. 2, pp. 121-3

4 This appears to be a more reasonable explanation than that of Labelle Prussin who believes that Hausa architecture is the result of Fulani influence. Labelle Prussin, 'Fulani-Hausa architecture', *African Arts*, Vol. 10, 1976, Pt.1, pp. 8-19

5 The change from curvilinear to rectilinear structures has been associated with the technological innovations that occurred at the time of the changes in the organization of the structure of society from a matriarchal to a patriarchal system. See Lewis Mumford, *The City in History*, Harmondsworth, Penguin, 1966

6 F.W. Taylor and A.G.G. Webb trans., *Labarun Al'Adun Hausawa Da Zantatukansu*, London, Oxford University Press, 1932, p. 179

7 Hugh Clapperton, *Journal of a Second Expedition into the Interior of Africa*, London, Murray, 1829, pp. 198-9; and Dixon Denham and Hugh Clapperton, *Narratives of Travels and Discoveries in Northern and Central Africa*

in the Years 1822, 1823 and 1824, London, Murray, 1826, Vol. 2, p. 103

8 K.A.C. Cresswell, *Early Muslim Architecture*, Oxford, Clarendon, 1932, Vol. 1, pp. 1-20

9 Clapperton, *Journal of a Second Expedition into the Interior of Africa* (see above note 7), p. 165

10 Clapperton, *Journal of a Second Expedition into the Interior of Africa* (see above note 7), p. 103

11 Mallam Mikaila, according to the current chief builder in Hausaland, Sarkin Maigina Haruna, was a devoted follower of Shehu 'Uthman Dan Fodio, being rewarded by him with a flag of office as chief builder after the successful jihad

12 J.C. Moughtin and A.H. Leary, 'Hausa Mud Mosque', *Architectural Review*, Vol. 137, February 1965, no. 816, pp. 155-8

13 Wilfred Thesiger, *The Marsh Arabs*, London, Longmans, 1964

14 Olive Macleod, *Chiefs and Cities of Central Africa*, London, Blackwood, 1912, pp. 212-37

15 Michael Crowder, 'The Decorative Architecture of Northern Nigeria', *African World*, February 1956, p. 9

Part 4
THE SJOBERG MODEL

Can any regularities or underlying similarities be read into the foregoing accounts of pre-industrial cities and their technologies? According to the author of reading 37, Gideon Sjoberg, an American sociologist, all pre-industrial cities, from Babylon to Kathmandu, are similar in their basic form and structure. He selects technology as the 'key independent variable' differentiating between societies, and therefore the main types of city, though he insists that this does not commit him to technological determin-ism. Nevertheless, the basic form of the pre-industrial city reflects its depen-dence on animal or human power, sparse tools, and, for most of the inhabitants, oral communication. Ultimately this explains why in all pre-industrial cities the centre is the hub of autocratic government and religion, and the elite's place of residence, why the lower classes live on the periphery, and why craftsmen concentrate in a particular street or quarter.

Many British historians have rejected outright sociological models such as Sjoberg's, though Peter Burke, a historian of popular culture in early modern Europe, has been more open-minded about the value of social theory in historical writing. His rather belated review of Sjoberg's book appears in edited form as reading 38. His main strategem is to point out exceptions to Sjoberg's generalizations: not all pre-industrial cities were fortified, not all had low levels of literacy, not all had poor amenities, not all were autocratic, and so on. He also doubts in principle that a clear-cut distinction can be drawn between pre-industrial and industrial cities; there seems to be no feature that is exclusive to either. Nevertheless, Burke does not reject urban models out of hand; he looks instead for a more 'developmental' one. Interestingly, it seems that Burke is not hostile to Sjoberg's privileging of technology as an explanatory variable; at one point, successive innovations in urban transport are seen as fundamental, at least in shaping the physical form of the city.

37

THE PRE-INDUSTRIAL CITY

by Gideon Sjoberg

Source: Gideon Sjoberg, *The Pre-industrial City: past and present*, Glencoe, The Free Press, 1960, pp. 1, 4–8, 10–12, 101, 104, 323–6, 328–9

The city and civilization are inseparable: with the city's rise and spread, man at last emerged from the primitive state. In turn, the city enabled him to construct an ever more complex and, we would like to believe, more satisfying way of life. Some scholars regard the city as second only to agriculture among the significant inventions in human history. We shall not quibble over the proper ranking due the urban community in man's storehouse of great inventions. It is sufficient to recognize that it is worthy of intensive treatment. [. . .]

Statement of purpose

Our aim is to describe and analyze the structure of the city, both in historical societies and in surviving literate non-industrial orders, before its transformation through industrialization. We also seek to provide background data on the origins of city life and the growth and spread of cities around the world.

Our principal hypothesis is that in their structure, or form, pre-industrial cities – whether in medieval Europe, traditional China, India, or elsewhere – resemble one another closely and in turn differ markedly from modern industrial-urban centers. Most writers have failed to distinguish the industrial and pre-industrial types. As a result, the data on pre-industrial cities negate many popular sociological generalizations based solely upon evidence from modern industrial American communities.

The most non-industrial cities today are those like Andkhui and Mazar-i-Sharif in Afghanistan and Kathmandu in Nepal, where the populace continues its pre-industrial mode of existence quite unaffected by industrial forms. Still largely pre-industrial cities abound in other parts of Asia, in North Africa, and in sections of southern Europe and Latin America.

Not only do pre-industrial cities survive today, but they have been the foci of civilization from the time of its first appearance in Mesopotamia in the fourth millennium BC. The 'ancient' cities of Athens and Rome, familiar to almost every school child, are in actuality relatively late creations and merely two out of a vast number scattered over much of Eurasia and North Africa around the beginning of the Christian era. Even when Europe entered the Dark Ages and city life waned over much of the continent, the Eastern Roman Empire and Spain experienced a vibrant urban life. Contemporaneously, cities were flourishing in Meso-America, North Africa, and Asia.

To return to our main thesis: pre-industrial cities everywhere display strikingly similar social and ecological structures, not necessarily in specific cultural content, but certainly in basic form. [. . .]

We seek to isolate for pre-industrial cities structural universals, those elements that transcend cultural boundaries. These cities share numerous patterns in the realms of ecology, class, and the family, as well as in their economic, political, religious, and educational structures, arrangements that diverge sharply from their counterparts in mature industrial cities.

The industrial–urban center is the standard against which we contrast the pre-industrial city [. . .]

For analytical purposes we distinguish three types of societies: the folk, or preliterate, society; the 'feudal' society (also termed the pre-industrial civilized society or literate pre-industrial society); and the industrial–urban society. Only the last two contain urban agglomerations: the pre-industrial and industrial cities, respectively.

To achieve this typology of societies, and consequently of cities, we take technology as the key independent variable – i.e., associated with varying levels of technology are distinctive types of social structure. Technology both requires and makes possible certain social forms. This viewpoint does *not* commit us to technological determinism, however, for recognized is the impact upon social structure of other variables – the city, cultural values, and social power – all of which can affect the patterning of technology itself. Nor do we, like sociologists of the 'ecological school,' conceive of technology as part of the 'biotic,' or subsocial realm.[1] Technology is not some materialistic, impersonal force outside the socio-cultural context or beyond human control; technology is a human creation par excellence.

Technology, as employed in this study, refers to the sources of energy, the tools, and the know-how connected with the use of both tools and sources of energy for the production of goods and services. Industrialism is that type of technology that utilizes inanimate sources of energy for driving its tools; associated with it are implements and know-how of a much more complex form than those in non-industrial systems. In industrial cities (subsystems of the industrial society) electricity, steam, nuclear fission, etc. supply the power that for so many millennia were provided by animals and human beings. We could consider energy itself as the key variable for distinguishing between the main types of cities and societies.[2] However, technology in the broader sense is a more satisfactory medium for explaining the structure of the pre-industrial city and its society.

The technological variable is highly useful for differentiating among folk, feudal, and industrial societies, as well as more specifically between pre-industrial and industrial cities. The folk, or preliterate, order's technology is an exceedingly simple one, and cities are absent. Only a few of the simpler folk societies – those that lack any stable food supply and utilize the crudest of implements – survive today; South America and New Guinea harbor some of these. Yet they were quite numerous only a few centuries ago. Most of the preliterate groups that have persisted into the twentieth century – some in West Africa, for example – are of the more advanced type that possess food production and storage techniques enabling them to support some full-time non-agricultural specialists. But even these can be distinguished from typical feudal societies by their lack of literacy and other key traits. [. . .]

The feudal society, vis-à-vis the typical folk order, has a more advanced agricultural technology that produces sufficient food surpluses to support large non-agricultural populations. In all instances this technology includes the cultivation of grains. It also embraces (except in ancient Meso-America) animal husbandry, large-scale irrigation works, the plow, metallurgy, the wheel, and other devices that multiply the production and distribution of agricultural surpluses. Nevertheless, the feudal society is almost entirely dependent upon animate, i.e., human and animal, sources of energy.

This more advanced technology operates in conjunction with (it both requires and makes possible) a complex social organization that is typified by a 'leisured,' literate elite or upper class. In all instances feudal orders contain cities. Relative to the total population, urban residents are few, but the presence among them of the elite makes the city's inhabitants significant far beyond their numbers. [. . .]

The industrial–urban society, in contrast to the feudal order, utilizes inanimate sources of energy, a complex set of tools, and specialized scientific know-how in the production of goods and services. As a result the greater portion of the industrial society's populace dwell in cities.

The industrial technology requires, and at the same time supports, a structural apparatus that diverges strongly from the feudal type in its ecological, class, familial, economic, political, religious, and educational aspects. The class system is a highly fluid one that emphasizes achievement rather than ascription, and social power is diffused throughout the city and society. Associated with the ill-defined class system is a small, flexible, conjugal family unit as the ideal norm. In turn these patterns are interwoven with a large-scale, rational economic structure characterized by a complexity in the division of labor far exceeding that in the feudal system and requiring highly trained specialists recruited according to universalistic criteria. The proliferation of skilled specialists implies the existence of mass literacy and an educational system that emphasizes science.[3] [. . .]

Differentiation of land use according to occupation is usual. [. . .] This localization of particular crafts and merchant activities in segregated quarters or streets is intimately linked to the society's technological base. The rudimentary transport and communication media demand some concentration if the market is to operate [. . .]

The temporal ordering of life within the city merits special comment. Activity proceeds at a slow pace, and the pre-industrial urbanite, compared to industrial man, does not think of time as a 'scarce commodity,' except within broad limits such as days or weeks. [. . .] All this is quite in keeping with the technology of the city, which typically lacks any precise measuring instruments.

The pre-industrial city in capsule form

Cities of this type have been with us, present evidence indicates, since the fourth millennium BC, when they first began their development in the Mesopotamian riverine area. Before long, in response to the growing technology and a variety of political forces, city life proliferated over a broader area. To an astonishing degree, pre-industrial cities throughout history have prospered or floundered, as the case may be, in accordance with the shifting tides of social power.

In terms of their population these cities are the industrial city's poor relations, few ranging over 100,000 and many containing less than 10,000 or even 5,000 inhabitants. Their rate of population growth, moreover, has been slow and variable as well, in accordance with the waxing and waning of the supportive political structure. Yet throughout the shifting fortunes of empire, and the concomitant oscillation in population growth and decline, certain persistent structural characteristics signalize pre-industrial cities everywhere.

As to spatial arrangements, the city's center is the hub of governmental and religious activity more than of commercial ventures. It is, besides, the prime focus of elite residence, while the lower class and outcaste groups are scattered centrifugally toward the city's periphery. Added to the strong ecological differentiation in terms of social class, occupational and ethnic distinctions are solemnly proclaimed in the land use patterns. It is usual for each occupational group to live and work in a particular street or quarter, one that generally bears the name of the trade in question. Ethnic groups are almost always isolated from the rest of the city, forming, so to speak, little worlds unto themselves. Yet, apart from the considerable ecological differentiation according to socio-economic criteria, a minimum of specialization exists in land use. Frequently a site serves multiple purposes – e.g., it may be devoted concurrently to religious, educational, and business activities; and residential and occupational facilities are apt to be contiguous. [. . .]

Economic activity is poorly developed in the pre-industrial city, for manual labor, or indeed any that requires one to mingle with the humbler folk, is depreciated and eschewed by the elite. Except for a few large-scale merchants, who may succeed in buying their way into the elite, persons engaged in economic activity are either of the lower class (artisans, laborers, and some shopkeepers) or outcastes (some businessmen, and those who carry out the especially degrading and arduous tasks in the city).

Within the economic realm the key unit is the guild, typically community-bound. Through the guilds, handicraftsmen, merchants, and groups offering a variety of services attempt to minimize competition and determine standards and prices in their particular spheres of activity. Customarily also, each guild controls the recruitment, based mainly on kinship or other particularistic ties, and the training of personnel for its specific occupation and seeks to prevent outsiders from invading its hallowed domain.

The production of goods and services – by means of a simple technology wherein humans and animals are almost the only source of power, and tools to multiply the effects of this energy are sparse – is accomplished through a division of labor which is complex compared to that in the typical folk order but, seen from the industrial city's vantage point, is surprisingly simple. Very commonly the craftsman fashions an article from beginning to end and often markets it himself. Although little specia-

lization exists in process, specialization according to product is widespread. Thus each guild is concerned with the manufacture and/or sale of a specific product or, at most, a narrow class of products.

Little standardization is found in prices, currency, weights and measures, or the type or quality of commodities marketed. In the main, the price of an item is fixed through haggling between buyer and seller. Different types and values of currency may be used concurrently within or among communities; so too with weights and measures, which often vary as well among the crafts. [. . .]

The expansion of the economy is limited not only by the ruling group's negation of economic activity, the lack of standardization, and so on, but very largely also by the meager facilities for credit and capital formation.

Turning from the economy to the political structure, we find members of the upper class in command of the key governmental positions. The political apparatus, moreover, is highly centralized, the provincial and local adminstrators being accountable to the leaders in the societal capital. The sovereign exercises autocratic power, although this is mitigated by certain contrary forces that act to limit the degree of absolutism in the political realm. [. . .]

Relative to the industrial–urban community, communication in the feudal city is achieved primarily by word-of-mouth, specialized functionaries serving to disseminate news orally at key gathering points in the city. Members of the literate elite, however, communicate with one another to a degree through writing. And the formal educational system depends upon the written word, the means by which the ideal norms are standardized over time and space.

Only the elite, however, have access to formal education. And the educational and religious organizations, with few exceptions, are interdigitated. The curriculum in the schools, whether elementary or advanced, is overwhelmingly devoted to predication of the society's tradi-tional religious–philosophical concepts. The schools are geared not to remaking the system but to perpetuating the old. Modern science, wherein abstract thought is coherent with practical knowledge and through which man seeks to manipulate the natural order, is practi-cally non-existent in the non-industrial city. The emphasis is upon ethical and religious matters as one is concerned with adjusting to, not over-coming, the order of things. In contrast, indus-trial man is bent upon revising nature for his own purposes.

Theoretical orientation

Inasmuch as pre-industrial cities in numerous divergent cultural milieu display basic similari-ties in form, some variable other than cultural values, in the broad sense, must be operative; regularities of this sort are not the result of mere chance.

Here technology - viz., the available *energy*, *tools*, and *know-how* connected with these - seems the most satisfactory explanatory vari-able.[4] This mode of reasoning should not com-mit us in any way to credence in tehcnological determinism or unilinear evolution; indeed we firmly reject these stands. In point of fact, we make frequent reference to social power in accounting for the fluctuating fortunes of cities and the fate of technology and give due recog-nition to its role in producing organization in the society. Nor do we ignore values. These, we have remarked in a number of contexts, account for certain divergencies from our con-structed type; [moreover] some values tend to be correlated with a specific level of technol-ogy; as a notable instance, the scientific

method has built-in values that must be dif-fused, along with the energy-sources, tools and requisite know-how, to underdeveloped areas if these are to industrialize. Further, we see the city per se as a variable to be reckoned with; rural and urban communities are in many ways intrinsically different. Although we lend priority to technology, we can not dispense with the other variables enumerated.

With these qualifications (and we hope the reader will keep them ever in mind), it seems clear that the transition from the preliterate to the feudal level, i.e., to the pre-industrial civi-lized order, or from the latter to the industrial-urban society is associated with certain crucial advances in the technological sphere. The very emergence of cities is functionally related to the society's ability to produce a sizeable sur-plus; and the orientation, quite late in history, to an industrial base made possible a kind of city never before imagined. To minimize tech-nology, as to ignore the value system, would be poor procedure. [. . .]

Notes

1 Amos H. Hawley, *Human Ecology*, New York, Ronald Press, 1950

2 Fred Cottrell, *Energy and Society*, New York, McGraw-Hill, 1955

3 Just as the mere existence of literacy constitutes a divid-ing line between folk and feudal societies, so too mass literacy, including lengthy formal education, serves as one convenient device for separating feudal and indus-trial societies

4 For other sociologists who have given attention to the technological variable see: Francis R. Allen et al., *Tech-nology and Social Change*, New York, Appleton-Cen-tury-Croft, 1957. Unfortunately, some sociologists tend to drift into a materialistic interpretation of technology, something we have tried to avoid.

38

SOME REFLECTIONS ON THE PRE-INDUSTRIAL CITY

by Peter Burke

Source: Peter Burke, 'Some Reflections on the Pre-industrial City,' *Urban History Yearbook*, 1975, pp. 13-19

In the last few years, a new word has gained popularity among historians: 'pre-industrial'. Specialists in the social and economic history of Europe before 1800 have become increasingly aware that the object of their studies is simply one case among others of what sociologists call 'traditional society', and that it is easier to understand traditional or pre-industrial Europe if it is compared and contrasted with other societies of this type. Thus Keith Thomas and Alan Macfarlane have illuminated English witchcraft by making comparisons with witchcraft in African tribal societies, while Frédéric Mauro and Witold Kula, among others, have compared the economies of early modern Europe with those of the developing countries today.[1] Even Richard Cobb, no great friend to the social sciences, has recorded that he came to understand eighteenth-century Paris better after visiting contemporary Calcutta.[2] In fact, the city is an obvious and splendidly tangible unit of comparison, and it is not surprising that the term 'pre-industrial city' is passing into general use.

Credit for defining this useful term and putting it into general circulation should go to the American sociologist Gideon Sjoberg, whose book on the subject presents a description of the main typical features of the pre-industrial city.[3] In other words, he has constructed a model. The author makes some sharp criticisms of historians who have come to conclusions about a particular city but remained blissfully unaware of the generality of their findings. He is of course absolutely right, but historians in their turn may have good grounds for criticizing his model in some respects. [. . .]

Such a historian cannot help noticing that the quality of Sjoberg's secondary sources, for his period, is not very high. On Venice, for example, he used only the reminiscences of W.D. Howells, a nineteenth-century American literary figure. For seventeenth-century France, he depends on G. Vanel's book on Caen, an entertaining antiquarian work but of little use if analysis is one's aim.[4] Writing in the late 1950s, Sjoberg could not easily have done better, but at just about that time more important and more relevant monographs began to appear, above all in France. Jean Delumeau published a book on Rome, and Jacques Heers one about Genoa. Pierre Deyon made a study of Amiens, and Bartolomé Bennassar one of Valladolid. [. . .] To juxtapose these and other monographs to Sjoberg's model awakes a certain uneasiness. There are serious discrepancies. To begin with points of detail, the historian of early modern Europe may well want to object to five features of the model.

1 The pre-industrial city is described as walled, fortified, with narrow streets, 'unpaved, congested, poorly lighted and poorly drained'.

Minority groups lived 'sealed off in quarters of their own'. It is perhaps only a minor qualification to remark that Venice was not fortified – nor, for that matter, were the great cities of Tokugawa Japan, Edō, Osaka, and Kyōtō. In France in the seventeenth and eighteenth centuries, the fortifications of many towns were razed because the king feared urban revolts. As for the state of the streets, Paris in the late seventeenth century is a famous example, but far from the only one, of a pre-industrial city whose streets were largely paved, lit and drained, and in a few cases widened to cope with a new kind of traffic – the coach. Permanent street lighting was introduced in a number of French towns in the course of the eighteenth century. Sjoberg is certainly right about minority groups. There were Jewish quarters in many towns of early modern Europe, including Venice, where the term 'ghetto' first came into use, and there were Moorish quarters in Seville and other Spanish towns. However, is there any important difference between these instances and 'Little Ireland' in Manchester in the 1840s? Chinatown in nineteenth-century London? Harlem in New York today?

2 From the economic point of view, the pre-industrial city is seen as a centre of trade and crafts. 'As a consequence of the low-level technology', Sjoberg writes, 'the categories of personnel engaged in service occupations . . . are relatively few.'[5] These points may seem no more than common sense, but all the same they are in need of a little qualification. Bennassar has produced figures to show that in Valladolid in 1570, the tertiary sector was the same size as the secondary one. As the seat of the Spanish court, Valladolid was untypical of the early modern city in some respects, but it was not unparalleled. In Rome in the early sixteenth century, the second most common occupation was that of innkeeper,

no doubt the result of the pilgrim trade. The primary sector should not be forgotten either. In Mediterranean Europe in particular, peasants frequently lived in towns and commuted to their fields. The 'journey to work' was thus not unknown in early modern times. Once more, Valladolid furnishes examples, and in Toulouse in 1695, 16 per cent of the population was employed in the primary sector.[6]

3 From the social point of view, the pre-industrial city is seen as having a rigid system of stratification, with differences in status expressed by ritual and by dress. If the point is simply to contrast pre-industrial and industrial societies, well and good; but Sjoberg writes as if bowing low to one's social superiors was a special characteristic of cities, which it was not.[7]

4 From the political point of view, the pre-industrial city, according to Sjoberg, is not democratic. It is ruled in an autocratic manner by a small group of aristocratic families.[8] He makes no mention of elections to municipal office in cities other than Athens in the age of Pericles. Yet such elections were important in a number of large cities in early modern Europe. In Venice the Great Council, the body which elected men to many important offices, was some two thousand strong in 1600. If this group is too small, in a city of 130,000 people, for us to describe Venice as a 'democracy' it is too large for us to talk of 'autocracy' either. In Renaissance Florence, some four or five thousand citizens (in a population of less than 100,000) were politically active, and important officials were chosen by lot in order to give all the citizens an equal chance. In France in the seventeenth and eighteenth centuries, there was universal male suffrage in some municipal elections, and officials were chosen by lot at Aix, Montpellier and Marseilles. Compared with the countryside around them, the towns of early modern

Europe were democratic indeed, unlike the towns of traditional China or Japan. Against this 'democratic' view it may be argued that the government of these European towns was usually in practice in the hands of a small group of families who intermarried and dominated, if they did not monopolize, the key offices, whether these offices were filled by co-option (as in Amsterdam, for example), or by election (as at Florence and Venice). However, the same point can be made about the industrial city. Despite the Municipal Corporations Act of 1835, the government of Birmingham, for example, remained in the hands of a small group of inter-related families.

5 From the cultural point of view, Sjoberg describes the inhabitants of pre-industrial cities as mainly illiterate, as unaccustomed to precise weights and measures and as lacking any sharp sense of time. On the contrary, they held the view that 'time is not a scarce commodity, something to be utilised to the fullest'.[9] These statements are in need of major qualifications. It is plausible to argue that nearly half the population of fourteenth-century Florence and seventeenth-century Amsterdam (to take only two examples), could read and write. The cities of early modern Europe were places where clocks regularly struck, night watchmen called out the hours and merchants wrote in their diaries reminders to themselves not to waste time.

These points of detail may add up to something more general. Sjoberg constructed his model in order to contrast the typical pre-industrial city with the typical industrial city. Naturally this procedure involves distortion – stressing some aspects of urban reality at the expense of other aspects. Well and good – such distortion is the price of the clarity and simplicity achieved by model-building. However, the price may become too high when students of

the city with purposes other than Sjoberg's take his model over. It is tempting to take it over because it is beautifully clear and explicit, but when our interests are different we need to construct different models with different emphases.

For example, we might be interested in contrasting the city and the country in pre-industrial societies. In that case, we would need to emphasise precisely what Sjoberg most neglects, the fact that in early modern Europe and elsewhere cities were often islands of *relative* social mobility, democracy and literacy, although the majority of the urban population did not change their social status, did not participate in the city government, and did not know how to read and write.

Alternatively, we might be interested in the characteristic features of the European city as opposed to cities elsewhere in the world before the Industrial Revolution. In that case it would be important to discuss the questions of political autonomy and town planning. The frequently self-governing communes of Europe, proud of their privileges, stand in obvious contrast to the bureaucratically-administered cities of China and of the Islamic world, and also to the tribal cities of West Africa. The frequently planned cities of Europe and China, whether laid out on the grid plan, for convenience, or designed as cosmic symbols, may be contrasted with the unplanned, relatively formless Islamic and African towns. Nineteenth-century Yoruba towns had virtually no public buildings. Sjoberg plays down these differences because to discuss them would have been a distraction from the essential purpose of his book.

[. . .]

This paper began by suggesting some revisions to Sjoberg's model on points of detail. Almost insensibly the discussion has had to widen, to consider whether this is the only useful model of the pre-industrial city or whether it is only one of several possible models. Such a discussion inevitably raises the further question whether a monolithic model of 'the' pre-industrial city is not more dangerous than it is useful. Is such a comment just an example of the professional myopia of the early modern historian? Or should 'pre-industrial city' be relegated to the scrap-heap reserved for models which do not perform what is required of them?

Before this question can be answered, there is one more objection to be raised. To my mind it is the most serious of all. It is an objection not simply to Sjoberg's model, but to any use of the concept 'pre-industrial city'. It is necessarily defined by contrast to the industrial city. But what is an 'industrial city'? Is it any city in a society where the take-off into industrialization has taken place? But does nineteenth-century Exeter (say) differ significantly from Sjoberg's pre-industrial city? In many respects it surely does not. Perhaps an industrial city should be defined as a city dominated by a particular industry, in the way that nineteenth-century Leeds and Manchester were dominated by textiles. However, Florence in the fifteenth century, Norwich in the seventeenth century and Lyons in the eighteenth century were also textile towns in this sense. In Lyons, 14 per cent of the total population, meaning something like half the working adults, were employed in textiles, and most of these in silk. Is the industrial city a city dominated by large factories? This would certainly exclude Florence and Norwich, but at the price of leaving out nineteenth-century Birmingham and Sheffield as well, since they were cities of small workshops. In other words, the 'post-industrial city' (as it may be more convenient to call it) was no more monolithic than the pre-industrial city. [. . .]

What is needed is a developmental model of the city; that is, a description of the most important changes which have occurred in the history of cities, presented as a sequence and related to one another. In what remains of this paper I do not propose anything so ambitious. I should simply like to clear the ground

for such a model by drawing attention to certain features of the larger cities of early modern Europe (in particular, cities of 100,000 inhabitants or more), and to ask exactly how they differed from post-industrial cities, and when and why the change from one type to the other took place.

It may be convenient to look in turn at four aspects of the pre-industrial European city. They are as follows: the city as artifact, the demographic and social structure of the city, its government, and finally, urban culture.

The city as artifact

Recent studies of the nineteenth-century city as artifact are among the best-known recent contributions to urban history. It has been emphasized that the coming of the railways involved clearing people out of the centre of the city and rehousing them at the periphery. In the late nineteenth century, from London to Boston, the streetcar and the train created the modern suburb. There were great opportunities for speculative builders, which some exploited to the full. The pedestrian city disappeared, the journey to work became common, and the middle classes were physically segregated from the working classes in different suburbs.

All this is true, but may call for two comments. In the first place, it should not be thought that social zoning and speculative building did not exist in early modern Europe. We are sometimes told that the pre-industrial city lacked social segregation. Aristocrats and poor men rubbed shoulders – or at least lived in the same street – in eighteenth-century London and Edinburgh. However, there is considerable evidence of the existence of rich quarters and poor quarters in some cities before 1800. In Paris, the Faubourg St-Germain was an aristocratic suburb in the seventeenth century, while the Faubourg St-Antoine was a suburb for the poorer artisans. [. . .]

As for the speculative builder, he made his

appearance on the urban scene long before the Industrial Revolution. In sixteenth-century Seville, a boom town, a certain Martín López de Aguilar built blocks of houses on speculation. In seventeenth-century Amsterdam, another boom town, entrepreneurs developed the ironically-named *Jordaan* area (the promised land) as a zone of cheap housing for poor immigrants. In Paris in the 1620s, Louis Le Barbier headed a syndicate of financiers who turned the fields opposite the Tuileries into the Faubourg St-Germain. It is well known how Nicholas Barbon became a successful property developer in London after the Great Fire.

A second point concerns urban transport. It is much too crude a picture simply to contrast the pre-industrial pedestrian city with the post-industrial city where transport is mechanized. There are at least four important stages to distinguish. There is the rise of the coach in the seventeenth century; of the railway in the early nineteenth century; of the tram in the late nineteenth century; and finally the rise of the motor car in our own century. A developmental model of the city as artifact would have to take into account the changes set in motion by all four modes of transport. Reyner Banham has suggested that the only way to see Los Angeles is out of the window of a moving car. Similarly, Lewis Mumford has argued that the way to see seventeenth-century Paris and Rome was on horseback or out of the window of a moving carriage. Changes in transport affected the shape of the city.[10]

The demographic and social structure

Historians of the post-industrial city, from Paris to Newburyport, rightly devote a good deal of attention to immigration, in particular to long-distance immigration. They emphasize the fact that unskilled immigrants often formed a pool of cheap labour and undercut local workers. It is easy to come away with the impression that

there was little immigration to pre-industrial cities, or that such immigration was mainly from neighbouring regions. This was not in fact the case. In eighteenth-century Lyons, over a thousand immigrants arrived every year, a figure equal to 1 per cent of the population. Thirty-three per cent of the men and 20 per cent of the women did not come from neighbouring regions. Already in the fourteenth and fifteenth centuries, immigrants to London seem to have come from all over the country rather than predominantly from the home counties, and in 1736 there were riots in London against the employment of cheap Irish labour. Was it the size of the city which determined its immigration pattern rather than the Industrial Revolution?

It is customary for social historians to make a fairly sharp distinction between the stratification systems of pre-industrial and post-industrial societies. Pre-industrial societies are seen as divided into 'orders' or 'estates', groups which were legally rather than economically defined. A noble tended to have high status even if he were poor. Group solidarity tended to be 'vertical' between a landlord and his peasants, a master-craftsman and his journeymen. About 1800, so the argument goes, this structure began to yield to another. Society was now divided into classes, groups which were defined by economic criteria, conscious of themselves and in conflict with one another. Horizontal solidarity replaced vertical solidarity. Edward Thompson and John Foster have provided some impressive evidence for these changes in the case of nineteenth-century England.[11] However, some pre-industrial cities on the Continent do seem to come nearer to the second model of the social structure than to the first. Fifteenth-century Florence was divided into three groups on essentially economic criteria. There were, as contemporaries put it, the 'rich' (*ricchi*), the 'middle class' (*mediocri*), and the 'poor' (*poveri*), and these groups were sometimes in conflict over political issues.

[. . .] Does the nineteenth century see a complete change in the social structure and in social consciousness, or simply the spread of social structures and mentalities which were already traditional in many large towns?

City government

It has already been suggested that the dominance of the city government by a small group of families who intermarried can be found after the Industrial Revolution as well as before. At the level of policy it is equally difficult to make a sharp distinction between pre-industrial municipalities and post-industrial ones. Historians of the nineteenth century have told us in detail about the politics of sanitation in Birmingham, Paris and elsewhere. Yet if Chamberlain was concerned about improving the water supply of his city, so was Pope Sixtus V. The Birmingham Water Committee or Drainage Board has its parallel in the *Congregazione delle strade* of sixteenth-century Rome. There may well be important differences between the sanitation of Renaissance Rome and Victorian Birmingham, but it is likely that these differences are at the level of technical means, not that of municipal policies. [. . .]

Urban culture

It is often suggested that the inhabitants of the post-industrial city differ from their urban predecessors in mentality. In a famous essay Georg Simmel, who has been followed in this by many sociologists, suggested that the large modern city is experienced by its inhabitants as a place of anonymity, freedom, privacy and loneliness. Yet the same points were made about cities in the early modern period. At the end of the eighteenth century Wordsworth described how in London people lived 'Even next door neighbours, as we say, yet still/Strangers, nor knowing each the other's name'. Descartes' praise of seventeenth-century Amsterdam for

its anonymity is celebrated. 'I could spend my entire life here without being noticed by a soul. I go for a walk here through the Babel of a great thoroughfare as freely and restfully as you stroll in your garden.' A seventeenth-century Parisian commented in similar terms that in Paris 'everyone lives so freely that it is normal for respectable people to live in the same lodgings without knowing one another.' Even in the sixteenth century, an English visitor could describe Venice as follows. 'No man there marketh another's doings or . . . meddleth with another man's living . . . No man shall ask thee why thou comest not to church . . . to live married or unmarried, no man shall ask thee why . . . And generally of all other things, so thou offend no man privately, no man shall offend thee.'[12] It seems reasonable to suggest that it is the large city, the city of 100,000 inhabitants or more, which engenders privacy and anonymity rather than the specifically post-industrial city.

Finally, the post-industrial city is often seen as the home of a particular ethos which it is convenient to describe as 'bourgeois' and which involves considerable stress on the virtues of industry, thrift, cleanliness, punctuality, temperance, self-control, achievement, deferred gratification and self-improvement through study; in a word, 'respectability'. This ethos is contrasted not only with the values of the 'dangerous classes', but also with those of the pre-industrial city. This latter contrast is surely a mistake. The urban sense of time in early modern Europe has been mentioned already. It is not difficult to find evidence of townsmen in the same period who believed in hard work, thrift, and achievement. [. . .]

Notes

1 K.V. Thomas, *Religion and the Decline of Magic*, London, Weidenfeld & Nicholson, 1971; A.D. Macfarlane, *Witchcraft in Tudor and Stuart England*, New York, Harper & Row, 1970; F. Mauro, *Le XVIᵉ siècle européen: aspects économiques*, Paris, Presses Universitaires de

France, Paris, 1966, p. 280f.; W. Kula, 'Il sottos viluppo economico in una prospettiva storica' in *Annali della fondazione Luigi Einaudi*, iii, 1969

2 R. Cobb, *The Police and the People*, Oxford, Clarendon, 1970, p. xxi

3 G. Sjoberg, *The Pre-industrial City*, Glencoe, Free Press, 1960. For criticisms of this book see esp. L. Mumford in *American Sociological Review*, xxvi, 1961, and O.C. Cox in *Sociological Quarterly*, v, 1964. The debt of these reflections to Mumford will be obvious

4 W.D. Howells, *Venetian Life*, rev. edn, Boston, Houghton, Mifflin & Co, 1907; G. Vanel, *Une grande ville aux 17ᵉ et 18ᵉ siècles*, Caen, 1910

5 Sjoberg, *The Pre-industrial City* (see above note 3), p. 203

6 P. Wolff, *Histoire de Toulouse*, Toulouse, 1958, p. 257

7 Sjoberg, *The Pre-industrial City* (see above note 3), ch. 5

8 Ibid., esp. pp. 220, 224

9 Sjoberg, *The Pre-industrial City* (see above note 3), pp. 286f, 209f

10 L. Mumford, *The City in History*, new edn, Harmondsworth, Penguin, 1966, pp. 421ff., 446

11 E.P. Thompson, *The Making of the English Working Class*, London, Victor Gollancz, 1963; J. Foster, *Class Struggle and the Industrial Revolution*, London, Weidenfeld & Nicolson, 1974

12 G. Simmel, 'Die Grossstädte und das Geistesleben', trans. in K.H. Wolff (ed.), *The Sociology of Georg Simmel*, Glencoe, Free Press, 1950, pp. 409ff.; W. Wordsworth, *The Prelude*, Book 7, quoted by R. Williams, *The Country and the City*, London, Chatto & Windus, 1973, p. 149; Descartes, letter to Guez de Balzac, 5 May 1631; H. Sauval, *Antiquités de la ville de Paris*, posthumously published, Paris, 1724, p. 62; W. Thomas (1549) quoted in K.V. Thomas, *Religion and the Decline of Magic*, (see above note 2), p. 527

ACKNOWLEDGEMENTS

PART 1

CARTER, Harold
An Introduction to Urban Historical Geography; 1983. Reprinted by kind permission of Edward Arnold Publishers.

MOOREY, P. R. S.
Ancient Mesopotamian Materials and Industries; 1994, Clarendon, Oxford.

HERODOTUS
The History of Herodotus, in *Great Books of the Western World*, 6: Herodotus, Thucydides; 1952. Reprinted from *Great Books of the Western World* ©1952, 1990 Encyclopaedia Britannica, Inc.

DAVID, A. R.
The Pyramid Builders of Ancient Egypt: a modern investigation of the Pharaoh's workforce; 1986. Reprinted with kind permission of Routledge.

BURFORD, A.
'Heavy Transport in Classical Antiquity' in *Economic History Review*, 2nd series, Vol. 18; 1960. Reprinted by kind permission of Blackwell Publishers, Oxford.

BURNS, A.
'Ancient Greek Water Supply and City Planning: a study of Syracuse and Acragas' in *Technology and Culture*, Vol. 12; 1974. Reprinted by permission of University of Chicago Press.

COULTON, J. J.
'Lifting in Early Greek Architecture' in *Journal of Hellenic Studies* Vol. 94; 1974. Reprinted by kind permission of The Council of the Hellenic Society.

ANDERSON, James C. Jr.
Roman Architecture and Society; 1997. Reprinted by kind permission of Johns Hopkins University Press.

VITRUVIUS
The Ten Books on Architecture, translated by Morris Hicky Morgan; 1960. Reprinted by kind permission of Dover Publications, Inc.

EVANS, Harry
Water Distribution in Ancient Rome: the evidence of Frontinus; 1994. Reprinted by kind permission of The University of Michigan Press.

MORLEY, Neville
Metropolis and Hinterland: the city of Rome and the Italian economy 200 BC-AD 200; 1996. Reprinted by kind permission of Cambridge University Press.

PART 2

WARD-PERKINS, Bryan
From Classical Antiquity to the Middle Ages: urban public building in northern and central Italy (AD 300-850), 1984, Oxford University Press, Oxford.

KENNEDY, Hugh
'From *Polis* to *Madina*: urban change in late antique and early Islamic Syria' in *Past and*

Present, Vol. 106; 1985, Oxford University Press, Oxford, pp. 15-23, 25-6.

WHITE, Lynn Jnr.
Medieval Technology and Social Change; 1962, Oxford University Press, Oxford.

GALLOWAY, James A., KEENE, Derek and MURPHY, Margaret
'Fuelling the City: production and distribution of firewood and fuel in London's region, 1290-1400' in *Economic History Review*, Vol. 49; 1996. Reprinted by kind permission of Blackwell Publishers, Oxford.

HERLIHY, David
Pisa in the Early Renaissance: a study of urban growth; 1958. Reprinted by kind permission of Yale University Press, New Haven, Conn.

EWAN, Elizabeth
'Town and hinterland in medieval Scotland' in *Medieval Europe 1992* (pre-printed papers for conference on medieval archaeology in Europe, University of York, 1992) Volume 1: *Urbanism* pp. 113-8.

GOLDTHWAITE, Richard A.
The Building of Renaissance Florence: an economic and social history; 1981. Reprinted by kind permission of Johns Hopkins University Press.

FONTANA, Domenico
Della transportatione dell'obelisco Vaticano, Rome, 1590', in *A History of Western Technology* by Friedrich Klemm, trans. Dorothea Waley Singer; 1959, George, Allen & Unwin.

FRIEDRICHS, Christopher R.
The Early Modern City, 1450-1750; 1995. Reprinted by kind permission of Addison Wesley Longman Ltd.

ISRAEL, Jonathan
'A Golden Age: innovations in Dutch cities, 1648-1720' in *History Today*, Vol. 45; 1995. Reprinted by kind permission of History Today and History Review.

PORTER, Stephen
The Great Fire of London; 1996. Reprinted by kind permission of Sutton Publishing.

BERNARD, Leon
The Emerging City: Paris in the age of Louis XIV; © 1970. Reprinted with kind permission of Duke University Press.

EVELYN, John
'Londinium Redivivum' in *The Writings of John Evelyn*, ed. Guy de la Bédoyère; 1995, Boydell Press.

HOBERMAN, Louisa Schell
'Technological Change in a Traditional Society: the case of the desagüe in colonial Mexico' in *Technology and Culture*, Vol. 21; 1980. Reprinted by kind permission of University of Chicago Press.

PART 3

The Book of Songs, trans. Arthur Waley, 1937, George, Allen & Unwin, London, pp. 231-34.

WU, Nelson I.
Chinese and Indian Architecture, 1963, Prentice-Hall International and George Braziller, New York and London, pp. 8-9, 11-12, 29-32, 34-5, 37-8, 42-3.

NEEDHAM, J. and WANG Ling
Science and Civilization in China, Volume 4, Part 2: *Mechanical Engineering*; 1965. Reprinted by kind permission of Cambridge University Press.

YING-HSING, Sung
T'ien-kung K'ai-wu: Chinese technology in the seventeenth century, trans. Sun E-tu Zen and Sun Shiou-chuan, 1966.
Re-issued by Dover Publications, 1997. Reprinted with their kind permission.

YING-HSING, Sung
Chinese Technology in the Seventeeth Century: T'ien-kung K'ai-wu. Translated from the Chinese and annotated by E-tu Zen Sun and Shiou-chuan Sun.

ROWE, William T.
Hankow: commerce and society in a Chinese city, 1796–1889; 1984. Reprinted with the permission of the publishers, Stanford University Press © 1984 by the Board of Trustees of the Leland Stanford Junior University.

ROWE, William T.
Hankow: conflict and community in a Chinese city, 1796–1895; 1989. Reprinted with the permission of the publishers, Stanford University Press © 1989 by the Board of Trustees of the Leland Stanford Junior University.

HUC, M.
The Chinese Empire: a sequel to recollections of a journey through Tartary and Thibet, trans. by Mrs. J Sinnett, 1859. PUBLIC DOMAIN.

MOUGHTIN, J. C.
Hausa Architecture, 1985, Ethnographica, London. Reprinted by kind permission of the author.

***

PART 4

SJOBERG, Gideon
The Pre-Industrial City: past and present, 1960. Reprinted with the permission of The Free Press, a Division of Simon & Schuster from *THE PRE-INDUSTRIAL CITY: past and present* by Gideon Sjoberg. © 1960 by The Free Press.

BURKE, Peter
'Some reflections on the Pre-industrial City' in *Urban History Yearbook*; 1975. Reprinted by permission of Cambridge University Press.

INDEX